Semiotics in Mathematics Education

SEMIOTIC PERSPECTIVES IN THE TEACHING AND LEARNING OF MATHEMATICS SERIES

Volume 1

Semiotics in Mathematics Education

Epistemology, History, Classroom, and Culture

Luis Radford
Université Laurentienne, Sudbury, Canada

Gert Schubring
Universität Bielefeld, Germany

Falk Seeger
Universität Bielefeld, Germany

SENSE PUBLISHERS
ROTTERDAM / TAIPEI

A C.I.P. record for this book is available from the Library of Congress.

ISBN 978-90-8790-595-8 (paperback)
ISBN 978-90-8790-596-5 (hardback)
ISBN 978-90-8790-597-2 (e-book)

Published by: Sense Publishers,
P.O. Box 21858, 3001 AW
Rotterdam, The Netherlands
http://www.sensepublishers.com

Printed on acid-free paper

Printed by Printforce, the Netherlands

TABLE OF CONTENTS

THE UBIQUITOUSNESS OF SIGNS

By way of introduction

In the past few years, semiotics has grown up at a startling pace. Such a fast development is not, however, the result of a generalized obstinate curiosity for understanding formal sign systems. On the contrary, current interest in semiotics results from the increasing awareness that we live in a world of signs and artifacts, and that the way we express ourselves and our manners of perceiving and acting in our world are deeply related to a tremendous variety of signs and sign systems (language included) that, ubiquitously, surround us.

In fact, it is this ubiquitousness of intricate webs of signs signifying signifiers which—while rendering virtually impossible a pure, non-mediated experience—often makes us miss the effect that signs have on us. Semiotics is—at least to a certain extent—a reflective step backward, an effort to understand the amazingly complex manners in which, through sign systems, individuals signify and are, in turn, signified.

But semiotics is more than a contemplative gesture: in contemporary semiotic perspectives the notions of culture and cultural praxis receive a new interpretation—interpretation which extends to history as well—making semiotics a form of practical understanding and social action (Thibault, 1991). This is why it does not come as a surprise that semiotics is increasingly considered as a powerful research field capable of shedding some light on what have traditionally been understood as self-contained domains of enquiry. It is not unusual, hence, to now find a sustained recourse to semiotics in contemporary studies about mass communication, economics, politics, literature, arts, history, human and non-human cognition, psychology, education, and so on.[1]

Semiotics and Mathematics Education

Mathematics education has not been an exception among the disciplinary fields which, one way or another, have drawn on semiotics. Of course, the question is: What exactly does semiotics have to contribute to mathematics education?

The answer is both simple and complex. It is simple to the extent that, obviously, mathematics is an intrinsic symbolic activity, that is to say, mathematics is something that we accomplish through written, oral, bodily and other signs. Semiotics, with its arsenal of concepts, appears well suited to help us understand the mathematical processes of thinking, symbolizing and communicating.

At the same time, the answer is complex, for processes of thinking, symbolizing and communicating are—as sociologists, anthropologists and literary critics found out several decades ago—subsumed in more general encompassing symbolic systems (see, e.g., Barthes, 1982; Bourdieu, 1994; Eagleton, 1983; Foucault, 1966; Lévi-Strauss, 1962). The inevitable embedded nature of our ways of thinking and doing into these ever-changing symbolic systems makes mathematical thinking and discourse not a mere personal affair, but something entangled with the cultural, historical and political dimensions of life. Semiotics, as a reflective step backward, offers a advantageous viewpoint—a fissure of the symbolic, a disturbance of the familiar, a bracketing of the quotidian—whence to investigate, resist and transform the signs and sign systems through which we breath and live.

About this book ...

Like actions, artifacts have a history. What about the history of this book?

This book is not a systematic exposition of the aforementioned problems. It can be better understood as a modest continuation of previous efforts undertaken by some scholars who, directly or indirectly, have shown the potential of semiotics in the field of mathematics education.[2] Thus, the papers included in this volume are in no way an attempt to deal with—let alone solve—all the problems we are facing in the field. It would be a vain and empty presumption to believe that semiotics —or any theory for that matter—could solve the complex problems surrounding the teaching and learning of mathematics. Semiotics is perhaps a symptom of what Canadian scholar Charles Taylor (2003) calls the "malaise of modernity", one of its symptoms being the awareness that reality is much more complex than we and the crafters of modernity previously thought. Contemporary semiotics is, in a sense, an avowal that an understanding of ourselves and our "reality" (in whatever sense we consider this term) cannot lie within the scope of a sole theoretical approach, regardless of how well conceptually equipped such an approach might be. It is in the nature of signs, indeed, to intercept several layers of reality—psychological, economical, political, and so on.

The success of the Semiotics Working Group—organized by Adalira Sáenz-Ludlow and Norma Presmeg from 2001 to 2004—of the International Group for the Psychology of Mathematics Education led us to think that it was important to continue to ensure a space where scholars could keep on thinking and exchanging about the use of semiotics in the field. It was in this context that, in 2005, Michael Otte and Luis Radford discussed the possibility of hosting a meeting to gather a small number of mathematics educators. The meeting took place from July 13 to 15 2006, in Germany, under the title *The promises and problems of a semiotic approach to mathematics, the history of mathematics and mathematics education.* It was organized by Falk Seeger, Gert Schubring and Michael Otte near Bielefeld and was attended by 14 individuals. The long standing concentration on research into C.S. Pierce's philosophy and semiotics at the IDM in Bielefeld made Germany in general, and Bielefeld in particular, a perfect place to meet. The fourteen participants from three continents met at the *Haus Ohrbeck*, a former Franciscan monastery, transformed into a nicely situated and quiet conference centre. The participants understood themselves as a working group and, at the end of their meeting, there was a sense that some progress had been achieved but there was also a clear felling that, to go further, the exchange had to be continued. It was then unanimously decided to continue the work the next year. Thus, a second meeting was organized in 2007. It took place from July 16 to 18, in Germany again, but this time at the *Landesturnschule* in Melle—a highly charming ambiance filled with old (but reconstructed) Westphalian farm buildings. Seventeen participants attended the meeting. The intensive work continued and enlarged the discussions of the previous years. During the closing discussion, it was decided that the results achieved by the group should be submitted to a larger public.

The spirit of the meetings was not to create a monolithic theorizing semiotic perspective. Since the beginning, the idea was to be respectful of the various semiotic traditions upon which mathematics educators had been drawing (e.g., Vygotsky's, Peirce's, Saussure's, etc.). This plurality is manifested in the papers included in this volume. Thus, instead of an all encompassing semiotic perspective, covering all possible semiotic issues, the reader will find here several problématiques dealing with questions about teaching and learning, epistemology, history and culture. The papers are not arranged in some specific order. Their arrangement is rather a path to be walked in the course of which one stops to see, on the horizon, a certain problem as posed and discussed from a certain perspective; one continues and stops again at another spot to now take a look at a different landscape, and so on.

Acknowledgments

We wish to thank the persons and institutions who made possible the preparation of this book, in particular The Social Sciences and Humanities Research Council of Canada, Heather Empey and Isaias Miranda.

NOTES

1 Thus, the latest issue of the famous journal *Semiotica*, founded in 1969 in Europe, is devoted to what has been termed "theater semiotics", while the journal based at University of Toronto, "Applied Semiotics", founded in 1996, has recently featured entire issues on a diversity of topics such as semiotics, religion and ideology; the visual in popular culture; semiotics and media; collective beliefs (see http://www.lulu.com/content/156896).

2 See, for instance, Anderson, Sáenz-Ludlow, Zellweger, & Cifarelli (2003); Goldin and Janvier (1998); Hoffmann, Lenhard & Seeger (2005), Hitt (2002); Janvier (1987), Radford & D'Amore (2006), Sáenz-Ludlow & Presmeg (2006).

REFERENCES/BIBLIOGRAPHY

Anderson, M., Sáenz-Ludlow, A., Zellweger, S., & Cifarelli, V. (Eds.). (2003). *Educational perspectives on mathematics as semiosis: From thinking to interpreting to knowing*. Ottawa: Legas.

Barthes, R. (1982). *Empire of signs*. New York: Hill & Wang.

Bourdieu, P. (1994). *Raisons pratiques*. Paris: Éditions du seuil.

Eagleton, T. (1983). *Literary theory*. (Second Edition, 1996). Minnesota, MN: The University of Minnesota Press.

Foucault, M. (1966). *Les mots et les choses*. Paris: Gallimard.

Goldin, G. A., & Janvier, C. (Eds.). (1998). *Representations and the psychology of mathematics education* (Vol. 17.1 & 17.2).: The Journal of Mathematical Behavior.

Hoffmann, M. H. G., Lenhard, J., & Seeger, F. (Eds.). (2005). *Activity and sign: Grounding Mathematics Education*. New York: Springer.
Hitt, F. (Ed.). (2002). *Representations and mathematics visualization*. Mexico: Departamento de matemática educativa, Cinvestav-IPN.
Janvier, C. (Ed.). (1987). *Problems of representation in the teaching and learning of mathematics*. Hillsdale, NJ: Lawrence Erlbaum.
Lévi-Strauss, C. (1962). *La pensée sauvage*. Paris: Plon.
Radford, L., & D'Amore, B. (2006). *Semiotics, culture and mathematical thinking, Revista Latinonamericana de Investigación en Matemática Educativa, Special Issue*. Available at: Retrieved from http://www.laurentian.ca/educ/lradford/).
Sáenz-Ludlow, A., & Presmeg, N. (2006). Semiotic perspectives in mathematics education. *Educational Studies in Mathematics. Special Issue*, *61*(1–2).
Taylor, C. (2003). *The Malaise of Modernity*. Toronto: Anansi.
Thibault, P. (1991). *Social semiotics as praxis*. Minneapolis, MN: University of Minnesota Press.

Luis Radford
École des sciences de l'éducation
Université Laurentienne
Canada

Gert Schubring
Institut für Didaktik der Mathematik
University of Bielefeld
Germany

Falk Seeger
Institut für Didaktik der Mathematik
University of Bielefeld
Germany

FALK SEEGER

INTENTIONALITY AND SIGN

A developmental perspective

Only one must not take a nominalistic view of Thought as if it were something that a man had in his consciousness ... if it is to mean Thought it is more without us than within. It is we that are in it, rather than it in any of us.
(Peirce, Letter to William James, *Collected Papers 8.256*)

INTRODUCTION

Ever since it became clear that an evolutionary perspective is an important ingredient of Peirce's semiotic approach (see, e.g., Burks 1997) it has become equally apparent that development is a particularly important cornerstone of any attempt to understand the relation between sign, mathematics, and learning. In apparent contrast, only few attempts can be found, which base a semiotic perspective not only on the developmental view of mathematics, but also on the developmental view of the learner.

My claim here is that it is fundamental to understand that the triadic sign-function is itself subjected to development.[1] The triadic sign-function itself is not some self-sufficient eternal structure but comes into being during the developmental processes of the child.

This is a new and important perspective for mathematics education because developmental research has shown that everything known from the Piagetian framework on time and developmental sequences has to be corrected. Very young infants can do things Piaget would have thought to be impossible. I will go into some detail of a critique of Piaget below. One key point: there has been considerable new research on how the triadic sign-function develops immediately after birth. Making this body of research relevant for embracing theoretical conceptions on mathematics education like the one presented by Dehaene (1997) seems promising. It is especially fascinating to see what infants know about the world of objects without ever having had the opportunity to manipulate the objects inhabitating this world.

Research on the ontogenesis (and the phylogenesis) of shared intentionality has shifted our attention to the early infant. And there is a corresponding shift of attention to the mathematical experiences of children before starting school, in order to find out what could be done to prevent the growing alienation from math in school. The importance of a semiotic perspective to early learning and

L. Radford, G. Schubring, and F. Seeger (eds.), Semiotics in Mathematics Education: Epistemology, History, Classroom, and Culture, 1–18.

mathematics education would support current initiatives to draw more explicitly on the development in mathematical cognition taking place before school, in order to preserve the positive attitude most kids have towards numbers, counting and arithmetic during their pre-primary and primary years.

The semiotic approach could also support a focus on the externality of mind when one is trying to make sense of how mathematical "competence" develops. It is extremely important, e.g., to conceive of "competence" not or not only terms of an internal capacity. The primacy of the exterior over the interior, of the social over the individual – or at least the development from the exterior to the interior, from the social to the individual – is of crucial importance for a critical conception of competence.

In earlier papers, I have emphasized that what seems most important for a semiotic perspective on the psychology of learning is that sign processes where the signs relate to objects are different from sign processes where the signs relate to people (Seeger, 2005, 2006; see also Hoffmann, 2007). This difference has been connected and discussed under many different names and concepts in the past: it has been called Learning II or Learning III by Gregory Bateson (1972); it has been discussed as the complementarity of content and social in mathematics learning (Otte, 1994); and it has been treated as secondary intersubjectivity (Trevarthen, 1979; Trevarthen & Hubley, 1978) in developmental psychology and infant research. Finally, the issue of intersubjecticty and intentionality is most interestingly related to questions of the self and the role of the self in learning and identity formation.

To look at semiotics in mathematics education from the point of view of development means to consider three developmental planes: the development of the individual human being, from infant to adult, which is the *ontogenetic* plane; the development of humans as a species, that is, the development from higher forms of behaviour in the great apes to human behaviour, which is the *phylogenetic* plane; and the level of development of human society from the beginning of human culture to current developments, which is the *historical* plane. These threee planes together form the approach to psychology known as the cultural-historical approach initiated by Lev S. Vygotskij. I will not be able to do justice to all three planes here – the historical plane in particular will not be given the attention it deserves.

In what follows, I will start from the semiotic perspective of Lev Vygostkij and take his famous example of the development of the pointing gesture in early infancy as a point of departure. In the next section, I will identify the problem of the two lines of development, namely, the biological and the cultural, which also goes back to Vygotskij and is basic and pertinent to the issue of semiotics and development. I will then discuss the pivotal concept of shared intentionality. In the ensuing section, I will briefly discuss the issue of biological and cultural development as the relation between psychological core functions and shared intentionality. I will close with a section on possible consequences for mathematics education.

THE DEVELOPMENT OF THE POINTING GESTURE

As a starting point for a developmental perspective on the sign-function, nothing could be better than Vygotskij's *The History of the Development of Higher Psychological Functions,* published in English in 1997 as part of the six volume edition of his works. His analysis of the development of the pointing gesture is paradigmatic for a semiotic approach focusing on the primacy of the exterior and the social.

> As an example, we will consider the history of the development of the pointing gesture; as we shall see, it plays an exceptionally important role in the development of speech in the child and is, to a significant degree, the ancient basis for all higher forms of behavior. Initially, the pointing gesture represents a simply unsuccessful grasping movement directed toward an object and denoting a future action. The child attempts to grasp an object that is somewhat too far away, his hands stretched toward the object are left hanging in the air, the fingers make pointing movements. This situation is the point of departure for further development. Here the pointing movement, which we may arbitrarily term a pointing gesture, appears for the first time. This is movement of the child objectively indicating an object and only an object.
>
> When the mother comes to help the child and recognizes his movement as pointing, the situation changes substantially. The pointing gesture becomes a gesture for others. In response to the unsuccessful grasping movement of the child there arises a reaction not on the part of the object, but on the part of another person. In this way, others carry out the initial idea of the unsuccessful grasping movement. And only subsequently, on the basis of the fact that the unsuccessful grasping movement is connected by the child with the whole objective situation, does he himself begin to regard this movement as a direction.
>
> Here, the function of the movement itself changes: from a movement directed toward an object, it becomes a movement directed toward another person by means of a connection; grasping is converted into a direction. Because of this, movement itself is reduced, is contracted, and that form of the pointing gesture is developed which we may rightly call a gesture for oneself. But movement becomes a gesture for oneself in no other way than being, at first, direction for oneself, that is, objectively having all the necessary functions for direction and gestures for others, that is, being thought of and understood by the nearby as a direction. In this way, the child is the last one to recognize his gesture. Its significance and function are initially made up of an objective situation and then by the people around the child. The pointing gesture most likely begins to indicate by movement what is understood by others only later becomes a direction for the child himself.

Thus we might say that through others we become ourselves, and this rule refers not only to the individual as a whole, but also to the history of each separate function. This also comprises the essence of the process of cultural development expressed in a purely logical form. The individual becomes for himself what he is in himself through what he manifests for others.

This is also the forming of the process of the forming of the individual. In psychology, the problem of the relation of external and internal mental functions is posed here for the first time in all its significance. Here, as has been said, it becomes clear why everything internal in higher forms was of necessity external, that is, was for others what it is now for oneself.

Every higher mental function necessarily passes through an external stage of development because each function is primarily social. This is the center of the whole problem of internal and external behavior. Many authors have long since pointed to the problem of interiorization, internalizing behavior. Kretschmer sees in this a law of nervous activity. Bühler reduces the whole evolution of behavior to the fact that the field of selection of positive actions is transferred inward from the outside.

But we have something else in mind when we speak of the external stage in the history of the cultural development of the child. For us to call a process "external" means to call it "social." Every higher mental function was external because it was social before it became an internal, strictly mental function; it was formerly a social relation of two people. The means of acting on oneself is initially a means of acting on others or a means of acting of others on the individual.

We can formulate the general genetic law of cultural development as follows: every function in the cultural development of the child appears twice, in two planes, first, the social, then the psychological, first between people as an intermental category, then within the child as an intramental category. This pertains equally to voluntary attention, to logical memory, to the formation of concepts, and to the development of the will.

We are justified in considering the thesis presented as a law, but it is understood that the transition from outside inward transforms the process itself, changes its structure and functions. Genetically, social relations, real relations to people, stand behind all the higher functions and their relations. From this, one of the basic principles of our will is the principle of division of functions among people, the division into two of what is now merged into one, the experimental unfolding of a higher mental process into the drama that occurs among people.

For this reason, we might term the basic result to which the history of the cultural development of the child leads us as sociogenesis of higher forms of behavior. (Vygotskij, 1997, 104-106).

The paradigmatic situation depicted in this rather long quotation can be looked at, e.g., from a semiotic-epistemological stance like Michael Hoffmann (1998) has done. He analyzed development, so to speak, as a circular reaction of deduction, induction and abduction in a Peircean sense. This could be called a classical "internalist" approach, insofar as the sequence of syllogisms explains the pointing gestures as some form of semiotic representation of cognitive processes. Even if the semiotic processes, as it were, are happening "outside"—visible between child and mother—the crucial processes as circular reactions reside "inside." The triadic structure of semiosis seems to dissolve into the circular reactions passing into a sequence of dyadic structures. In a strict sense, such a picture appears as non-social.

Among many other attempts, Michael Tomasello has recently come up with some interesting research on non-human primates and human infants in the Max Planck Institute of Evolutionary Anthropology in Lepzig. His research on the development of shared attention towards the end of the first year nicely fits those questions arising from Vygotkskij's ideas on the development of the pointing gesture. Tomasello (1999) presents pointing and declarative gestures and their relation to "shared attention" from a developmental psychological perspective. At nine months, human infants start, for the first time, to transform dyadic interaction with other humans and humans plus objects into triadic interaction, which means that they begin to see interaction in a triangle—in other words, as reflexive interaction. The transition from deictic pointing to declarative pointing in particular seems to be very important here (see Figure 1), the basic difference between the two forms of pointing being reflexiveness. While "deictic pointing" means simply pointing to something the child wants to get hold of, "declarative pointing" occurs for its own sake. When the child utters: "Look ... (what a sweet puppy)!" the purpose of declarative pointing is nothing but to share the same experience. Declarative pointing, thus, appears to be a good indicator for a basic, more or less elaborated, triadic structure of the sign-process (see also Camaioni et al., 2004).

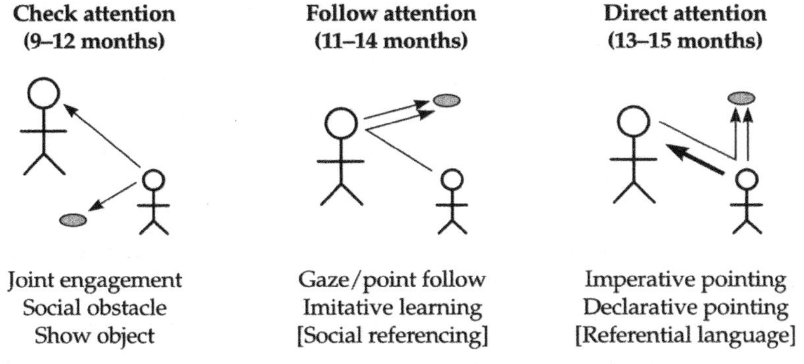

Figure 1. Three main types of shared attention (Tomasello, 1999)

Tomasello states that, starting from the ninth month, the infant is capable of perceiving – and also influencing – the intentions of other people. As an example, he cites the shared direction of attention. At the age of nine months, the infant can actively construct a triadic relationship between herself, an interesting object and a second person also relating to the interesting object. At this age, the infant notices, e.g., if the mother directs her attention towards an interesting object or not. The child can follow the mother's gaze and she tries to direct the mother's gaze herself or call the mother's attention. At the age of 12 to 14 months, the child can intentionally call the mother's attention by using communicative means which direct her attention towards the intended object. The mother understands the sign or interprets the child's behaviour like in Vygotskij's example, directing her attention towards the interesting object.

In contrast to Vygotskij's example of the development of pointing out a grasping movement, while sharing attention, the child and her communication is from the start directed towards the mother and her gaze. The child notices if the mother is directed towards her with communicative intention, and by using diverse means, actively makes this turn. This is also true if the child does not use any conventionalized signs, e.g., through pulling a sleeve or nudging.

Sociogenesis, as postulated by Vygotskij, necessitates, as it were, the active communicative participation of the child from the beginning. The child brings her definition of communication and communicative means into this situation – and this starts from the point of development where she is capable of addressing others as intentional agents having purposes that can be perceived and influenced. It follows from this view that the communicative means (the signs) of having influence entail a major reorganization of the psyche – as has been described above in the long Vygotskij quote. Furthermore, only the active role of the child in the communicative process is making clear why there is something like a feedback of a social means onto the self's psyche. Because, from the moment in which the child uses signs actively and successfully, the world is at her disposal in two ways: sensuously, emotionally and cognitively on the hand, and, on the other hand, symbolically in the form of signs and their meanings – at first only relating to certain aspects of everyday life and later expanding into all realms of reality and of the psyche. As this process progresses, the two ways with signs gradually merge into a single one.

The results of research exhibit that infants at the age of nine months are already capable of understanding the intentions of other agents. Just as this capacity is born out of interaction, it is at the same time stimulating further interactions of shared attention with other agents. Here, Vygotskij seems to underestimate children's capacities to understand: understanding does not develop behind children's backs, but it is in its early forms already well rehearsed, fine tuned and advanced. Vygotskij here shares Piaget's underestimation of children's capacities – though not with the same sharpness.

The incredible extent to which the capacities of infants have been underestimated in the past can be shown by using the example of the capability of putting oneself in the position of an observer, an example which was investigated by Piaget. Piaget used the so-called "three-mountain task" (see Figure 2) to ask children to put themselves in the position of an observer watching a landscape. They were asked, e.g., what a person would see from position 2 or position 3, while the child was observing from position 1. Piaget found that up to approximately the age of 8 years, the children were not capable of telling what an observer would see from a position different from the one they were currently taking. And this happened even if the children had been put previously in the other positions. Piaget concluded that the children were still entangled in an egocentric perspective. For Piaget, decentering was an enormously important concept: without decentering, children would not be in the position to grasp the essential features of the environment (for an extended discussion see Kesselring, 1981).

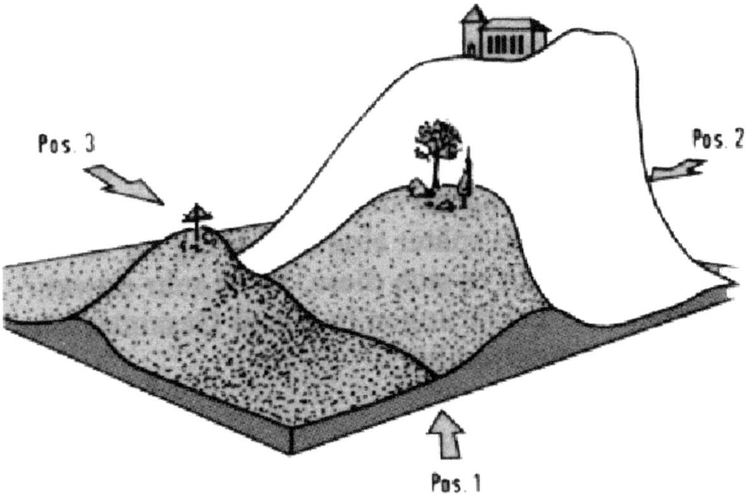

Figure 2. Piaget's Three-Mountain task (from Montada, 2002)

Piaget's experiment engendered ample discussion and criticism as early as 1975 (see Borke, 1975). In her book, Margaret Donaldson (1978) took the critique of Piaget to a new level. She demonstrated, like Borke, that children are very clearly capable to decenter when they are confronted with tasks they can consider meaningful and relevant. In these tasks, children easily and at an early age put themselves in the position of an observer.

The failure to decenter is, apart from that, responsible for many "errors" Piaget had found with the children in experiments where the number, the volume or the mass of objects did no change, but children perceived changes because the appearance had changed. "Non-conserving", as Piaget called it, was explained as a failure to decenter from certain external properties of the objects in question.

Now, it is extremely interesting that Hare et al. (2000, 2001) were able to demonstrate with chimpanzees that the animals knew very well what a conspecific could see from his position and what he could not see.[2] Chimpanzees, as it were, dispose of exactly the kind of capability that Piaget was identifying with children only at an age of around 6 and even later.

Now, what have the results of research on the ontogenesis of intersubjectivity and mutual understanding shown so far? They have given strong support to the perspective starting from the mind as exterior. We have indicated that the exterior perspective might be an essential complement of an "internalistic" view of thinking and learning. There are more than just the structures of the working brain and the local distributed process-structures. From the beginning, there is an "external" cognitive structure between humans. One could really say that what makes Peirce's dictum "Man is a sign" so plausible is not so much the fact that we can now better visualize how semiotic structures are created within the brain. It seems plausible not because men simply *are* signs, but because semiotic processes function according to the primordial process of interactive mutual human understanding, of shared intentionality. From the beginning, human infants learn to view conspecifics as intentional beings similar to the self (Tomasello, 1999). It is this understanding of intentionality, the insight that the behavior of others has to be perceived and interpreted in relation to a third instance, which basically provides the key to understanding the transition from a mere social determination of thinking, learning and behavior to a culturally specific determination.

Table 1. Some domains of social activity trasformed into cultural activity (Tomasello, 1999)

Domain	*Social*	*Cultural*
Communication	Signals	Symbols (intersubjective, perspectival)
Gaze of others	Gaze follow	Joint attention (intersubjectivity)
Social learning	Emulation, Ritualization	Cultural Learning (Reproducing intentional acts)
Cooperation	Coordination	Collaboration (role taking)
Teaching	Facilitation	Instruction (mental states of others)
Object manipulation	Tools	Artifacts (intentional affordances)

Table 1 from Tomasello (1999) can be read as a presentation of the differences between dyadic and triadic social processes. This table summarizes the form in which shared intentionality is realized in diverse cultural activities – in contrast to merely social activities. Dyadic processes appear as simple building blocks of the interaction with persons or objects, they are social without having any further triadic cultural implication. The table should not be read as an attempt to clearly define what is *social* and what is *cultural*. Rather, the table elicits a couple of interesting questions for semiotic research in mathematics education. Outstanding questions would be related to the difference between social, dyadic and cultural triadic processes in mathematics education. Some discussion has already gone on in the past, e.g., in relation to cultural learning and the apprenticeship model of learning and teaching (Lave, 1988; Lave & Wenger, 1991; Brown et al., 1989).

SHARED INTENTIONALITY AND THE TWO LINES OF DEVELOPMENT

One of the unique features of Vygotskij's approach to development is known to be the "development of higher psychological functions." Higher psychological functions can be understood as specifically human types of psychological functions compared (1) to the psychological functions of our animal forebears in phylogenesis, and (2) to the psychological functions of the youngest children in ontogenesis.

At the beginning of *The History of the Development of Higher Psychological Functions,* Vygotskij (1997) makes clear that there is another dimension, in juxtaposition, so to speak:

> The concept of "development of higher mental functions" and the subject of our research encompass two groups of phenomena that seem, at first glance, to be completely unrelated, but in fact represent two basic branches, two streams of the development of higher forms of behaviour inseparably connected, but never merging into one. There are, first, the processes of mastering external materials of cultural development and thinking: language, writing, arithmetic, drawing; second, the processes of development of special higher mental functions not delimited and not determined with any degree of precision and in traditional psychology termed voluntary attention, logical memory, formation of concepts, etc. Both of these taken together also form that which we conditionally call the process of development of higher forms of the child's behaviour (Vygotskij, 1997, p. 14).

In the following quotation, Vygotskij painstakingly explains that he sees a big difference between a biogenetic version[3] of human development, viewing ontogeny as a repetition of phylogeny, and his idea of an *analogy* of ontogenesis and phylogenesis:

> In the development of the child, two types of mental development are represented (not repeated) which we find in an isolated form in phylogenesis: biological and historical, or natural and cultural development of behavior. In ontogenesis both processes have their analogs (not parallels). This is a basic and central fact, a point of departure for our research: differentiating two lines of mental development of the child corresponding to the two lines of phylogenetic development of behavior. This idea, as far as we know, has never been expressed; nevertheless it seems to us to be completely obvious in the light of contemporary data from genetic psychology, and the circumstance

> that it has this far stubbornly escaped the attention of the researchers seems completely incomprehensible (Vygotskij, 1997, p. 19).

After reading Vygotskij's careful presentation of the concepts, we still need to know more about how exactly the second line of development, the cultural and historical line, evolves. We know that the two lines continue to coexist or that they merge into one single line, depending on what part of Vygotskij's work we refer to.

In a first attempt, one could say that the first line of development is the typical development of the basic biological equipment of humans, while the second line is development through and into the culture. The demarcation criterion between these two developmental lines is not exactly the point when the human is given birth to – at least as far as learning is concerned. If a feature of biologically grounded functions is that they do not have to rely on learning, learning does not begin with the child leaving the mother's womb. Accordingly, learning can be understood as a typical feature of the culture.

INTENTIONALITY AND SHARED INTENTIONALITY: TOWARDS A DEFINITION

Intentionality has recently come to be a research topic of considerable interest. The problems and ideas around the development of intentionality promise to give fuel to attempts at formulating Vygotskij's idea of the two lines of development in greater detail – at least as far as the "higher" psychological functions are concerned. It has been the decisive feature of a higher psychological function in Vygotskij's sense, in that it defines the difference between humans and non-humans. In the light of a cultural-historical perspective on "intentionality" (Tomasello, 1999) and "shared intentionality" (Tomasello & Carpenter, 2007), it becomes evident that, even though the great apes share many of the features of human intentional understanding, they fail to incorporate reflexive shared intentionallity into their practical life. A couple of highlights may illustrate what we are talking about here:

- chimpanzees can fully understand what another chimpanzee can see,[4] e.g. when food is presented in a place right in between the two animals, it is always the one with the higher rank taking the food. A chimpanzee with a lower rank would never take the food in the presence of a higher-ranking chimpanzee. When a blind is used in such a way that the animal with the lower rank can see the food and the one with the higher rank cannot see it, the lower ranking chimpanzee takes the food – knowing that the other cannot see (see Hare et al., 2001). This provides evidence for the extent to which chimpanzees are consciously acting;
- in contrast, chimpanzees cannot perform a relatively trivial gesture like pointing at a certain object. Referential pointing (we will come back to that later) is not an option for the great apes, and Tomasello gives an answer to the question: "Why don't apes point" (Tomasello, 2006);
- chimpanzees have tremendous problems working together cooperatively if the social ranking does not fit. Melis et al. (2006) have demonstrated that chimpanzees are quite successful as pairs in a task requiring cooperation and coordination, if they are tolerant and have no problems with the social rank of the cooperating partner. If they do have problems, cooperation is practically impossible.

Intentionality can be regarded as an individual or as a social construct depending on whether the context of activity is a more solitary one or a group situation. The type of *shared intentionality* we are dealing with here goes, however, beyond the mere sharing of goals in a social or a group situation. Shared intentionality, as we discuss it here, is defined by Tomasello as "two people experiencing the same thing at the same time *and knowing together that they are doing this*" (Tomasello & Carpenter, 2007, p. 121, emphasis source). In a new article on infant pointing, Tomasello et al. (2007) come back to our point of departure: infant pointing. Here, they distinguish between a lean and rich interpretation of protoverbal forms of communication (like pointing). The lean position sees the relation between child and adult as merely instrumental: the child wants the adult to do something. The rich interpretation, however, sees that the child is acting on the basis of understanding the adult as a social being trying to influence her intentional/mental state. The rich interpretation requires that we take into account the shared context and its power to define relevance: a *common ground* is absolutely necessary for that. It is equally necessary for an account of what it means to "know something together."

Shared intentionality, as it were, emerges on the basis of the belief that the other human being is "like me" (Meltzoff, 2007), that the other human being is a conspecific sharer in what I have in common with others[5].

Obviously, one is not born with fully-fledged shared intentionality. Shared intentionality has to develop in the early years – and it is a surprise how early certain functions are at the child's disposal – functions like empathy or helping being basic elements in the development of shared intentionality.

What we are going to do now is to rephrase the problem of the two lines of development in two different parts: one part on cognitive core functions and their relation to shared intentionality, and one part on language and the two lines of development.

CORE FUNCTIONS AND THE RELATION TO SHARED INTENTIONALITY

In this section, we would like to present some ideas about how one can imagine a relevant biological line of development. With "relevant" we mean that we are not talking about the necessity of eating and having water to drink and oxygen to breathe. We mean core psychological functions which we share with our animal forebears. A recent example of an attempt to synthezise research on these basic psychological functions is presented by Spelke and Kinzler (2007). While Spelke's first publications (see Spelke, 2000) focused on one single system, e.g., the number system, Spelke and Kinzler have identified four core knowledge systems: object representation, representation of agents and their actions, a core number system, and a system of representation of the geometry of the environment. We will briefly sketch the basic features of these four systems.

The core system which has received the widest attention is the system of object representation. In a sense, research on visual perception has spelled out most of the principles pertinent to this field. If one takes Gestalt psychology and Gibsonian psychology on principles like the cohesion of objects, the boundaries of objects, the constancy and illusion of object perception, most of what defines the field of object representation is covered. It is interesting to note that Spelke argues that the features of this system are basically the same whether one looks at a newborn infant or a newly hatched chick – and they remain virtually unchanged during the life of the adult human being, so that the core system of object representation remains constant over human development. This feature of object representation is in line with findings from very young children who seem to possess a rather vast amount of physical knowledge (like in the experiments of Karen Wynn, 1992, 1995) without having had a chance to handle and manipulate objects – because their motor development does not enabled them to manipulate objects freely. This is in sharp contrast to a Piagetian view on the ontogenesis of physical knowledge, which would presuppose the extensive manipulation of objects.

The core system representing *agents and their actions* seems at first glance to be overlapping the system of object representation – if one thinks only of causation of movements and the like. Very young infants, however, interpret social and non-social actions, e.g., infants do not interpret the movement of object as goal-directed and do not try to imitate those movements. All these representations of agents can also be found with newly hatched chicks, newborn monkeys and so on. Mirroring behavior with a corresponding activity of mirror neurons seems to be an important and ancient part of adaptive behaviour.[6]

The core *number system* is perhaps the most abstract of the core systems. We have to bear in mind, however, that we are talking here about numbers up to five, plus or minus two. For human infants, children and adults, and for adult non-human primates number representations have the same basic threefold property: First, number representations are imprecise and they become less precise the larger the cardinal value (scalar variability); second, number representations are abstract insofar they apply to diverse entities, from stones, to apples, to cars and horses, to sequences of sounds or sequences of light, and sequences of actions: third, number systems can be compared and combined by operations of addition and subtraction. The above properties of number systems seem to prevail in all human cultures – whether they have number words for three, four or five or not.

The last core system has to do with orientation in the geometry of the environment. When infants, children, or adults lose orientation, they start to reorient according to magnitudes of the geometry of the surface of the environment like distance, angle and sense directions. Landmarks play a much stronger role for adults in case of disorientation.

If we add a fifth, tentative, element, i.e. identifying and reasoning about potential social partners, we would get the situation depicted in Figure 3, which attempts to illustrate the fact that the further development of these core systems will not take place if it is not based on the common ground of shared intentionality.

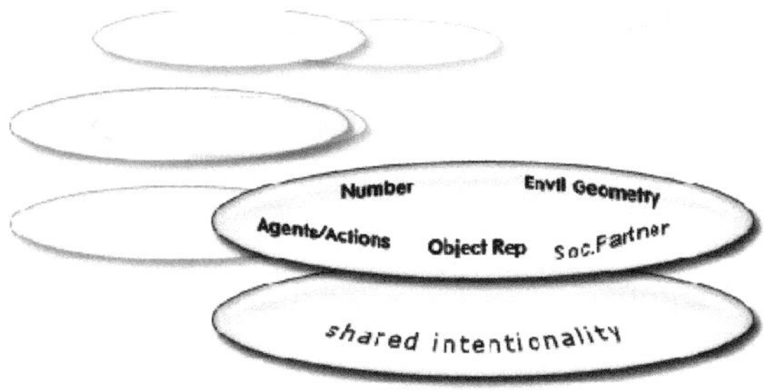

Figure 3. Core functions and shared intentionality

The question now arises about how shared intentionality succeeds in influencing the core functional systems. A tentative answer is given in Figure 4: the social semiosis of the cultural *praxis* of living and upbringing can be understood as mediating shared intentionality and the core functional system leading to a full development of those systems – turning it into the *cultural* line of development.

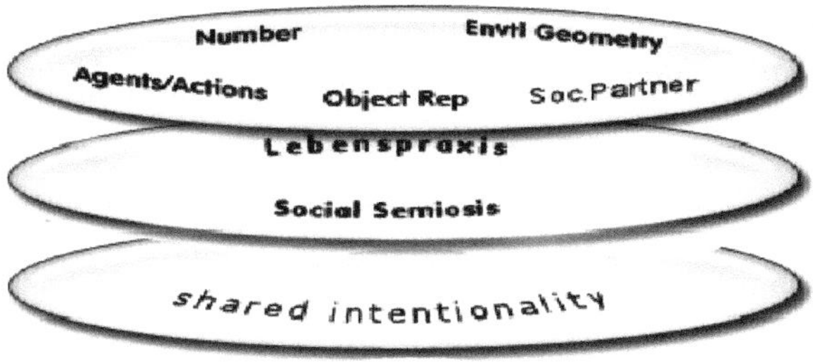

Figure 4. Core functions, shared intentionality and social praxis

CONCLUSION FOR MATHEMATICS EDUCATION

Thinking about the effective and potential role of a semiotic perspective in mathematics education[7] makes one realize that here, just as in other scientific disciplines, there is an enourmous diversity of approaches where one would rather see differences rather than similarities – but for the common denominator "semiotic." The diversity of perspectives also means an additional handicap for the disciplinary anchoring of a semiotic perspective in mathematics education. Reflecting on this issue might not lead to very reliable perspectives because the territory of disciplinary orientation is currently undergoing major modifications and restructuring. So we had better turn to possible outcomes within mathematics eduaction.

The creative and critical potential of a semiotic perspective for mathematics education will unfold in full, given what has been said above, in the realms of the general goals and objectives for mathematics education, as we gain a deeper understanding of what the standard formulations of goals, like mathematizing, exploring, argumenting, formalizing, actually mean in a framework based on shared intentionality.

From an "internal" semiotic perspective, much can be said about general goals and objectives for math education – and there are already a good many volumes devoted to these themes. Heuristic strategies and formalization in particular seem to lend themselves to being applied. The goal area of heuristic strategies seems like an appropriate field of application for Peirce's idea of abduction (see Hoffmann, 1998; Voigt, 2000), while formalization calls up his idea of diagrammatic thinking. Michael Hoffmann (2003) has specified what Peirce's idea could mean in the mathematics classroom: there must be *representations* of the problems at hand, second, there must be *experiments* with these "diagrams" and third, the results of these experiments must be *observed.* Hoffmann particularly emphasizes the fact that "semiotics praxis" has diverse formats, which arrange themselves in temporal sequences – very much like the well-known steps in problem solving. But his examples do not speak of the social nature of these formats – just like the social and cultural dimensions do not play a significant role in the discourse on learning objectives and standards.

Thus, an important conclusion for mathematics education would be to revise its goals and standards in light of the outstanding importance of the sociogenesis of learning, especially in relation to the shared intentionality of teacher and student in the math classroom. If it is correct to say with Vygotskij that the development of learning primarily is the movement of something between people to the inside of a person, it is of decisive importance to know how students and teachers interpret each other continuously and permanently, how they perceive and understand the other person's intentions and how they can act on the basis of such an understanding. To understand these interrelations as semiotic means specifically to address one's attention to the fact that much is recognized, acquired and learned that is not consciously recognized and acquired. Implicit learning is simply forming a large part of the learning process as a whole. Even if it is seen largely only as an irritation to official mathematics teaching and learning, it seems necessary to accept it as as an important element for the learning of mathematics. For very practical reasons alone, this seems a timely issue if one acknowledges how strong the emotional framing of mathematics learning has been and is today.

Apart from being relevant for the more general discussion on the learning objectives of mathematics education, I would like to emphasize three fundamental themes exhibiting the strong meaning of semiotically inspired mathematics education.

– A semiotically inspired approach to mathematics education is a clear alternative to any form of structural mathematics. Now, it seems that a structural approach to mathematics is far from being a really hot topic in math education, considering its demise together with set theory. I feel, however, that under the guise of an approach focusing on "competency", many old ideas of a structural origin have come back to life (see Wittmann, 2005) – without being explicitly perceived as such. If, for the definition of any mathematical competency, there is nothing else to do but choose the tasks that have to be solved in order to be "competent," the quintessential issue will be avoided, namely, precisely defining how this competency can be developed within teacher-student interaction – respectively, how it must necessarily be developed.

– A cultural-historical view of semiotics in mathematics education would put equal emphasis on the means and tools of math thinking as well as on the emphasis on discourse and intentional understanding in the classroom. It turns out, however, that the power of tools in mathematics only unfolds in the hands of those who know about this power and have access to it – the analogy of tools like a hammer and SimCalc[8] is, after all, misleading. The benign knowledge-producing use of means and tools does not only demand an introduction into their application and potential – it demands continuous discursive embedding and framing. In addition, it holds that all cultural tools beyond mere hammers need a cultural framing – means called artifacts with intentional affordances in Tomasello's list in Table 1 above.
– A semiotic approach, furthermore, would make it possible to deal with the issue of implicit learning in a fresh and promising way. In the last decade, it turned out that imitation as a specific form of implicit learning is not easy to understand because of its own quite complicated structure – and is, at the same time, quite fundamental for learning (see, e.g., Meltzoff & Prinz, 2002; Meltzoff, 2002). It should soon be possible to understand the relation between imitation and autonomy in such a way that imitation is no longer seen as a defective, so to speak, "unofficial" form of learning. The exclusive emphasis on learning mathematics autonomously and the identification of learning mathematics and conscious mental construction have not produced a very friendly balance over the years. The fate of constructivism in mathematics education could be instructive here: it is not enough just to underline the unlimited constructive spirit of learners and make the classroom the center of discourse. Imitation as a social process presupposes understanding others and their intentions – and this promotes mutual understanding as well as the understanding of one's own, individual, form of learning. On this basis, it can be understood why it is better, in the development of competence, to go from "performance" to "competence" (see Cazden, 1981) – and, it is only later that mathematical competence can be seen as a necessary ingredient for mathematical activity. Another example regarding the relevance of emotion is the well-documented difference between girls and boys in mathematics: while girls score best on tests throughout the primary grades and up to age 13 or 14, after that time, the picture changes completely. Now, the boys begin to get better test scores than the girls. It seems obvious that this has nothing at all to do with mathematical competence, but is obviously related to emotions, attitudes related to mathematics and to emerging conceptions of the self and the role of mathematics within that self-picture.

In our above presentation of develeopmental psychological research on infants and shared intentionality, we have seen that learning from subjects and persons, in a certain sense, precedes learning from objects.[9] We have started from the assumption that it makes a big difference whether learning starts from the objects or from the subjects. A further assumption has been that imitation and understanding the intentionality of others is a decisive force for introducing and acquiring cultural practices.

Also, within semiotics, we can distinguish between an "object-semiotic" and a "subject-semiotic." Object-semiotics would primarily put an emphasis on factual-empirical and content-related questions, while a subject-semiotics could focus on what has long been called a "critical" perspective. In that way, semiotics would gain a voice in the discourse on mathematics education, e.g., in the discussion on the meaning of the results of international assessment tests like TIMSS and PISA and the enourmous amounts of data generated by those studies. The structuring effect of a subject perspective on semiotics would scrutinize whether or not it is viable to make shared intentionality the basis of teaching and learning in mathematics.

With this in mind, a semiotic perspective of shared intentionality would also check the claims of various interpretative horizons and chains of meaning. The current emphasis on the "object" in the content of math learning often implies talking about the pressure one feels to simply accept that the subject of mathematics is necessarily tough and even strange. From the point of view of shared intentionality, this can only be a half-truth and is certainly not constitutive for mathematics education. It seems irrefutable that, especially with a tough subject, there should be a mutual and shared understanding of goals and motives.

NOTES

1 I use "sign-function" here instead of sign in order to underline that this is not about the specific form of a sign but about the processes of signification, of meaning, and so on.

2 Hare et al. (2000, 2001) placed a dominant and a non-dominant individual chimpanzee in a competitive situation in relation to food. In such a situation, the dominant individual always comes first. Some pieces of food were visible to both individuals, while others were only visible to the non-dominant chimpanzee. The fact that the non-dominant now preferred to take the food not visible to the dominant individual showed that he knew very well what the other, dominant, individual could see and not see.

3 The biogenetic approach claimed that each successive stage in the development of an individual represents one of the adult forms that appeared in its evolutionary history. Haeckel formulated the slogan: "Ontogeny recapitulates phylogeny." The recapitulation theory has also been used as a model for arranging curricula in teaching and learning according to development. Correspondingly, lower grades had to be preoccupied with earlier stages of cultural development, while higher grades had to deal with more recent accomplishments. In the words of Herbert Spencer (1885), "If there be an order in which the human race has mastered its various kinds of knowledge, there will arise in every child an aptitude to acquire these kinds of knowledge in the same order.... Education is a repetition of civilization in little."

4 As it were, chimpanzees would do well in the three-mountain task Piaget presented to children in order to find out whether they were still in an egocentric phase or whether they could de-center and put themselves in the position of an observer.

5 There is also a relation to the distinction between I-mode and We-mode shared intentionality introduced by Tuomela (2007).

6 We cannot go into the fascinating details and role of research on mirror neurons here. For an interesting account see Rizzolatti et al. (2002), Rizzolatti. & Sinigaglia (2007), and Ramachandran (2008).

7 See, e.g., the volumes edited by Anderson et al. (2003), Hoffmann (2003), Sáenz-Ludlow and Presmeg (2006).

8 For information on SimCalc see: http://simcalc.umassd.edu/ and Hegedus & Moreno-Armella (this volume)

9 Elsewhere (Seeger 2005), I have tried to describe this difference related to sign-processes as inter-semiotic and intra-semiotic processes, and I have discussed the concepts of "mediation" and "networking" as being relevant for the semiotics of teaching and learning. The difference between mediating and networking largely corresponds to the difference between "social" and "cultural", dyadic and triadic processes discussed above (see Table 1).

REFERENCES

Anderson, M., Sáenz-Ludlow, A., Zellweger, S., & Cifarelli, V. V. (Eds.). (2003). *Educational perspectives on mathematics as semiosis: From thinking to interpreting to knowing*. Toronto: Legas.

Bateson, G. (1972). *Steps to an ecology of mind*. San Francisco: Chandler Publishing Company.

Borke, H. (1975). Piaget's mountains revisited: Changes in the egocentric landscape. *Developmental Psychology, 11*, 240–243.

Burks, A. W. (1997). Logic, learning, and creativity in evolution. In N. Houser, D. D. Roberts, & J. Van Evra (Eds.), *Studies in the logic of Charles Sanders Peirce* (pp. 497–534). Bloomington, IN: Indiana University Press.

Brown, J. S., Collins, A., & Duguid, P. (1989). Situated cognition and the culture of learning. *Educational Researcher, 18*, 32–42.

Camaioni, L., Perucchini, P., Bellagamba, F., & Colonnesi, C. (2004). The role of declarative pointing in developing a theory of mind. *Infancy, 5*(3), 291–308.

Cazden, C. B. (1981). Performance before competence: Assistance to child discourse in the zone of proximal development. *Quarterly Newsletter of the Laboratory of Comparative Human Cognition, 8*(1), 5–8.

Dehaene, S. (1997). *The number sense: How the mind creates mathematics*. Oxford: Oxford University Press.

Donaldson, M. (1978). *Children's minds*. London: Croom Helm.

Hare, B., Call, J., Agnetta, B., & Tomasello, M. (2000). Chimpanzees know what conspecifics do and do not see. *Animal Behaviour, 59*, 771–786.

Hare, B., Call, J., & Tomasello, M. (2001). Do chimpanzees know what conspecifics know? *Animal Behaviour, 61*, 139–151.

Hegedus, S. J., & Moreno-Armella, L. (2008). Analyzing the impact of dynamic representations and classroom connectivity on participation, speech and learning (*this volume*).

Hoffmann, M. H. G. (1998). Erkenntnistheoretische Grundlagen des Lernens: Lernen als Verallgemeinerung. In M. Neubrand (Ed.), *Beiträge zum Mathematikunterricht* (pp. 311–314). Hildesheim: Franzbecker. (enlarged version here: http://www.uni-bielefeld.de/idm/personen/mhoffman/papers/GDM.htm)

Hoffmann, M. H. G. (Ed.). (2003). *Mathematik verstehen – Semiotische Perspektiven*. Hildesheim: Franzbecker.

Hoffmann, M. H. G. (2007). Learning from people, things, and signs. *Studies in Philosophy and Education, 26*(3), 185–204.

Kesselring, Th. (1981). *Entwicklung und Widerspruch – Ein Vergleich zwischen Piagets genetischer Erkenntnistheorie und Hegels Dialektik*. Frankfurt/Main: Suhrkamp.

Lave, J. (1988). *Cognition in practice. Mind, mathematics and culture in everyday life*. Cambridge: Cambridge University Press.

Lave, J., & Wenger, E. (1991). *Situated learning – Legitimate peripheral participation*. Cambridge: Cambridge University Press.

Melis, A., Hare, B., & Tomasello, M. (2006). Engineering cooperation in chimpanzees: Tolerance constraints on cooperation. *Animal Behaviour, 72*(2), 275–286.

Meltzoff, A. N. (2002). Elements of a developmental theory of imitation. In: A. N. Meltzoff & W. Prinz (Eds.), *The imitative mind* (pp. 19–41). Cambridge: Cambridge University Press.
Meltzoff, A. N. (2007). 'Like me': A foundation for social cognition. *Developmental Science, 10*(1), 126–134.
Meltzoff, A. N., & Prinz, W. (Eds.). (2002). *The imitative mind.* Cambridge: Cambridge University Press.
Montada, L. (2002). Die geistige Entwicklung aus der Sicht Jean Piagets. In R. Oerter & L. Montada (Eds.), *Entwicklungspsychologie* (5th ed., pp. 418–442). Weinheim: Beltz.
Otte, M. (1994). *Das Formale, das Soziale und das Subjektive.* Frankfurt/Main: Suhrkamp.
Ramachandran, V. S. (2008). *Mirrors in the mind.* London: Baker & Taylor.
Rizzolatti, G., Fadiga, L., Fogassi, L., & Gallese, V. (2002). From mirror neurons to imitation: Facts and speculations. In A. N. Metzoff & W. Prinz (Eds.), *The imitative mind* (pp. 247–266). Cambridge: Cambridge University Press.
Rizzolatti, G., & Sinigaglia, C. (2007). Mirrors in the brain: How our minds share actions, emotions, and experience. Oxford: Oxford University Press.
Sáenz-Ludlow, A., & Presmeg, N. (Eds.). (2006). Semiotic perspectives in mathematics education – A PME Special Issue. *Educational Studies in Mathematics, 61*, Whole No. 1–2.
Seeger, F. (2005). Vygotskij und Peirce – Anmerkungen zur Philosophie und Psychologie der Semiose. In M.-C. Bertau, A. Werani, & G. Kegel (Hrsg.), *Psycholinguistische Studien 2.* Aachen. Shaker-Verlag.
Seeger, F. (2006). Ein semiotischer Blick auf die Psychologie des Mathematiklernens. *Journal für Mathematik-Didaktik, 27*(3/4), 265–284.
Spelke, E. (2000). Core knowledge. *American Psychologist, 55*, 1233–1243.
Spelke, E. S., & Kinzler, K. D. (2007). Core knowledge. *Developmental Science, 10*(1), 89–96.
Spencer, H. (1885). *Education: Intellectual, moral, and physical.* New York: Appleton.
Tomasello, M. (1999). *The cultural origins of human cognition.* Cambridge, MA: Harvard University Press.
Tomasello, M. (2006). Why don't apes point? In N. J. Enfield & S. C. Levinson (Eds.), *Roots of human sociality: Culture, cognition and interaction* (pp. 506–524). Oxford: Berg.
Tomasello, M., & Carpenter, M. (2007). Shared intentionality. *Developmental Science, 10*(1), 121–125.
Tomasello, M., Carpenter, M., & Liszkowski, U. (2007). A new look at infant pointing. *Child Development, 78*(3), 705–722.
Trevarthen, C. (1979). Communication and cooperation in early infancy. A description of primary intersubjectivity. In M. Bullowa (Ed.), *Before speech: The beginning of human communication* (pp. 321–347). London: Cambridge University Press.
Trevarthen, C., & Hubley, P. (1978). Secondary intersubjectivity: Confidence, confiding and acts of meaning in the first year. In A. Lock (Ed.), *Action, gesture and symbol* (pp. 183–229). London: Academic Press.
Tuomela, R. (2007). *The philosophy of sociality: The shared point of view.* New York: Oxford University Press.
Voigt, J. (2000). Abduktion. *Beiträge zum Mathematikunterricht – Vorträge auf der 34. Tagung für Didaktik der Mathematik vom 28.2.–3.3. 2000 in Potsdam* (S. 694–697). Hildesheim: Franzbecker.
Vygotskij, L. S. (1997). The history of the development of higher mental functions. In R. W. Rieber (Ed.), *Collected works* (Vol. 4). New York: Plenum Press.
Wittmann, E. Ch. (2005). Eine Leitlinie zur Unterrichtsentwicklung vom Fach aus: (Elementar)-Mathematik als Wissenschaft von Mustern. *Der Mathematikunterricht, 50*(2/3), 5–22.
Wynn, K. (1992). Addition and subtraction by human infants. *Nature, 358*, 749–750.
Wynn, K. (1995). Origins of numerical knowledge. *Mathematical Cognition, 1*, 35–60.

Falk Seeger
Institut für Didaktik der Mathematik
University of Bielefeld
Germany

CRISTINA SABENA

ON THE SEMIOTICS OF GESTURES

Learning and teaching mathematics requires the activation of *a variety of resources*, which can be grouped around what I consider to be two chief sources of mathematics knowledge: the *body* and its activity with artifacts, and the activity with *signs*[1]. The cognitive significance of the body has become one of the major topics in current psychological and cognitive studies. In particular, the "Embodied Mind" paradigm (Lakoff & Núñez, 2000) has situated the origins of all human knowledge—mathematics included—in bodily experiences and perceptions. On the other hand, knowledge formation is embedded in cultural and social contexts, and the use of signs becomes crucial in cognition. Such a polarity results in being of particular interest in the context of the teaching and learning of *mathematics*, a discipline traditionally considered "abstract" on the one hand, and extensively based on perceivable signs on the other.

Many studies in mathematics education research have framed teaching and learning activity within semiotic perspectives, focusing on written semiotic systems such as algebraic symbolism. Only recently has attention been dedicated to considering bodily means of expression as *semiotic resources* in the learning process, and to looking at their relationship to written mathematical symbolism (see for instance Radford, Bardini & Sabena, 2006 on the role of rhythm in the context of algebraic generalization).

In this chapter, I focus on the contribution of gestures to the mathematics teaching-learning processes in the classroom context. The role of gestures will be outlined by situating them as semiotic resources and looking at how they are closely intertwined with more traditionally studied semiotic systems (such as language and written signs).

The chapter is divided into three parts: theoretical reference points, an analysis of examples from the classroom context, and a final discussion based on the analysis provided.

THE STUDY OF GESTURES

When people talk, they gesture. Gesturing is a widespread phenomenon: it is found in all cultures that have been observed, occurring across a wide range of tasks and ages, even when the listener is not physically present, or cannot be seen. Gestures are part of what is called "nonverbal communication", which includes a wide-ranging array of behaviors such as the distance between people in conversation, eye contact, voice prosody, body posture, and so on. According to the traditional view of communication, all these acts, although important in framing a conversation,

L. Radford, G. Schubring, and F. Seeger (eds.), Semiotics in Mathematics Education: Epistemology, History, Classroom, and Culture, 19–38.

have little to do with the conversation itself, which is instantiated in uttered speech ("verbal communication"). In this perspective, gestures are interpreted as nothing more than hand waving, embellishments, means for letting out excess energy, bids for the listener's attention, or regulators in the communicative exchange. Such a dichotomist view has been widely challenged in the last three decades, starting from the work of Kendon (1980), who was among the first to propose that at least one form of nonverbal behavior—gesturing—cannot be separated from the conversation itself[2].

Nowadays, research in a number of disciplines (such as psychology and all its branches, cognitive linguistics, and anthropology) is increasingly showing the fundamental importance of gestures not only in communication, but also in cognition. Curiously, it has been the interest in cognition prompted by Chomsky's view of linguistics as a kind of purely "mental science" that has led to the vigorous investigation of gestures by those interested in language:

> If language is a cognitive activity, and if, as is clear, gestural expression is intimately involved in acts of spoken linguistic expression, then it seems reasonable to look closely at gesture for the light it may throw on this cognitive activity (Kendon, 2000, p. 49).

What is "Gesture"?

"Gesture[3]" is used in various ways in ordinary language and also in literature. It sometimes refers to any movement that people perform while talking, including movements of the hands and arms, adjustments to posture, self-adaptors (e.g. touching one's hair), functional actions (such as picking up something), nervous tics, etc. It is clear that not all such body movements play a part in learning processes; therefore, it is reasonable to look for definitions and operative distinctions.

In Kendon's research, gesture is assigned a general meaning, which he organizes in a continuum ranging from spontaneous gestures (which he calls "gesticulation") at one edge, to sign languages at the other edge (Kendon, 1988):

gesticulation → pantomimes → emblems → sign languages.

At the center of the continuum, we find *pantomimes* that are namely significant gestures without speech (like twirling a finger around in a circle to describe a vortex without using words), and *emblems*, which are endowed with a culturally specific standard form and are used in an intentional way to communicate precise meanings (e.g., the OK gesture). McNeill (2000) analyses this distinction with respect to the relationship to linguistic properties and to conventions, and the character of semiosis. Gesticulation and pantomime[4] are described as being unconventional and not seeming to obey any system constraints, and therefore lacking all linguistic properties. Sign language, on the contrary, is itself a fully developed linguistic system and is purely conventional. Emblems are in between, showing evident conventionality but not forming a linguistic system[5]. McNeill concludes that "nothing about the visual-manual modality *per se* is incompatible with the presence of linguistic properties" (McNeill, 2000, p. 4).

Analyzing conversational settings, McNeill (1992) defines gestures as "the movements of the hands and arms that we see when people talk" (McNeill, 1992, p. 1). In this approach, widely adopted in successive research studies in psychology, gestures end up being viewed as distinct but inherently linked with speech utterances[6]. Indeed, in the context of mathematical activities in the classroom, where the focus is not communication *per se* but knowledge formation, phenomena accounted for as gestures may occur in a wider range of actions beyond mere conversation, such as handling tools or "simply" thinking in silence. A working definition suitable to mathematics learning settings can therefore include in the term "gesture" all *those movements of hands and arms that subjects (students and teachers) perform during their mathematical activities and which are not a significative part of any other action*[7] *(i.e. writing, using a tool, ...).*

This definition has to be intended in an inclusive way, and is certainly not clear-cut. In any case, whatever definition is taken, it so happens that the notion of gesture remains quite fuzzy, in the sense that there is no hard-and-fast line between what is "gesture" and what is not. Hence, how can one identify gestures and distinguish them from other body movements? Gestures are usually characterized as follows (Kendon, 1996; McNeill, 1992): they begin from a position of rest, move away from this position, and then return to rest. The central part of the movement, generally recognized as expressing the conveyed meaning, is called *stroke* or *peak*; it is preceded by a *preparation* phase (hand/arm moving from its resting place, and usually to the front away from the speaker), and symmetrically succeeded by a *retraction* phase (hand/arm back to the quiescence). Speakers of European languages usually perform gestures in a limited space in the frontal plane of the body, called *gesture space*, which goes roughly from the waist to the eyes, and includes the space between the shoulders. However, differences have been detected according to age (with children the space is larger) and different cultural settings[8].

Types of Gestures

Even narrowing the meaning of "gesture" *à la* McNeill, the term comes to include a variety of behaviors that do not form a single category. In McNeill's study, the following basic gesture types are identified[9]:

- *iconic gestures*: bear a relation of resemblance to the semantic content of discourse (object or event);
- *metaphoric gestures*: similar to iconic gestures, but with the pictorial content presenting an abstract idea that has no physical form;
- *deictic gestures*: indicate objects, events, or locations in the concrete world;
- *beats*: the hands move along with the rhythmical pulsation of speech, lending a temporal or emphatic structure to communication;
- *cohesives*: tie together thematically related but temporally separated parts of the discourse.

Iconic, metaphoric and deictic gestures have a main imagistic and pictorial component, whereas beats and cohesives are kinds of discourse-structuring gestures.

Each type uses the gesture space in a different way—i.e. the iconics are mostly in the center part, the deictics more in peripheral regions—and each category has a different story of development in children. However, a same gesture can show features belonging to more than one category. This follows from the fact that the taxonomy is not based on the mere shape or kinesthetic characteristics of the gesture, but on its relationships with external elements, such as the accompanying speech and the task at hand. Gesture interpretation is therefore context-dependent and needs information about the action schemes provided by our experience with objects in the world (Cassel, 1998). *Iconic gestures*, also referred to as *representational* (Kendon, 1988; Kita, 2000), through their imagistic features provide information not only on some object that is being represented, but also on the particular point of view the speaker is taking towards it. *Metaphoric gestures* provide an image of something invisible, an image of an abstraction. For instance, the idea of a topic may be presented as a bounded container supported by the hands: a certain portion of space comes to be used to present something else that is inherently nonspatial. *Deictic gestures* are pointing movements and are prototipically performed with the index finger. They have the function of focusing joint attention on a shared reference and are the most context-dependent, since they derive their interpretation essentially from the context in which the listener-speaker interaction takes place[10]. In pointing, the movement may not be simply linear, but can follow various patterns, which can have semantic implications and iconic features. Apparently simple, pointing is on the contrary a complex, typically human act. A special case of pointing is called *abstract pointing*, when there is no actual physical pointed object, rather the pointed empty space houses an introduced reference, for instance an abstract concept about where the speaker has been at a certain moment he is talking about. McNeill recognizes a certain meaning to such pointed space: "the speaker appears to be pointing at empty space, but in fact the space is not empty; it is full of conceptual significance. Such abstract deixis implies a metaphoric use of space in which concepts are given spatial forms" (McNeill, 1992, p. 173).

As it emerges in the discussion, the taxonomy is not to be intended in a rigid way, rather it is meant to suggest some of the various functional dimensions gestures can contribute to communication. In the next paragraph, we shall see that the strict link recognized between language and gesture has implications that go beyond communication and involve cognition itself.

Gestures, Speech and Thought

In psychological studies, gestures and speech are generally considered as complementary sides of the same coin: *linguistic systems* are described as conveying meaning in a *segmented, analytic, linear, and hierarchically structured* way; *gestures*, on the contrary, are characterized as *global, synthetic, multidimensional and never hierarchical* (McNeill, 1992, p. 19). In his semiology, Saussure (1916/1967) attributes the linear-segmented character of spoken language to its unidimensionality, contrasted to the multidimensionality of meanings. This characterization is taken up by Goldin-Meadow, who contrasts language and gesture features:

> Language can only vary along the single dimension of time. [...] This restriction forces language to break meaning complexes into segments and to reconstruct multidimensional meanings by combining the segments in time. But gesture is not similarly restricted. Gestures are free to vary on dimensions of space, time, form, trajectory, and so on, and can present meaning complexes without undergoing segmentation or linearization (Goldin-Meadow, 2003, pp. 24–25).

Language is made of elementary constituents and its rules of production involve both atomic (single) and molecular (compound) signs. On the contrary, gestures are described as endowed with holistic features that cannot be split into atomic components. The modes of production and transformation are often idiosyncratic[11] to the subject who produces them (McNeill, 1992, p. 19 ff.). Furthermore, one gesture can combine many meanings (synthetic character) and two gestures produced together do not combine to form a larger, more complex, hierarchically higher gesture[12], as language, in contrast, does: gesture is free to vary on dimensions of space, time, form, trajectory and can present meaning complexes without undergoing segmentation or linearization, nor being constrained to precise standards of forms (McNeill, 1992).

Being very different forms of expression, each with its own specificity, gesture and speech appear, however, to be intimately related: they are semantically and pragmatically co-expressive, they are essentially synchronous[13] in time and meaning, and they develop together in children. From cognitivist perspectives, the benefit gesture brings is often interpreted in terms of *lightening the cognitive burden*, in an analogue way that writing a problem down can reduce the effort needed to solve the problem. Research findings seem, in fact, to show that gesturing confers a significant cognitive benefit on verbal and visuospatial memory, by reducing the load of working memory and thus allowing it to organize the cognitive resources to solve a given task better (Goldin-Meadow, Nusbaum, Kelly & Wagner, 2001). Gesture can therefore constitute an embodied form for off-loading cognitive work into the environment (Wilson, 2002). Concerning the gesture-speech relationship, McNeill puts it in a clear way: "[Gestures] are tightly intertwined with spoken language in time, meaning, and function; so closely linked are they that we should regard the gesture and the spoken utterance as different sides of a single underlying mental process" (McNeill, 1992, p. 1).

This quotation, coming from the very beginning of McNeill's cornerstone book, highlights an important dimension in gesture studies in psychology, namely their role in cognitive processes. Though differing in the specific positions they assume on the issue, the most recent studies in cognitive psychology and linguistics agree in recognizing the importance of the function of gestures in cognition, up to upholding that gestures and language should be viewed within a unified conceptual framework (Goldin-Meadow, 2000; Kita, 2000; McNeill, 1992)[14].

In McNeill's discussion, gesture is related to the important dialectics of individual and social planes:

> The gesture supplies the idiosyncratic, the personal, and the context-specific aspects of thought, to be combined with the socially regulated aspects that come from the conventions of language. Such a combination implies a dialectic of gesture and language in which the gesture provides the momentary context of speaking and language carries this individuality to the social plane where it is categorized, segmented, reformatted, and dressed up for the world. Putting these themes together, we can conceive of thought as fundamentally an inner discourse in which gestures play an intrinsic part (McNeill, 1992, p. 2).

Going in the direction of extending Vygotsky's claim that "thought is not merely expressed in words; it comes into existence with them" (Vygotsky, 1934/1986, p. 218), McNeill (1992) interprets the essential speech-gesture cognitive unit as supporting an active *constitutive* role of gestures on thought: "gestures do not just reflect thought but have an impact on thought. Gestures, together with language, *help constitute thought*" (p. 245, emphasis in the original).

In other words, according to a Vygotskian perspective, we can frame gesture and language in a unifying semiotic mediating[15] function: *gestures together with language provide the materiality in which thought can arise and unfold.*

TAKING PEIRCE'S SEMIOTIC APPROACH

To analyze the role of gestures as semiotic resources in the classroom, they have to be inserted into a semiotic approach allowing a very general definition of "signs". This is well accomplished in Peirce's pragmatic theory.

Peirce defines a *sign* as a triad composed by the sign or *representamen* (that which represents), the *object* (that which is represented), and the *interpretant*:

> A sign, or representamen, is something which stands to somebody for something in some respect or capacity. It addresses somebody, that is, creates in the mind of that person an equivalent sign or perhaps a more developed sign. That sign which it creates I call the interpretant of the first sign. The sign stands for something, its object. It stands for that object, not in all respects, but in reference to a sort of idea, which I have sometimes called the ground of the representamen (C.P., 2.228).

The object is the referent, that which the sign represents. Before being interpreted, the representamen is pure potentiality. A sign can "stand to somebody for something in some respect or capacity" only because the relation "to stand for" is mediated by an interpretant. The interpretant is not the interpreter of the sign; rather it is what guarantees the validity of the sign even in the absence of the interpreter. The interpretant is another representation referred to the same object. It can be an equivalent significant—but Peirce would speak of "sign"—in another semiotic system (i.e. a drawing to explain a word meaning), an index to the single object, implying an element of universal quantification ("all the objects like this"), another definition in the same semiotic system (i.e. salt for sodium chloride), an emotive association (i.e. dog for fidelity), the use of synonyms. The list does not have the effect of reducing the interpretant to the intensional properties of a content and hence to the series of denotations and connotations of an expression: for Peirce, an interpretant can even be a complex discourse, inferentially developing all the logical possibilities implicated in the sign; an interpretant can even be a syllogism deduced from a regular premise. Furthermore, it can be a behavioral response, a disposition and many other things as well. What is striking and innovative in Peirce's introduction of the interpretant in the very definition of sign is that the sign is thereby endowed with an intrinsically dynamic character. In fact, the interpretant is a sign that translates and explains the previous one, and this other sign in its turn requires another sign as interpretant, and so on in a chain of infinite interpretation, establishing a process of *dynamic unlimited semiosis*. Peirce's idea of semiosis involves the relation Sign-Object-Interpretant as irreducibly triadic. He often insists that the triadic relation is genuine, in the sense that it cannot be reduced to any of the couples formed by its terms.

The notion of sign introduced by Peirce is very general: every phenomenon, simple or complex as it may be, can be interpreted as a sign entering into a semiotic process[16]. The essential feature for something to be a sign is to be able to represent something else to somebody, where "to represent" means "to stand for, that is, to be in such a relation to another that for certain purposes it is treated by some mind as if it were that other" (C.P., 2.273). Adopting a Peircean semiotic perspective, everything entering into a semiotic process is a sign, and therefore a larger variety of phenomena, included body movements and gestures, can be considered as semiotic resources.

Besides the general characterization of signs, a relevant aspect of Peirce's theory with respect to the present research is its pragmatic character. In fact, it considers the *context* of sign production and reception and defines the sign through its action on the interpretant. This is a main difference from Saussure's program for linguistics, where there is no provision for the study of the actual contexts in which speakers communicate with one another.

A third theoretical tool gained from Peirce's theory is its insightful classification of signs. Peirce's theory identifies three[17] basic types of signs, according to the manner in which they are capable of signification. A sign refers to its object in an *iconic* way if it looks like it, resembles it. The equal sign "=" is an example of an iconic sign: it is an icon of an idea. Introducing it, Recorde commented that "nothing could be more equal". A sign refers to its object in an *indexical* way if it is materially affected by it. Indices involve an element of contextuality, as smoke with respect to fire. Finally, a sign is a *symbol* if it refers to its object by means of a rule, such as language words or algebraic formulas. Symbolic rules can be formulated *a priori*, due to conventions, or *a posteriori*, due to cultural habits. The classification is to be intended as stemming from an observative process, rather than an abstract definition of formal categories, and the categories do not have sharp and fixed boundaries[18]. On the contrary, every sign can be characterized by a prevailing element, but there do not exist signs relegating a category to its pure state.

A BROAD SEMIOTIC APPROACH

In mathematics education, the analysis of the contribution of gesture has only recently appeared in the literature and this according to different approaches: see the studies of Radford and his collaborators on the contribution of gesture in the objectification process (Radford, 2003; Radford, Bardini & Sabena, 2007), Roth's research on speech-gesture (Roth, 2001, 2002), as well as Edwards' and Arzarello & Robutti's studies on the gesture-inscription relationships (Edwards, 2003; Arzarello & Robutti, 2004). In a short time, the topic has nevertheless engendered a certain interest and become the focus of study groups at international conferences, such as the PME Research Forum (Arzarello & Edwards, 2005).

My contribution on this issue stems from *considering gestures as semiotic resources* that students and teacher use in the classroom. Accordingly, the resulting semiotic analysis of teaching-learning phenomena has to be broad enough to include standard signs such as language and mathematical symbols, in addition to embodied resources such as gestures. To tailor such a "broad semiotic approach" to mathematics activity, I will use the notion of *semiotic bundle* pointed out recently by Arzarello (2006).

Arzarello builds his construct by widening the notion of semiotic system (see Ernest, 2006)[19] to the notion of *semiotic set*. A semiotic set[20] is characterized by the three following features:

- a set of signs, which might be produced through different actions such as uttering, speaking, writing, drawing, gesticulating and handling an artifact;
- a set of modes for producing and transforming these signs; such modes may possibly be rules or algorithms, but also more flexible actions or modes used by the subject[21];
- a set of relationships between the signs and their underlying meaning structure.

Gestures are the most striking example of signs that constitute a semiotic set (in Arzarello's sense) but not a semiotic system (in Ernest's sense). Gestures in fact can be considered as signs, in the Peircean sense, since they can "stand to somebody for something in some respect or capacity"; furthermore, they are endowed with underlying meaning structures, which are highly context-bound; finally, they are produced by bodily enactments, in a balance between idiosyncratic features and conventional character that does not follow precise rules nor algorithms.

Semiotic sets constitute the ingredients of the semiotic bundle. In fact, a *semiotic bundle* is defined as composed by:

- a collection of semiotic sets;
- a set of relationships between the sets of the bundle.

A semiotic bundle is to be intended not as a juxtaposition of semiotic sets, rather as a unitary system, whose components are distinguished for the sake of analysis. The most striking example is represented by the speech-gesture enactment: a tight unity between them has been generally advocated in recent psychological studies (see the discussion above). From a "broad semiotic" point of view, they can be seen as a bundle formed by two semiotic sets, one of which (speech) is also a semiotic system.

In a general sense, the semiotic bundle can be analyzed by considering the enactment of different semiotic sets at the same time, or the evolution of their interplay unfolding in time. In a terminology evocative of Saussure, they can be called *synchronic* and *diachronic* analyses[22] (Arzarello, 2006). But as the semiotic bundle comprises an extended field of observation with respect to traditional ones, looking at the relationships and dynamics in the bundle involves broadening the view to include phenomena other than those studied in the literature, such as, for instance, the transformational processes of semiotic systems (e.g., the notions of treatments and conversions pointed out by Duval[23]). Such a hypothesis has provided the basis for a study carried out within the research group of Torino, and reported in an extended way by Sabena (2007).

EXAMPLES FROM THE CLASSROOM CONTEXT

Using the frame of the semiotic bundle, I analyze two episodes drawn from the video-recording of classroom activities. As in psychological studies, videos provide the corpus of data on which qualitative analysis is carried out. The methodology includes the selection of significative episodes, which undergo a detailed analysis, parceled at levels of seconds of time (micro-analysis).

The two episodes come from a classroom discussion during a teaching experiment that followed the students and the teacher over a period of three years. The teaching experiment proposed an early introduction to the basic concepts and ideas of calculus, from the first years of high school onward (grades 9-11). Students were involved in a number of group activities with the use of different kinds of software, and in classroom discussions. The teacher was a teacher-researcher who had been collaborating with the research group for many years. In the present case, during a grade 10 lesson, he has asked a good student, Andrea, to explain to the class how they had solved the group work activity, which consisted in finding out the graph and equation of the derivative of a given function. The activity was intended to have the students work in the symbolic register, and specifically on the application of the formula for the differential ratio, which had been presented by the teacher in one of the preceding lessons.

First Episode

Andrea is at the blackboard, where the teacher has traced the graph of the given function. He has written the formula of the differential ratio and is about to explain its link to the graph.

[3:54 – 4:29][24]

29. **Teacher:** So, you've written, f(x+h) minus f(x) divided by h, what does it represent?

30. **Andrea:** Practically it represents the slope at this point here (*he is pointing to the graph*)

31. **T:** Uh

32. **A:** Because <u>h is a wonderful invention</u> [04:07] that, practically, is a… it is a… [04:09] it is an infinitesimal; not, not a true number, since it is something that is not zero, but at the

end, we can exclude it (*pointing gesture toward the formula*), treating it almost as if it were zero, but it is not zero, otherwise, otherwise we would nullify

[4:07 *pointing to h*]

[4:09 *pointing to abscissas in the graph*]

33. **Luigi** (overlapping): It is so small [4:26-1], that it tends [4:26-2] to zero

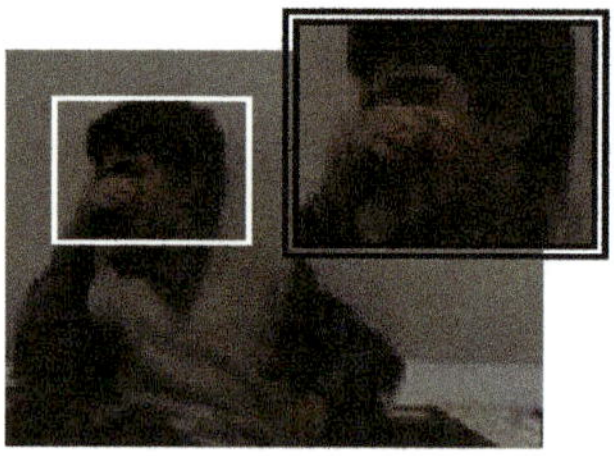

[4:26-1 *Δ-gesture*]

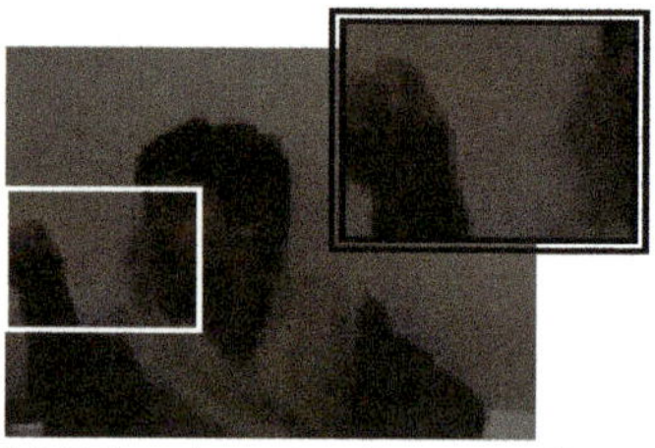

[4:26-2 *hand moving horizontally from left rightwards, with index and thumb touching each other*]

34. **T:** Sorry?
35. **L**: It is so small [4:27-1&2] that it tends [4:28] to zero

[4:27-1 *Δ-gesture*]

[4:27-2 *fingers get closer*]

[4:28 *hand moving horizontally, from left rightwards*]

The teacher (#29) is asking about the formula of the incremental ratio that Andrea has written on the blackboard near the graph. Specifically, he is asking for an interpretation of the symbolic algebraic formula in terms of the graphic-geometric and numeric frames ("what does it represent")[25].

Addressing this question, Andrea at first deictically refers to the graph ("here" coordinated with a pointing gesture, #30), incorrectly relating the incremental ratio to the slope of the function (this passage, due to limited spaced, has been omitted). Then, to justify the link between the two registers, he focuses on the nature of "h", with respect to the numerical register (#32)[26].

Whereas Andrea uses mainly speech, Luigi (#33) uses speech-gesture synchronies. He performs them twice (again in #35), keeping the same multiple semiotic character: the same language and gesture structure, as evident in the transcription and in the pictures. Let us enter into the details of the semiotic bundle. The utterances leave the subject h unexpressed and are divided by a pause into two parts, separated in the transcription by a comma. The two parts are matched with two different gestures and express two different (linked) meanings. The first part refers to h as both the width of an interval (numerical register) and the length of a segment (graphic-geometric register). The first meaning (related to the numerical evaluation) is mainly expressed through language ("It is so small"), whereas the second one (related to the graphic-geometric setting) is depicted through a gesture that is recurrent in the video recordings of the classroom activities, and which I have dubbed the *Δ-gesture* ([4:26-1], [4:27-1&2]). Adopting McNeill's classification, this gesture can be considered an *iconic* gesture with respect to a segment-line in the Cartesian plane, and a *metaphoric* gesture, with reference to a numerical interval. These two referents come to be *blended* (or *condensed*) in the same gesture. Furthermore, the *dynamic* character of the gesture performed during the second utterance[27] ([4:27-1&2], #35) makes apparent one specific aspect of the segment-interval at stake in the current discussion: the fact that h is considered as not only small, but indeed as getting smaller and smaller. This idea is at the base of the limit process to which the derivative function has been linked in the previous lessons. It is then more explicitly expressed in the second part of the sentence, which shifts to refer to h as a symbol in the formula of the differential ratio (with its underlying numerical instantiation). Here, words and gestures are referring to the operation of the passage to limit, usually expressed in the inscription "$h \to 0$" (which is not yet written on the blackboard). Luigi's hand, which has just embodied the numeric-graphic register in a condensed way, now shifts to depict the inscription of an arrow referring to the symbolic register[28].

However, I want to stress that the two parts of each sentence (#33 and #35), just analyzed separately, do indeed form a global contribution, something that emerges by analyzing both speech and gesture. In fact, linked by the construction "so…that", the two propositions form a whole unique sentence. And, in both cases (#33, 35), the latter gesture [4:26-1, 4:28] is performed in continuity with the former one [4:26-2]: the hand keeps the configuration, with index and thumb close to each other, and starts the horizontal movement exactly where the previous Δ-gesture has completed its dynamics.

The given analysis shows how, in the semiotic bundle, each resource takes on a peculiar role, allowing pointing out and linking different aspects involved in the mathematical tasks, in particular the graphical, symbolic and (underlying) numerical ones. This happens through modalities that escape the traditional descriptions, given in terms of transformation of semiotic systems or registers: no transformation of this kind is performed and no semiotic system is in fact acted upon. The mathematical objects at stake are being made apparent–*objectified* in Radford's

(2003) sense—within the semiotic bundle, involving the semiotic *set* of gesture and its close relationship to speech in mechanisms that I call *blending* or *condensation* because inherently different aspects of mathematical concepts are displayed in them in a multi-semiotic way and merge, as a result, in the bundle.

Second Episode

The second episode starts one minute later than the previous one, with an intervention by the teacher:

[5:42 – 6:60]

51. **T:** Ok, the thing I'm interested in […], however, is to try to understand this fact: this h is a small interval, small, but anyway a well defined small interval; in this case: for now, you haven't yet made it tend to zero, have you?

52. **A:** Yes, well [*he writes a sign for the limit*, 5: 50]

53. **T:** For now it is a number

54. **A student:** In this case, it is there, still there

[5:50 *Andrea has added the symbol for the limit*]

55. **T:** It is. What does f(x + h) minus f(x) divided by h represent? Is it the slope of the tangent or of what?

56. **A:** When h tends to zero?

57. **T:** No, at this moment, I recall that you have said: we take an, an increment h and so we pass from x to x + h, right?

58. **A:** Uh, uh…

59. **T:** And indeed h is small, but it is not equal to zero [*pointing his glasses at "x" and then "x+h"*, 6:10 and 6:11] in this case, is it? Nor it is tending to zero: it is a very precise object, it is this here… [6:16]

[6:10 *the teacher is pointing to the abscissa x*]

[6:11 *the teacher is pointing to the abscissa x+h*]

[6:16 *the teacher is pointing two fingers at the two abscissas*]

61. **A:** Well…

62. **T:** What does this object represent (*pointing at the incremental ratio formula*) when h (*performing the same gesture as in 6:16*) represents this distance, this small interval?

63. **A:** If h represents that, hence…it represents this… [6:30] it is this…I know the slope of this line passing through these two points here [6:38] […]

[6:30 *tracing the secant line segment*]

[6:38 *pointing to the two points of abscissas x and x+h*]

68. **T:** Is that the tangent?

69. **A:** No, it's not even a tangent: it is a…secant. The more, the more I reduce, it is small [6:49], when I arrive at saying that h is supersmall [6:51] then it represents the slope [6:53] of the tangent at that point there, at that point there (*repeating the gesture*)

[6:49 *extending the two forefingers*]

[6:51 *approaching the two forefingers*]

[6:53 *lowering his hand, with thumb and forefinger closed*]

The teacher wants to clarify the fundamental steps of the procedure that is being discussed in a blended way. He guides the students in proceeding step by step and focuses their attention on the links between the differential ratio formula and the graphical and numerical settings before the limit passage. In line 57, he is echoing and reformulating Andrea's account, addressing not only this particular student, but also all of the students in the classroom, to make the situation clear. Hence, while talking, he points at different inscriptions on the blackboard, suitably correlating them: first, he makes explicit the object which is the focus of his intervention, i.e. the increment h (#57); then (#59) he stresses the fact that h is small but different from zero: he clearly states this in speech and concurrently performs two distinct acts of pointing on the graphical setting. His gestures ([6:10], [6:11]), highlight that x and x+h correspond to two different abscissas in the Cartesian plane, thus, they determine a certain (non zero) interval. Finally, he specifies that he is considering h not as tending to zero, but as a "very precise object" (#59) corresponding to a certain distance in the Cartesian plane.

The reference to the distance, i.e. to the graphical-geometrical meaning of h, is not given by the teacher's speech, but by means of a gesture ([6:16]). Index and thumb are extended to touch exactly the abscissas "x" and "x+h". The strongly *deictic* features of the gesture are endowed with *iconic* features as well. In fact, the fingers are pointing at two parts of the x-axis in a very precise way, but are also, at the same time, representing the distance between the abscissas in an iconic way.

The configuration of the hand is the same as the Δ-gesture, but, in this case, the gesture is strongly indexically related to the given inscription, in which the fixed nature of h is well expressed. By means of his gesture, the teacher is *grounding* the students' processes of knowledge production in the inscription of the didactic setting. In particular, the gesture is making apparent a certain feature of h (i.e. that it is fixed and not zero) before the limit passage. With its sharp reference to the inscription, this gesture therefore invokes a different function than the Δ-gesture. In fact, the physical dynamicity of the Δ-gesture endows it with the characteristic of representing in a unique sign both a certain small distance, different from zero, and the process of making it smaller and smaller (involved in the limit of the differential ratio). The two meanings are not clearly distinct in the gestural act, neither when the fingers are kept fixed nor when they are made to get close to each other. In this sense, the Δ-gesture can be considered as a *condensing* gesture. Accordingly, to completely *disambiguate the meaning*, we see that the teacher is using different gestural resources: either two well-separated pointings ([6:10], [6:11]), or a gesture similar to the Δ-one but with the function of pointing at the inscription on the blackboard ([6:16]).

Having made clear that he is making h different from zero, the teacher then directs the students' attention to the geometrical meaning of the incremental ratio (#62). Andrea shows his answer by deictically referring to the graph (#63). He focuses on the limit process, whose result provides the slope of the tangent line (#69). Andrea's words suggest that he is thinking of the resultant stage of h in the limit process as an actual infinitesimal ("when I arrive at saying that h is supersmall"). Leaving aside epistemological observations, what is interesting from our point of view is that Andrea's utterance is accompanied by gestures that are now detached from the blackboard inscriptions and performed in his *gesture space*. The student's body is now turned from the blackboard towards the interlocutors (see pictures [6:45-53] contrasted to [6:30-38]). The spatial location of the gesture mirrors the actual *detachment from the concrete* that the limit process requires. In fact, since it involves an infinite number of steps, the limit process needs to be *imagined.* More generally, processes or objects related to the infinite carry with them a certain level of *abstraction* from the concrete. Accordingly, we observe that the reference to material inscriptions, characterizing Andrea's semiotic bundle in previous lines (made striking by observing the pictures), is here abandoned, and the utterance is accompanied by *gestures performed in the air* (see [6:49], [6:51], [6:53]). I have also noticed and described the same phenomenon in the different context of algebraic generalization (see Sabena, Radford & Bardini, 2005). In both cases, the students' semiotic activity is characterized by gestures that are detaching from the concrete materiality provided by the setting, and expressing, in an embodied way, abstract features of the mathematical objects that are being objectified. In the present case, Andrea is performing two gestures. The former ([6:49-51]) is formed by two pointed forefingers that approach each other and represents the process of getting h smaller and smaller in a *dynamic* way, as expressed in the co-occurring utterance ("the more I reduce, it is small"). The latter ([6:53]) is formed by the right hand lowered down with index and thumb seeming to trace something. It expresses, in a condensed way, both a feature of the function (its slope, i.e. the slope of its tangent line), and the fact that it is computed at a certain, given point (see the co-occurring speech: "it represents the slope of the tangent in that point there").

DISCUSSION

In the course of mathematical activity in the classroom, gestures appear as providing specific ways of carrying out the semiotic process. The analysis has specified how these ways differ from those provided by semiotic systems such as language and symbols.

A first characterization concerns *dynamism*. Gestures allow reference to mathematical objects, intrinsically endowing them with certain dynamic characteristics. Cognitive linguistic studies have pointed out that speaking of static objects in terms of dynamic features is a typical metaphoric cognitive mechanism that can be interpreted as reflecting the fundamental embodied nature of human thinking (Lakoff & Núñez, 2000). Gesture, as a semiotic resource inherently linked to the body, allows the display of such characteristics in a rather direct manner. Surely, a certain dynamic character is typical of calculus as well. Calculus can, in fact, be considered essentially as the culturally established mathematical form of thinking about change, specifically modeling variation and accumulation. Furthermore, from an historical point of view, one of the roots of calculus, dating back to medieval times, is the problem of motion. With regard to gestures, a certain dynamic character is present in specific kinds of gestures that I have pointed out in the course of data analysis: an example in the protocols above is the *Δ-gesture*, i.e. the configuration formed by index and thumb partly extended as if holding a little stick. The two fingers can be kept more or less closed, or may be made to approach each other. In the data, the Δ-gesture usually co-occurs with linguistic expressions referring to increments of variables, or distances on the Cartesian plane[29]. The dynamic possibility allowed by this gesture appears functional to the development of the mathematical meanings of calculus. In fact, the two fingers constituting the gesture can be easily approached until they touch each other, offering an embodied semiotic resource for depicting the dynamic metaphor underlying the limit process.

Another interesting feature shown by the Δ-gesture is described in terms of *condensation* or *blending*. In fact, the iconic features of the gesture, with respect to a small line-segment, induce one to identify graphic-geometric references; however, the speech contribution in which the gesture shows up may refer to the numerical register, or to a symbolic representation ("h"). Looking at the semiotic bundle as an integrated system, *the condensed or blended character of the semiotic bundle* is therefore a second characterization of the semiotic activity (in the broad sense outlined), coming from focusing on the role of gestures. The multi-semiotic character of the semiotic bundle allows us to deal with different aspects of the mathematical objects involved in the activity in a very integrated way. "Condensed" is a term of Vygotskian derivation (Vygotsky, 1934/1986), whereas "blended" comes from cognitive linguistics (Lakoff & Núñez, 2000). As concerns the Δ-gesture, the condensed-blended character of the bundle (and of the gesture itself) parallels the complex nature of representations in the Cartesian plane, described in terms of "conceptual blending" in the embodied cognition approach (Lakoff & Nuñez, 2000). Thanks to this character, the semiotic set of gestures therefore results as resonating with the semiotic system of the Cartesian reference, crucial in calculus. I remark, however, that by drawing such a parallel, I do not intend to underestimate the differences in the semiotic character of gestures and of the written graphical representations in the Cartesian plane, nor the important differences in the semiotic potentiality they provide to the mathematical activity.

Finally, a third contribution of gestures to the semiotic mathematical activity is seized by looking at their relationship with the displayed inscriptions in the bundle. In fact, gestures provide a peculiar way of *grounding* the process of knowledge production in the perceptual *contextual dimension* of the unfolding mathematical activity, but also of *detaching* from it to embody a *certain character of generality*. In this shifting, the spatial location of gestures with respect to the subjects' body and to the physical configurations of the inscribed resources seems to play a particular role. Looking at hours of videos of classroom activities, in different classes and contexts (see also Sabena, 2004; Sabena et al., 2005), I have observed various examples in which the students' or the teacher's gestures shift from grounding the reference on the given inscriptions, to embody some aspects of the current mathematical activity in the gesture space, or vice versa. The shifting appears to be related to the focus of the semiotic process: if the focus relates to the *contextual* character of the activity, gestures show marked deictic features and often refer to some written semiotic system; on the contrary, if the focus involves *imagination* in an important way or a certain *generality* in the discourse, gestures appear to be detached by the materiality in which the inscriptions are embedded and are often performed "in the air". Generally, in this latter case, gestures exhibit an iconic or metaphoric character. Typical examples have concerned the Δ-gesture, in which the iconic features are endowed with a condensed-blended character (as discussed above), reflecting the complexity of the Cartesian representation. Adopting Peircean semiotics, my interpretation is that beyond *indexical* and *iconic* features, this kind of gesture also acquires a certain *symbolic* character with respect to the *general* and *abstract* process of differentiating a function. The iconic features are interpreted as constituting the link between deictic and symbolic references. The contextual/general polarity can therefore be associated with the index/symbol polarity, as shown in the diagram (figure 1).

The figure represents identified correspondences between the features of signs (gestures, in particular), as integrated in the semiotic bundle, and the nature of mathematical meanings that are emerging in the activity. Indexically-featured signs correspond to a strong contextual dimension of the involved meanings, whereas, at the opposite edge, symbolically-featured signs are linked to a higher level of generality and abstraction. Iconic features may allow for the passage from the former to the latter.

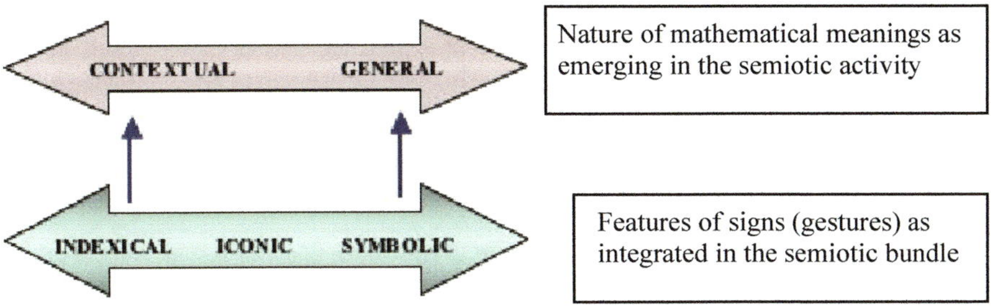

Figure 1. Relationship between gestures and mathematical meanings

The overall analysis that has allowed us to identify the outlined results has been carried out by framing gestures as signs, according to Peirce's account, and by looking at their contribution to the teaching-learning processes in classroom. I am aware that only a glimpse at this fascinating issue has been provided in this paper. The scope and soundness of the results, together with their implications at the educational level, are to be investigated in future research.

AKNOWLEDGMENTS

This paper stems from my doctoral study. I want to thank my supervisors, Ferdinando Arzarello and Luis Radford, for their passion for research and their constant and valuable support.

NOTES

1 For the time being, I take the term "sign" in a naïve sense.
2 For an overview of the history of gesture studies, see Kendon, 1982; McNeill, 1992.
3 The term comes from the Latin "gestus", for action, for carrying out an activity and for performing.
4 I'm not considering theatrical pantomime.
5 Emblems are certainly cultural-dependent. For instance, the set of emblem gestures that southern Italian speakers use have been found to be different from the set used by English speakers (Kendon, 1995).
6 Quite a more radical approach would be defining gesture, following Studdert-Kennedy, as "an equivalence class of coordinated movements that achieve some end" (Quoted in Kendon, 2000, p. 47). On this basis, it would be possible to argue for a fundamental identity between speech and gesture.
7 "Action" is distinct from "movement" in that the former is intentionally goal-oriented.
8 The question of the dependence of gestures on languages and culture is largely open. Evidence for language specificity of representational gestures seems to be emerging (Kita, 2000; McNeill & Duncan, 2000).
9 A number of classification schemes for nonverbal behavior, including gestures, have been developed over the years: for a discussion, see McNeill, 1992, pp. 75–77, or Goldin-Meadow, 2003, p. 3 ff..
10 The same is true for the deictic function in language, which is carried out by temporal and spatial adverbs and adjectives, such as "this", "there", "here", "now" and by pronouns such as "I" and "you". *Deixis* means "display" or "reference" in Greek.

[11] The idiosyncratic character of gesture seems to have been overestimated, as Kendon comments: "It is often said that gesticulation is idiosyncratic, each speaker improvising his own forms. So far as I know, no one has ever really tested this claim. My own experience in gesture-watching suggests to me that people are far more consistent in what they do gesturally than this 'idiosyncrasy' claim would lead one to imagine" (Kendon, 1996, p. 10).

[12] Actually, there is no general consensus among researchers about the possible "compositional" or "hierarchical" characters of gestures. Kendon (1996), in particular, seems to be more inclined to see these features in gestural acts as well.

[13] McNeill pinpoints synchronies at the phonological (i.e. stroke phase of gesture coincides with the phonological peak of the utterance), semantic (i.e. content) and pragmatic (i.e. function) levels (McNeill, 1992, p. 26 ff.).

[14] Two well-argued theories are the "Growth Point Theory", brought forward by McNeill's group (McNeill, 1992; McNeill & Duncan, 2000), and the "Information Packaging Hypothesis" sustained by Kita's group (Alibali, Kita & Young, 2000; Kita, 2000). Both assign a prominent role to gestures in thinking processes, to the contrary of some earlier theories, such as the "Lexical Retrieval Hypothesis", which restricts the role of gesture in giving lexical access to generating sentences.

[15] In his book, however, McNeill never mentions *semiotic mediation*, nor does he speak of gestures in terms of *mediation*.

[16] The theory undoubtedly appears more comprehensive than previous dyadic ones, such as Saussure's linguistic one (Saussure, 1916).

[17] Peirce's classification of signs is very elaborated. In fact, each of the three terms constituting the sign is subdivided into three categories. The presented classification refers to the trichotomy of the *object*.

[18] This is analogous to the classification of gestures, as discussed above.

[19] In Ernest's words:

> The term semiotic system is here used to comprise three necessary components. First, there is a set of signs, each of which might possibly be uttered, spoken, written, drawn, or encoded electronically. Second, there is a set of rules of sign production, for producing or uttering both atomic (single) and molecular (compound) signs. [...] Third, there is a set of relationships between the signs and their meanings embodied in an underlying meaning structure (Ernest, 2006, p. 69–70).

[20] As Arzarello (2006) remarks, the word "set" is to be interpreted in a very broad sense, e.g. as a variable collection.

[21] It is this feature that distinguishes Arzarello's semiotic *sets* from Ernest's semiotic *systems*, the former being produced and transformed according to modes that are not necessarily *rules*. As a consequence, Arzarello's notion can be considered as a loose version of that of Ernest.

[22] Striking examples of the diachronic analysis of signs from a psychological perspective can be found in Vygotsky's work. See, for instance, the pioneering observations concerning the *genetic relationships* between gestures and written signs (Vygotsky, 1997), and the account of the genesis of inner speech from external speech (Vygotsky, 1934/1986). Inner speech can, in fact, be seen as a semiotic set, arising out of a semiotic system (language).

[23] According to Duval's framework, "treatments" are transformations within a same semiotic system, or register (e.g. computations in the algebraic domain); "conversions" are transformations between two different "registers", e.g. graphing a function from its formula (Duval, 1995). In this paper, I use the terms "register" and "semiotic system" synonymously.

[24] For the transcript, I use the following notations. [00:00] – [00:00] indicates the time-interval of the episode. The numbers stand for minutes:seconds. [00:00] is also used to refer to the pictures illustrating the gestures. The speech production is parsed according to the subject's participation; the numbering starts at the beginning of the activity. Punctuation marks are used to characterize utterances in terms of inflections as normally used; in particular, dots ... indicate pauses in speech. Underlined words correspond to the main phase (including the stroke) of the co-timed gesture. Text

in parentheses () contains additional information, mainly synthetic descriptions of gestures. [...] indicates omissions.

25 In fact, the modeling activities on which the meaning of the mathematical objects at stake (i.e. the behavior of functions in terms of slope) had been established in previous lessons and involved mainly geometrical and numerical semiotic systems.

26 Andrea's account is obviously not rigorous and shows intrinsic contradictions. An early approach to calculus, such as the one presented in this research, necessarily requires that certain notions be addressed before the students have the theoretical and technical instruments to deal with them at a rigorous level. It is not the goal of the present contribution to discuss the topic from epistemological and didactical points of view. I just observe that such didactic choices are in line with the ideas that (i) learning is not a process accomplished once and for all, but to the contrary, it unfolds over time; (ii) learning does not follow fixed hierarchical, linear paths, rather it is a more complex, multidimensional phenomenon.

27 Since the camera was filming Andrea, and moved to Luigi only slightly after he began speaking, we do not know if the gesture accompanying the first part of line 33 is similar to the corresponding one in line 35 (shown in picture 4:27-1).

28 Adopting the classification of Edwards (2003), this is an example of *iconic-symbolic gesture*, i.e. a gesture which, given its iconic features, resembles a symbolic inscription.

29 Given the available data, it is not possible to infer whether or not this gesture was used by the teacher in his explanation, and therefore may be considered as having been "introduced" by his didactic practice. This may be the case and the same may be true for other gestures as well. The genetic aspect of gestures therefore emerges as an interesting but delicate issue and one that has not been addressed in the analysis.

REFERENCES

Alibali, M. W., Kita, S., & Young, A. J. (2000). Gesture and the process of speech production: We think, therefore we gesture. *Language and cognitive processes, 15* (6), 593–613.

Arzarello, F. (2006). Semiosis as a multimodal process. *Revista Latinoamericana de Investigacion en Matematica Educativa*, Numero Especial: Semiotics, Culture, and Mathematical Thinking, pp. 267–299.

Arzarello, F., & Robutti, O. (2004). Approaching functions through motion experiments. In: R. Nemirovsky, M. Borba, & C. DiMattia (Eds.), *Bodily activity and imagination in mathematics learning, PME Special Issue of Educational Studies in Mathematics* 57.3, CD-Rom, Chapter 1.

Arzarello, F., & Edwards, L. (2005). Gesture and the construction of mathematical meaning. *Proceedings of PME 29*, Melbourne (Australia), Research Forum 02, Vol. 1, pp. 22–145.

Cassell, J. (1998). A framework for gesture generation and interpretation. In *Computer vision in human-machine interaction* (pp. 191–215). Cambridge University Press, 191–215.

Duval, R. (1995). *Sémiosis et pensèe humaine*. Bern: Peter Lang.

Edwards, L. D. (2003). A natural history of mathematical gesture. *American Educational Research Association Annual Conference*, Chicago, April 2003.

Ernest, P. (2006). A semiotic perspective of mathematical activity. *Educational Studies in Mathematics, 61*, 67–101.

Goldin-Meadow, S. (2000). Beyond words: The importance of gesture to researchers and learners. *Child Development*, Vol. *71*, n° (1), pp. 231–239.

Goldin-Meadow, S. (2003). *Hearing gesture. How our hands help us think.* Cambridge, Massachussetts, and London, England: The Belknap Press of Harvard University Press. Cambridge, Massachussetts, and London, England.

Goldin-Meadow, S., Nusbaum, H., Kelly, S. D., & Wagner, S. (2001). Explaining math: Gesturing lightens the load. *Psychological Science*, *12*, 516–522.

Kendon, A. (1980). Gesticulation and speech: Two aspects of the process of utterance. In M. R. Key (Ed.), *The relation between verbal and nonverbal communication*, (pp. 207–227). The Hague: Mouton.

Kendon, A. (1982). The study of gesture: Some observations on its history. *Recherches Sémiotiques/Semiotic Inquiry, 2*(1), 45–62.
Kendon, A. (1988). The significance of gesture: how it is achieved. *Papers in Pragmatics, 2,* 60–83.
Kendon, A. (1995). Gestures as illocutionary and discourse structure markers in Southern Italian conversation. *Journal of Pragmatics, 23*(3), 247–279.
Kendon, A. (1996). An agenda for gesture studies. *Semiotic Review of Books, 7*(3), 8–22.
Kendon, A. (2000). Language and gesture: Unity or duality? In D. McNeill (eEd.): *Language and gesture* (pp. 47–63). Cambridge University Press, 47–63.
Kita, S. (2000). How representational gestures help speaking. In D. McNeill (Ed.), *Language and Ggesture: Window into thought and action,* (pp. 162–185). Cambridge, UK: Cambridge University Press.
Lakoff, G., & Núñez, R. (2000). *Where mathematics comes from: How the embodied mind brings mathematics into being.* New York: Basic Books.
McNeill, D. (1992). *Hand and mind: What gestures reveal about thought.* Chicago: University of Chicago Press.
McNeill, D. (2000). Introduction. In D. McNeill (Ed.), *Language and gesture: Window into thought and action,* (pp. 1–10). Cambridge, UK: Cambridge University Press.
McNeill, D., & Duncan, S. D. (2000). Growth points in thinking-for-speaking. In D. McNeill (Ed.), *Language and gesture: Window into thought and action,* (pp. 141–161). Cambridge, UK: Cambridge University Press.
Peirce, C. S. (1931–1958). *Collected papers, Vol. I–VIII.* Edited by (C. Hartshorne, P. Weiss, & A. Burks, Eds.). Cambridge, MA: Harvard University Press, Cambridge, Massachussetts. (CP, Volume number. Paragraph number).
Radford, L. (2003). Gestures, speech, and the sprouting of signs: A semiotic-cultural approach to students' types of generalization. *Mathematical Thinking and Learning, 5*(1), 37–70.
Radford, L., Bardini, C., & Sabena, C. (2006). Rhythm and the grasping of the general. *Proceedings of PME 30,* (Vol. 4, pp. 393–400). Prague (Czech Republic), Vol. 4, pp. 393–400.
Radford, L., Bardini, C., & Sabena, C. (2007). Perceiving the general: The semiotic symphony of students' algebraic activities. *Journal for Research in Mathematics Education,* Vol. *38,* N° (5), pp. 507–530.
Roth, W.-M. (2001). Gestures: Their role in teaching and learning. *Review of Educational Research, 71,* 365–392.
Roth, W.-M. (2002). From action to discourse: The bridging function of gestures. *Journal of Cognitive Systems Research, 3,* 535–554.
Sabena, C. (2004). The role of gestures in conceptualisation: An exploratory study on the integral function. *Proceedings of PME 28,* (Vol. 4, pp. 145–152). Bergen (Norway), Vol. 4, pp. 145–152.
Sabena, C. (2007). *Body and signs: A multimodal semiotic approach to teaching-learning processes in early Calculus,.* PhD Dissertation, Turin University.
Sabena, C., Radford, L., & Bardini, C. (2005). Synchronizing gestures, words and actions in pattern generalizations. *Proceedings of PME 29,* (Vol. 4, pp. 129–136). Melbourne (Australia), Vol. 4, pp. 129–136.
Saussure, F. (1916). *Cours de linguistique générale.* Paris: Payot.
Vygotsky, L. S. (1934/86). *Thought and language.* Cambridge, MA: MIT Press. (Revised Edition., translated and edited by A. Kozulin, Trans. and Ed. Original work published in 1934).
Vygotsky, L. S. (1997). *Collected works, Vol. 4* (R. Rieber, Ed.). New York: Plenum.
Wilson, M. (2002). Six views of embodied cognition. *Psyconomic Bullettin & Review, 9* (4), 625–636.

Cristina Sabena
Department of Mathematics
University of Torino (Italy)

RAYMOND DUVAL

EIGHT PROBLEMS FOR A SEMIOTIC APPROACH IN MATHEMATICS EDUCATION

INTRODUCTION

Mathematics, whether in its teaching or in its more advanced practices, is the field where the use of signs is the most complex and the heterogeneous range of signs used is the most extensive. This has proven to be so intrinsic to mathematical activity that we feel we can assert: there is no mathematical thinking without using semiotic representations to change them into other semiotic representations. This raises big issues, not only about mathematics learning and the specific cognitive processes involved in mathematical comprehension, but first and foremost about which kind of semiotic approach is relevant for analysing mathematical thinking and knowledge. Do the different classical semiotic theories allow for an analysis of the use of images, diagrams, language and symbols in mathematics?

These theories suffer from the limitations of the particular kind of signs they make representative of all kinds of signs (Duval, 2006b). The result is a considerable gap between the existing semiotic tools they provide and the semiotic complexity found in any given mathematical production. A semiotic theory relevant for the problems that mathematics education must tackle remains still to be found. It must, at the same time, meet the mathematical demands necessitating the use of various kinds of semiotic representations, the need for a tool that can accurately distinguish all the kinds of representations used, and the methodological requirements for making verifiable analyses of students' productions.

We will present eight problems that are inevitably encountered when we analyse mathematical activity, from the only aspect where it is objectively given and observable, its multiform productions. We will follow a progression: first of all, problems raised by the wide variety of semiotic representations used from the elementary level onward; then, problems about the links between mathematical processes and the various creative changes of semiotic representation. These links cannot be reduced to concepts. Finally, the methodological and theoretical problems about modelling the functioning of mathematical thinking will be dealt with.

Each problem does, in fact, require a detailed presentation in itself. But they are all connected. The complexity of the semiotic approach to mathematics and cognitive processes underlying their comprehension would be missed if we

L. Radford, G. Schubring, and F. Seeger (eds.), Semiotics in Mathematics Education: Epistemology, History, Classroom, and Culture, 39–61.

confined ourselves to looking at one or the other. An overview of the eight problems is the best introduction to each.

I. HOW CAN WE ANALYSE THE RELATIONSHIP BETWEEN THE VARIETY OF SEMIOTIC REPRESENTATIONS USED IN MATHEMATICS AND WHAT THEY STAND FOR?

This problem is the preliminary issue that is never asked, because we stick to a common response mainly based on a functional definition of signs or *representamens*. The relationship between semiotic representations and what they represent should be more or less the same for all the kinds of representations. Therefore, the structural differences between language and images, or between symbols and icons, would have mainly a psychological and educational significance. But also, and this seems more incomprehensible, the relationship is supposed to be the same in mathematics and outside mathematics. This is not true. To understand the problem, it is important to take two points of view into account, namely, both the semiotic and the epistemological.

The semiotic issue arises when we cope with the wide range of semiotic representations used in mathematics teaching from primary school to university, and not with the students' local and transitory productions. The semiotic issue comes first of all from mathematics, and not from students. Thus, in order to carry out any given mathematical activity, it is necessary to deal with images, which are non-iconic visualisations and not only iconic representations, with an exponential increase of symbolic notations and also with linguistic statements, at least for expressing problems and wording properties (Figure 1 below). Let us take a typical problem used to bring out the meaning of multiplication from concrete situations: "Jane bought 5 slabs of chocolate for 7 shillings each. How much did she spend?"

In order to solve this problem, some of the following semiotic representations must be mobilised.

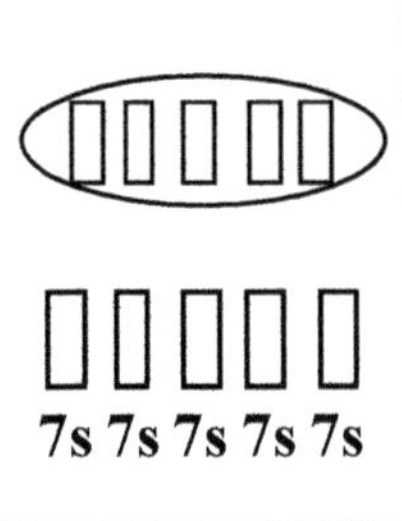	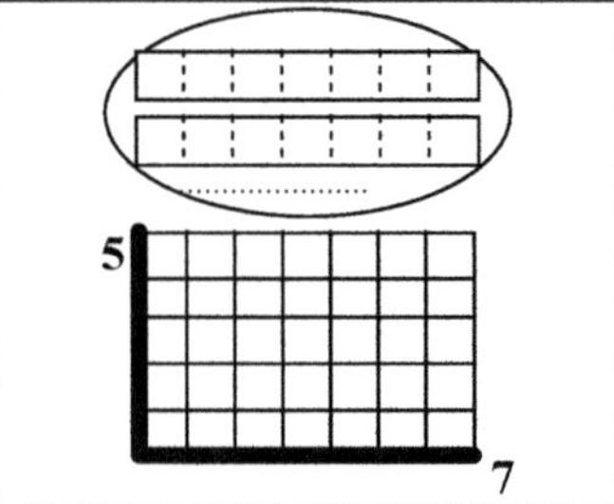	7 +7 +7 +7 +7 or 7 × 5	**ONE ... IS WORTH seven...** **or** ONCE seven, TWICE seven, THREE TIMES seven ...
1. Two Iconic visualisations of five slabs, with or without their price label	2. Two Non-Iconic visualisations open to two possible views, focusing either on surface units or on a two-dimensional lines network	3. Two Symbolic (decimal) descriptions	4. Two kinds of descriptive wordings of the visual or symbolic representations.

Figure 1. Eight possible semiotic representations

Semiotics is not concerned with the objects represented in this range of representations, but with the way these various representations can stand for something else. Thus, in Figure 1 above, are we facing two kinds of representations (iconic and symbolic) or three (visualizing, symbolic, linguistic) or four (visual iconic, visual non-iconic, symbolic and linguistic)? (Q1). The identification of the cognitive gaps between every kind of semiotic representation usable in a mathematical activity is an essential prerequisite for analysing the problems of comprehension in the learning of mathematics (Duval, 2006a, p. 110).

However, from the point of view of knowledge acquisition, what matters first is the represented objects and not how they are represented; concrete objects such as slabs of chocolate, and knowledge objects such as numbers and two operations on integers. From this point of view, the eight representations above raise a crucial question, regardless of the fact that some seem easier or more spontaneous than others. Among the eight representations possible, how do we determine which are representing the same thing? More generally, how can we recognise whether two different representations stand for the same **object or not? (Q.2)**

How any representation presents an object varies greatly depending on whether it is iconic, symbolic or linguistic, etc. This is obvious. But this should not distract us from the most important fact. The relationship of any semiotic representation to what its stands for involves an intentional operation of designation. For each kind of semiotic representation, we have to distinguish two levels in the relationship between representations and the represented objects:

- direct designation by using only basic units proper to the kind mobilised
- constructed designation by using the rules of composition specific to the kind of representation chosen

Thus, for the representation of the number "thirty five" in Figure 1 above, we are coping with four different constructed representations:

Direct designation of an integer	Iconic visualisation 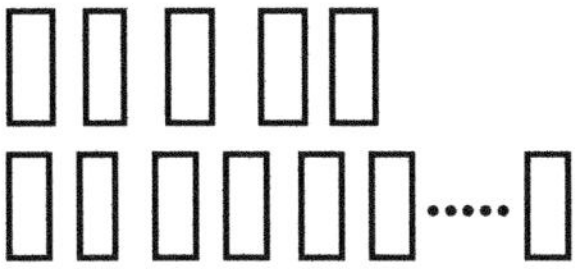	Decimal notation (5) 35
Constructed designation of an integer	The non-iconic visualisations in Figure 1	7 + 7+ 7+ 7+ 7 7×5

Figure 2. The two levels of the relationship of signs to a represented object

Here, for iconic visualisation, each rectangular shape can refer either to a slab or to any one-unit mark. Juxtaposition is the only rule of composition. For the symbolic notation of numbers, both number designation and rules of composition depend on the decimal basis.

For analysing a process of knowledge acquisition, we must also take into account how we objectively get access to the knowledge objects. In most knowledge fields, it is through direct perception, or by collecting samples or discovering remnants (fossils), or using physical devices (microscope, spectroscope) and the representations they produced. But, do we get access to numbers, functions, etc. through this experimental way or, on the contrary, by using semiotic representations, whatever the kind used? In other words, is the relation of semiotic representations to the represented object the same in and outside mathematics, regardless of the interpretant? (Q.3)

The questions (Q.2) and (Q.3) above must lead us to reconsider all the classical definitions given for the signs and also question their relevancy, at least as regards their use in mathematics, because they are limited to the more general function that signs meet. They only tell us about the purpose of any given signs' use: to stand for something else, to evoke what is absent, or to communicate to the others what we are thinking of (Duval, 2006b). However, such conceptions of the roles of signs lack what characterises the specific way signs are used in mathematics. Semiotic representations are interesting only insofar as they can be changed into other representations (Duval, 2006c).

It is this function of change that matters and it leads to the development of very precise structural analysis of the possibilities for change, which are specific to each kind of representation. Thus, the semiotic question (Q1) must be raised in a different way: Which possibilities for changing any given representation into another does each semiotic system provide? (Q4). For instance, in Figure 1 above, iconic, non-iconic visualisations, symbolic and linguistic representations do not provide the same power of internal change.

II. HOW CAN WE ANALYSE THE CHANGES FROM ONE GIVEN SEMIOTIC REPRESENTATION TO ANOTHER?

Mathematical activity is always based on some sequence of successive changes from one semiotic representation into another, without any supply of information from external sources. This can be observed not only in the most elementary use of numbers and geometrical figures but also in algebra and calculus.

In order to analyse the semiotic changes which are mathematically relevant, we have to start from the following crucial fact about semiotic analysis as the relationship of signs to object: **there are always many possible semiotic representations of the same object**. The higher processes of thought, and especially mathematical activity, are based on this plural semiotic object representation (Duval, 1995a). Frege faced this when he wondered how to resolve the Kantian dilemma between *a priori* logical reasoning, illustrated by the tautology a = a, and any reasoning developing new knowledge, illustrated by the equivalence a = b, often without an *a priori* basis (Frege, 1971, p. 102). Frege found the answer by distinguishing sense from reference: a and b have different senses but refer to the same object. This distinction is powerful because it allows one to analyse the substitution processes that constitute calculation and valid reasoning. However, it cannot be used as it is to analyse all forms of mathematical activity, because it suffers from three limitations:

- It was used mainly for symbolic expressions. But it raises paradoxical difficulties when it is applied to linguistic statements. Besides, such a distinction has no meaning when it comes to symbols, since their sense is only in the reference, be it decided or chosen. This is not the case with words, whose sense is made by semantic oppositions.
- The difference in sense is the result of various constructed designations possible within the same semiotic system, and not the result of the use of various semiotic systems.
- The deep link between the sense of a semiotic representation and the particular system used for producing it is missed. Each representation provides a different content according to the semiotic system used for producing it, while the represented object becomes the representations' invariant (Figure 2 above).

This is why we prefer to use the following triplet:

{{ representation **content**, semiotic register used}, represented **object** }

Thus, we get two semiotic sources for generating the cognitive differences in sense: on the one hand, variations with the constructed designation and, on the other hand, variations with representation registers. Then, for analysing the change from one given representation to another, we can combine both semiotic and cognitive viewpoints, without subordinating the one to the other (Figure 3 below). It is easy to notice that we have three and only three kinds of semiotic change of representation, two of them only being mathematically relevant, because they are made in relationship to the represented objects: conversions and treatments.

The problems of comprehension in learning mathematics are due either to the cognitive gap between two representation contents (Figure 3 below, (1)) or similarities/differences not mathematically relevant between them (*ibidem* (3)). It is question (2) *(ibidem)* which conveys the recurrent and insurmountable trouble for most students. We also get a better comprehension of Frege's distinction (*ibidem* (4)); through successive local reference invariance, we can globally construct or discover a new object.

In order to analyse the cognitive processes underlying any mathematical activity, and problems of students' comprehension in learning mathematics, we must carefully distinguish these two types of semiotic change: conversion and treatment. But, the necessity of this distinction is almost always ignored or denied. This is for two reasons:

First, at a surface level, conversions often appear transitory when it comes to choosing the most powerful register for treatment. What matters is treatment in the chosen register, because it corresponds to a mathematical processing.

Next, conversions (Figure 3 (1) (2)) are considered as nothing more than encoding/decoding processes (*ibidem* (5) (6)). Change from an equation to its graphical representation, or the use of literal expression for setting up any word and numerical description into an equation are the examples put forward to justify this conviction. Even in these cases, is this right?

<table>
<tr><td colspan="3"></td><td colspan="2">COGNITIVE POINT OF VIEW</td></tr>
<tr><td colspan="3"></td><td>The representation's CONTENT</td><td>The denoted OBJECT as the invariant</td></tr>
<tr><td rowspan="3">SEMIOTIC POINT OF VIEW: THREE KINDS OF CHANGE</td><td rowspan="2">The denoted object is taken into account and there are rules of combination</td><td>DIFFERENT KINDS OF SIGNS
Statement→Image
Statement→ Equation
Figure →Statement :
CONVERSION</td><td>(1) Nothing in common between the two representations' contents</td><td>(2) Is it the same object with other properties made explicit or a different one?</td></tr>
<tr><td>SAME KIND OF SIGNS
Statement→Statement
Equation → Equation
Figure→Figure :
« TREATMENT »</td><td>(3) Similarities and differences, not all mathematically relevant, even if meaningful</td><td>(4) A new object is found or constructed (Frege's situation)</td></tr>
<tr><td>The denoted object is not taken into account</td><td>RULES of ONE-TO-ONE MAPPING for discrete units
Phonetic → Graphic
Analogical →Digital
(alphabet, binary codes, Morse)
ENCODING</td><td>(5) Matching characters to a previously segmented content, without selecting any information.</td><td>(6) No denoted object can be recognised without decoding to come back to initial rep.</td></tr>
</table>

Figure 3. The two sides of the three kinds of semiotic representation change

Any setting up into an equation requires three quite different discursive operations (Figure 4 below). The first, a direct (re)designation, is like an encoding operation. But this is not the case for the second, constructing a designation. This is not obvious for students. The third requires choosing two different designations for the same denoted object, something that is out of reach for most students. Here is a very simple problem to highlight the gap between these three discursive operations.

"A newspaper and its supplement cost together 1,10 €. The newspaper A costs 1 € more than its supplement. What is the price of the newspaper ?".

Conversion is neither an encoding/decoding process, nor a secondary process for doing mathematics. It is a specific semio-cognitive process underlying any mathematical activity with which students must cope by themselves. Not distinguishing conversion and treatment means leaving them with these two questions always unanswered. How does one come to recognise that two representations whose contents have nothing in common are however two representations of the same object (Figure 3 (2))? How does one come to recognise what is mathematically relevant in two representations whose contents are visually, linguistically, symbolically, etc. similar, apart from some small differences? (Figure 3 (3)).

THREE DISCURSIVE OPERATIONS	Various linguistic expressions	Literal expression (**lexicon reduction**)
1. Direct or individual designation	newspaper supplement	b
2. Syntactically constructed or functional designation	The newspaper costs **1 more than...** Cost of the supplement, etc.	b + 1
3. Finding out **two designations for the same object** (*according the Frege's distinction*) for getting a relation of equivalence between them	— The cost of the newspaper **and** the cost of its supplement — They cost together 1,10	b + (b+1) = 1, 10

Figure 4. Setting up into an equation: conversion or encoding?

III. WHAT ENABLES US TO RECOGNISE THAT TWO SEMIOTIC REPRESENTATIONS HAVE NOTHING IN COMMON, YET STAND FOR THE SAME KNOWLEDGE OBJECT?

This issue is the reversal of the Frege's problem: how can there be continuity and progress of the same mathematical processing through the substitution of a new expression for another given? Frege's distinction works under two assumptions:

- We already know that the two representations stand for the same object.
- Expressions are produced in the same register (mainly the symbolic one).

The first assumption is obviously at the opposite side of the learner's situation. The second one amounts to forgetting the complexity of mathematical activity, which uses various kinds or representations. Then, the reversal of Frege's problem is: what continuity is there in the conversion of semiotic representations? Unlike the other fields of knowledge, this is the big cognitive issue for comprehension in mathematics, because the epistemological situations of access to knowledge objects are opposite.

Outside mathematics, each representation can be associated to the represented object itself, because we can get a perceptive or instrumental access to the object, apart from semiotic representations. For instance, we can juxtapose a physical chair with a drawing of the same chair and a description of the chair, as Kosuth did in a famous picture (Duval, 2006c). So, from these associations, a drawing and a description can be recognised as being two representations of the same thing. Outside mathematics, we always get the possibility of this double juxtaposition: {object, one representation of this object} and {one representation, another representation of the same object}.

In mathematics, such a double juxtaposition is impossible, because there is no access to knowledge objects by perception or by instruments. We have only juxtapositions of semiotic representations whose contents have nothing in common: {equation, graph, diagram, statement, etc.}. How do we find out whether we are facing the same knowledge objet, or not, if we are not able to change from one to another?

There are two widely shared answers, which turn out to be dead ends. The first relies on one of the two characteristics of mathematical "concepts": they have to be independent of any semiotic representation. But one misses the other as the result of the epistemological situation of mathematical objects: they are only accessible by using semiotic representations. This answer comes up against the cognitive paradox of mathematics. Namely, how can one distinguish the represented object from the content of the representation used, if there is no access to mathematical objects apart from semiotic representations? The second answer relies on the psychological process of association: any recognition would be based on previous situations in which various representations of the same knowledge object would have been given together. But the trouble is still the same: how can we recognise the same thing in representations when a double juxtaposition is not possible?

The only solution lies in the following requirement: some correspondences have to be discriminated between the relevant units, which constitute the respective contents of the two representations. Take the example of the linear function, the introduction of which mobilises several semiotic representations. Then, to change from one to the other, learners are expected to make two kinds of connections between graphs and their equations:

- one numerical by local focusing on intersection points and reading a couple of numerical values, required for plotting any graph
- the other qualitative by global focusing on visual oppositions mathematically relevant in relation to the equation's semantic characteristics, required for a comprehensive approach

From the numerical change, which is a coding procedure, to the qualitative one, which is a conversion, there is a gap. Conversion requires discriminating the visual features of the graph and the symbolic designations of the equation, and making links between the respective units of these two representation contents. That requires specific tasks, which are like an experiment in which the visual variation of line, curve, etc is taken as an independent variable and where the changes in the symbolic designation are the dependent variable (Figure 5 below). One must focus attention not on each isolated representation, but on variations in one register and the co-variations in the other.

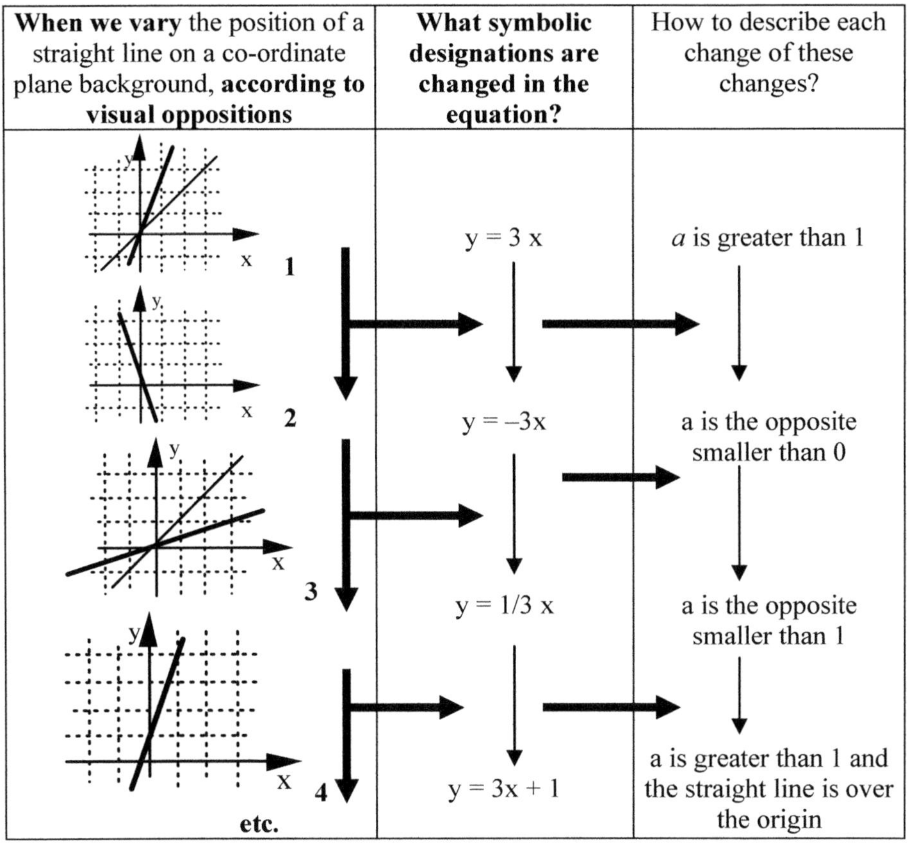

Figure 5. Organisation of a discriminating task

Thus, through this kind of task, students are in a position to discriminate visual features which are independent of each other but always visually merged, and can articulate them with the specific symbolic units in the equation: here the slope of the straight line and its position relative to the origin of the axes. From this example, we can gain an insight into four important points:

- Unlike for encoding procedures and treatments, there are no rules of conversion (Duval, 2007a).
- Conversion in one direction can be without any cognitive link with this in the reverse direction (Duval 2006a, p. 123).
- Correspondences between visual units and symbolic units are set without starting by taking into account the represented object. In other words, each representation content denotes only one object but includes many units merged or explicitly organised into a whole.

- Teaching is a source of misunderstanding about representation and knowledge object. Cartesian graphic representations are introduced for learning what a function is (that is, knowledge object) and are subordinated to this goal alone, whereas, the functioning of this kind of representation is completely independent of it and allows the representation of geometrical objects that are not functions.

IV. CAN THE SEMIOTIC COMPLEXITY OF MATHEMATICS CAN BE AVOIDED BY MAKING STUDENTS WORK ON CONCRETE OR REAL PROBLEMS?

This is the assumption underlying the rhetorical oppositions between “formal” or “theoretical” mathematics and “intuitive” or “practical” mathematics. Acquisition of mathematical knowledge should be easier in contexts of solving real problems, which involve bringing into play some mathematical procedure.

The issue is twofold. The question is whether solving these problems does not require that students use several kinds of semiotic representations and be able to discriminate all their relevant connections already. It is also whether the embedding of mathematical procedures in real situations does not hide a misleading levelling of iconic and non-iconic visualisation. Consider two examples.

A problem in order to develop multiplicative procedures

This problem presents two points of interest for teaching. Physical items can be used for understanding the described situation, and an iconic representation can be drawn as a complete support for counting. The crucial point is that students realise the two connections marked by arrows below.

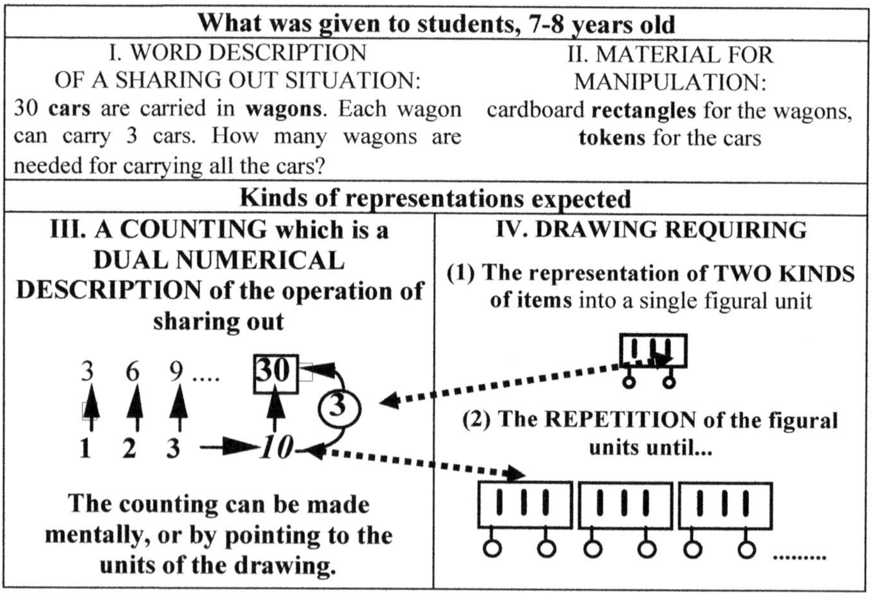

Figure 6. Working field for solving a concrete problem of sharing

To enter into this "concrete" (?) working field, students must be able to come and go between two or three of these various representations. The drawings produced by students show failures:

- about connections between the drawing and the statement of the problem: the wagons are without cars or too many cars are drawn
- about connections between a relevant drawing and the counting

Figure 7. Insufficient connections

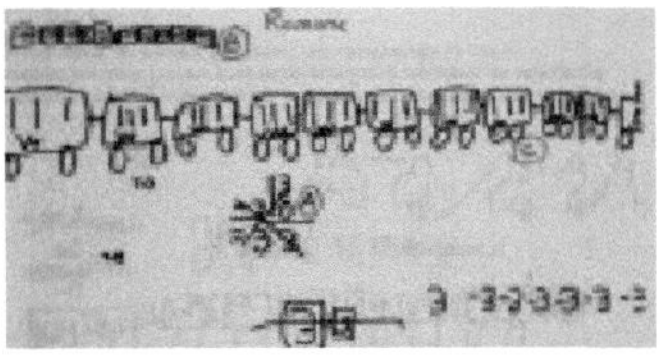

Figure 8. No connection between 3 AND its repetition in each wagon

Romane (Figure 8) had started writing a sequence of 3 and gave up. What is most significant is the choice made by the teacher. She did not pay attention to the troubles caused by these elementary conversions implicitly expected, but considered that these iconic representations were the reason for the failures. So she forbade the students from making drawings. Three months later, some students who had managed and succeeded at this problem by additive strategy, failed.

An equivocal non-differentiation between iconic and non-iconic visualisation

One classical application of mathematics to the real world is the use of Thales' Theorem to obtain measurements of distant objects. This requires two kinds of visualisation: iconic and non-iconic (Duval, 2006b). A visualisation is iconic when, for instance, it represents positions or shapes of real-world objects. It is non-iconic when it is organised according to internal constraints and gives access to all cases possible.

<table>
<tr><th colspan="2">ICONIC VISUALISATION</th><th>NON-ICONIC VISUALISATION</th></tr>
<tr><th>I) Image or photo</th><th>II) Diagram of the situation of the aiming at ... for the measurements to be taken (bold segments)</th><th>(III) Geometrical configuration</th></tr>
<tr><td>Showing the respective positions of two landmarks (buildings, lighthouse and ship, shades of pyramid and stick...) or two points on them (top, bottom)</td><td>ONE aiming at the height of... from the ground
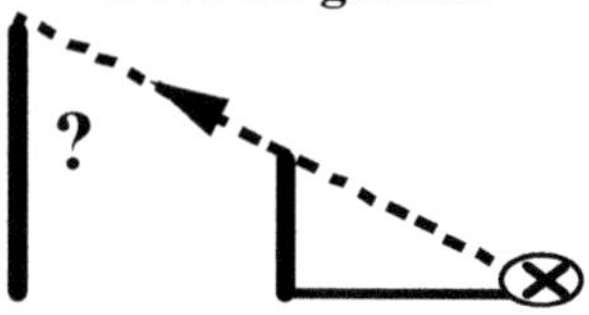

ONE at a far object from the top of....
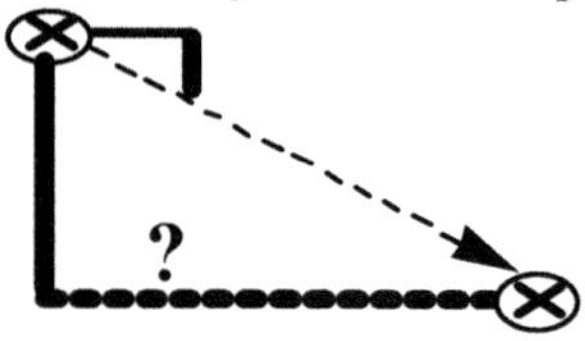

TWO aiming at the same object from two distant points
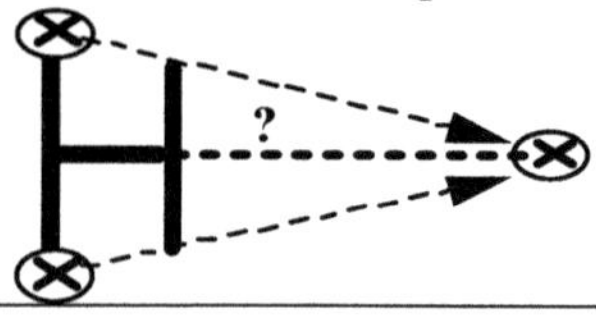
</td><td>— two similar triangles
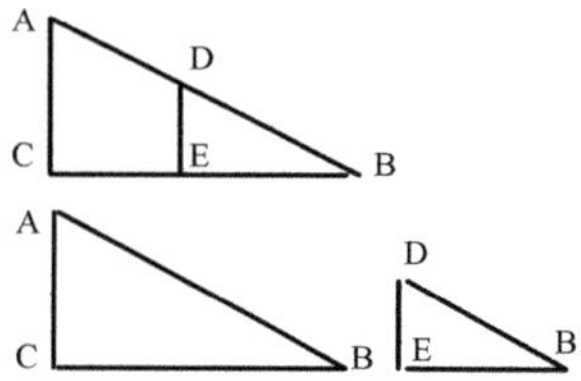

— with the given “the segments in bold are parallels”; we have seven identities of ratios
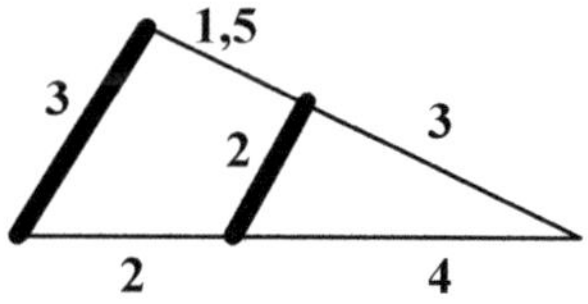
</td></tr>
</table>

Figure 9. Working field for application to real situations

There is a gap between the diagrams and the geometrical figures. Several diagrams are needed to show how to take the measurements required for calculating the length not measurable with an instrument. Only one geometrical figure gives all the ways of calculation. What matters in the diagrams is the line of sight and its anchorage, but in the geometrical figure, it is the intersection of straight lines by two parallel lines. In order to jump from one to the other, a qualitative condensation of the different diagrams into one figure topologically similar to the one of Thalés' Theorem has to be implicitly made. More precisely, the three diagrams must be mapped with the seven identities of ratios. This semiotic gap and the cognitive jump required are always hidden and ignored in teaching.

In textbooks, (I) and (III) are most often given separately or superimposed on one another, and likewise sometimes (I) and (II). But the variation of real situations is omitted, and it is assumed that superimposition of the geometrical figure on various diagrams should be obvious.

The same non-differentiation between iconic and non-iconic visualisation can be highlighted about the introduction of functions, etc.

V. WHY ARE SOME KINDS OF SEMIOTIC REPRESENTATIONS MORE NEEDED AND MORE SYSTEMATICALLY DEVELOPED IN MATHEMATICS THAN ALL OTHERS?

Whatever the curriculum, the fundamental fact of mathematics education from primary schools to the most advanced levels is that some symbolic systems become more systematically used than others. The mathematical need for generalisation is not the only explanation. We have pointed out that the most important mathematical use of semiotic representations is their internal potential to be changed into other representations, and we have distinguished two kinds of relevant change.

From a mathematical point of view, treatments are in fact the only important kind of semiotic transformation, because it is only through treatment that explanation is produced or a proof achieved. So, the semiotic systems needed are those that have the greatest potential for treatment, and above all, those whose treatment procedures can be made into algorithms, such as within symbolic systems. The choice of any conversion is always subordinated to the treatment to be run.

We must examine the semiotic side of treatment, not primarily about symbolic systems used as in algebra or in calculus, but about the commonly shared systems such as figures and native language. Are they obvious or spontaneous, or are they at the opposite extreme of the familiar practice of interpreting and reasoning?

Visual treatment on discrete sets

In the following very well known example, the visual treatment is enclosed in two conversions. An open sequence of iconic representations of integers to be added is changed into two equal triangular parts of a rectangle. The linking of three changes carries this out.

SYMBOLIC REPRESENTATIONS	**VISUAL TREATMENT**	
1 + 2 + 3 +... + n = ? ***Conversion from decimal to iconic representation and shift from 1D to 2D—the setting out of marks*** →	**1** **1 1** **1 1 1** **1 1 1 1**	
	1 0 0 0 **1 1** 0 0 **1 1 1** 0 **1 1 1 1**	**1. Change the triangular setting out into a square table.**

<table>
<tr><td></td><td>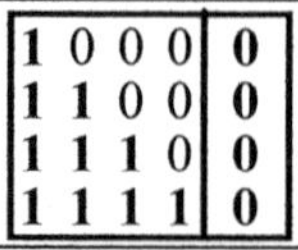
</td><td>2. Change the square setting out into a rectangular one by adding a new column.</td></tr>
<tr><td>$\frac{n(n+1)}{2}$ □ ←
Conversion that is a literal DESCRIPTION of triangular setting out.</td><td>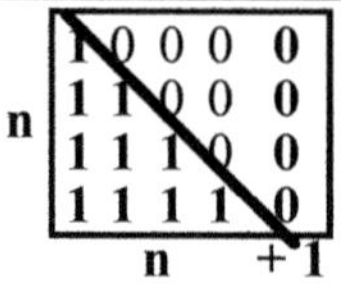
</td><td>3 Re-configure the rectangular setting into two equal triangles.</td></tr>
</table>

Figure 10. Splitting up of visual treatment into steps

Whatever the function given to this visual treatment for solving the problem—intuition, heuristic, explanation, proof—continuity and progress through the four successive visual representations must implicitly or explicitly recognised in order for visualisation to work and meet any of these functions.

The atypical case impossible to dismiss: treatment within native language

We need, on this subject, a more precise definition of what constitutes a mathematical treatment. The above example has already highlighted two striking features: a change made in the same kind of representation and producing something new through a continuous derivation.

But this is not sufficient to account for how any valid reasoning involved in proofs works. This change in mathematics consists of a substitution operation as in any calculation, although it is a little more complex, because two levels have to be separated (Duval, 2007b):

- A level of deduction step in which a conclusion is substituted to premises according to the rule of detachment *(modus ponens*). This can be done only in reference to a third statement, which is a theorem, a definition or an axiom.
- A level of step organisation is necessary, in which the conclusion of the last deduction step is substituted to the one(s) of the previous step(s).

Although it is less easy to operate and check than in any symbolic system, it can and must be done also in the native language, at least for teaching elementary geometry. The trouble is that such a treatment based on substitution operation is an unnatural practice of speech in one's native language, even in debates. Here, linking sentences is based on a composition process as in painting. What gives meaning to a new statement, in speech, is its contextual relations to all others statements, recent or remote, taken as a whole. Natural reasoning also works in this way.

Thus, the use of theorems in mathematics is diametrically opposed to the use of arguments outside mathematics (Duval, 1992). We do not use theorems in the same way as we use arguments. Several convergent arguments are often required to justify a thesis, whereas only one theorem is sufficient for a necessary conclusion at the level of step organisation. The difficulty of discriminating opposite ways of functioning there seems all the more insuperable, since their common use of the same logical connectors is completely equivocal.

If symbolic systems fit treatments better by a twofold discursive substitution than native language, then should teaching abandon any introduction to proving through the native language, and prefer diagrammatic and more computational presentations? In this case, how can students be made to understand the specific mathematical requirements for defining, since there is no definition without using the native language?

VI. WHAT ARE THE SEMIOTIC SYSTEMS IN THE FIELD OF GEOMETRY?

Geometry is the area where the visual representation of shapes and solids, the conceptual analysis of their intrinsic spatial relationships, and the creative power of deductive reasoning seem completely merged together. This is why geometry is the touchstone for a semiotic approach to mathematics. Let us take the "geometric figures" that are like symptoms of this cognitive merger. How to analyse them? To manage an efficient method we have to carry out three necessary dissociations.

Separating the visual aspect from the properties stated

The so-called "geometric figures" are visual representations that are coded with letters or marks indicating given properties. But letters and marks refer to a statement. Also, we always have, be it explicitly or implicitly, a dual representation.

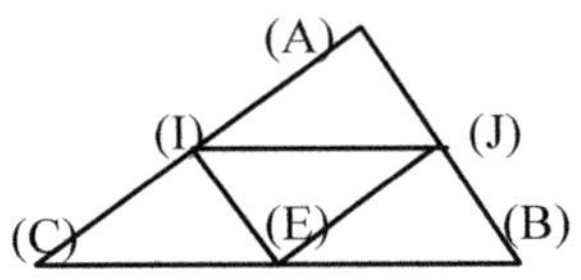

AC and EJ are parallels. AB and EI are parallels. CB and IJ are parallels.
Prove that E is the middle of CB

Figure 11. A dual representation articulated by anchor marks

In any dual representation, letters and marks anchor the reference points of the statement in the visual representation, but they belong to the statement and not to the visual representation. The visual representation works in a way completely independent of what the given encoded imposes, because for the same statement we have several possible visual variations.

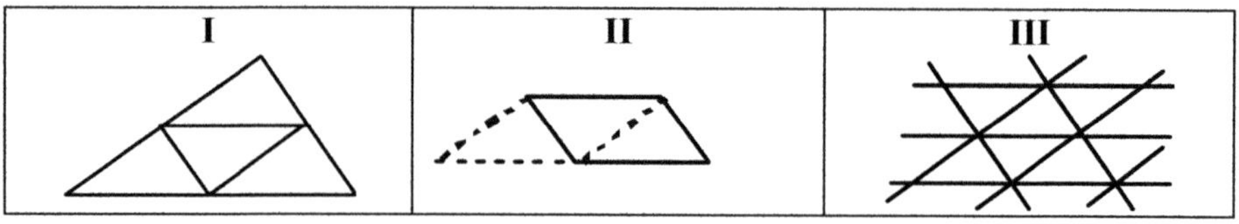

Figure 12. Visual variations of a geometric figure

If (I) is given as the starting dual representation then (II) must be recognised as a subfigure of (I) for solving the problem, and (II) (III) can also be given as starting dual figures within one dual representation. Analysing any geometric figure requires that we analyse the visual variations in which it appears as a transitory foreground amongst many others.

Discriminating figural units in any geometric figure

It is the requirement whose complexity is seriously misunderstood because, without this discrimination, any figure, even the most simple, becomes misleading or opaque. This complexity is due to the possibility of quite different ways to split up the content of a geometric figure into figural units. For distinguishing them, we have to take into account the number of dimensions. Thus, we get these two basic ways of splitting up.

The figural units have a number of dimensions equal to that of the starting figure: 2D → 2D in the framework of plane geometry. In this case, the figural units correspond to the shapes, which are closed outlines. For example, in Figure 12 above, we can discriminate in (I) two figural units (the two parallelograms (II) or three triangles). This is the spontaneous perceptive interpretation.

The figural units have a number of dimensions less than the one of the starting figure: 2D → 1D (or 0D). In this case, the figural units are straight lines free of any closed outlines. For example, in Figure 12, we can discriminate in (I) six figural units which are the straight lines making up the network (III) underlying (I) and (II). This discrimination goes against the immediate perceptive organisation.

It is through this double possibility of splitting up that geometric figures constitute a particular kind of semiotic representation with a powerful potentiality for visual treatment (Duval, 2005a).

Two visual treatments mutually incompatible

The visual treatments depend first on the kind of figural units with which you work.

The treatment by reconfiguration works with figural units with the same number of dimensions as the one of the starting figure. It is this one which is highlighted in most teaching studies, because it provides a non-formal justification for formulas or geometric properties. Its function is mainly heuristic. But it is less obvious and natural than most studies assume, because it depends on several factors triggering or inhibiting the various reconfigurations (Duval, 1995b).

The treatment by dimensional deconstruction works with figural units with a smaller number of dimensions than in the starting figure. This is the relevant visual treatment from a mathematical point of view. The reason is obvious. The geometric properties being relations, they can be visually represented only by two visual units. For instance, the definitions of polygons, polyhedra, are based on this dimensional deconstruction of shapes and solids. More generally, this amounts to work with the network of straight lines underlying a geometric figure, in which many other different figures can be recognised.

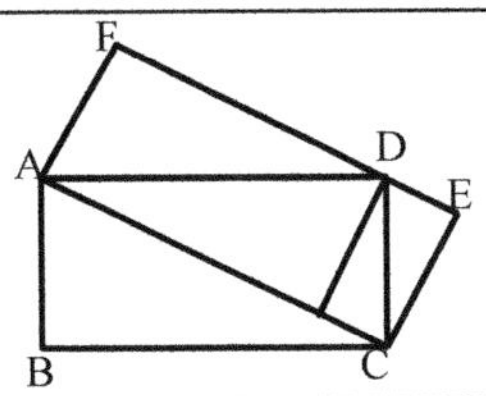	**Let ABCD be a rectangle. Is the area of the rectangle ACEF greater, equal to or less than the area of ABCD? Explain your answer.**
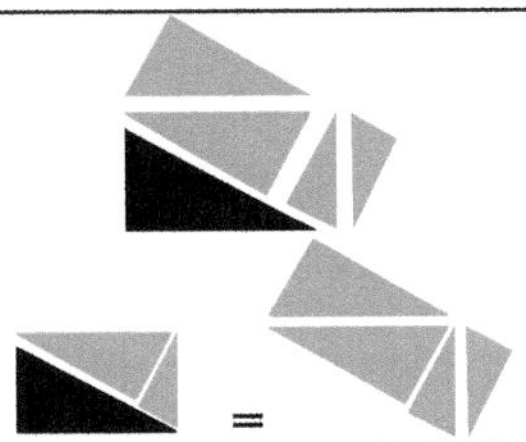	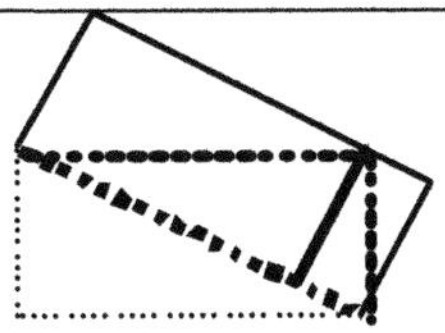
Treatment by reconfiguration: overlapping of 2D figural units	**Treatment by dimensional deconstruction, focusing on dotted lines as diagonals and not as edge of 2D shapes.**

Figure 13. Two visual treatments

This poor example can help us to get at the semiotic and cognitive complexity of geometry. We can condense it into three proposals:

(1) Dimensional deconstruction is a semiotic operation which goes against the spontaneaous perceptive identification (Duval & Godin, 2005). Most students have no inkling of this. No deep progression can be expected as long as geometry teaching misunderstands this.

(2) Each of these two visual treatments turns specifically towards one of the two kinds of discursive justification: either description/explanation or deduction.

(3) The ability to manage these two visual treatments is a prerequisite for analysing a geometric figure according to the given encoded.

VII. DOES THE SEMIOTIC APPROACH LEAD TO DEVELOPING A METHODOLOGY PROPER TO RESEARCH IN MATHEMATICS EDUCATION?

First, the main thing is not to confuse two questions. One concerns the identification of variables that determine the development of comprehension in mathematics. The other concerns the analysis and interpretation of students' productions that are collected either in classrooms, through individual observation or experiments. The first is too often ignored or reduced to the second, and yet, it is crucial because it involves a model of the functioning of mathematical thinking.

What are the variables that need to be taken into account for organising a learning situation?

The core variables must be determined from the unique epistemological situation of mathematics (above I (Q.3)), the cognitive paradox it creates (III) and the nature of mathematical activity as a change of semiotic representations (II, V). This means that we cannot start first from the particular mathematical contents to be taught, but through them, we must focus on the conditions that will allow students to understand and use them as expected from a mathematical point of view.

We have seen that there are two fundamental cognitive processes intrinsically linked to the mobilisation of multiple semiotic systems of representation: the conversion and treatment. This means that we must start by separating the variables related to the conversion and those related to the treatment. But to make this separation operational, we must have a precise classification of the various registers of semiotic representation, i.e. systems used not only for communication or objectivation, but also for the purpose of treatment.

The principle of classifying all possible kinds of representation, both the semiotic and non-semiotic, is very simple: there are as many kinds of representations as systems producing representations. If we confine ourselves to semiotic systems, there is the classic distinction between languages and all the various kinds of pictures (drawings, figures, diagrams, etc.). But this distinction must be crossed with another essential for mathematics: those whose the treatments can be made into algorithms and those for which this is not possible when it comes to the native language or geometrical figures (V above). In this way, we get the four classes of registers used in mathematics (Duval, 2006a, Figure 1, p. 110).

Now, it is easy to identify the variables related to conversion. The cognitive distance underlying any conversion depends on the kinds of source and target registers. So, there are as many kinds of conversions to investigate and take into account, as there are various couples of registers.

The variables related to treatment depend on two things specific to each register:

- the means of producing quite new representations (for instance, by constructed designation for the discursive registers)
- the means of changing semiotic representations in a continuous and creative way by composition, substitution, reconfiguration, dimensional deconstruction, according to the possibilities specific to the register chosen

These two kinds of cognitive variables must be taken into account to organise, as well as experiment, devices as activities for learning in the classroom.

How can we analyse and interpret the students' productions?

This issue is the main challenge and a flimsy point in the current research in mathematics education. The vast majority of studies are validated by reproductions *in extenso* or much too lengthy quotations of what very few students have told, written or drawn. We are in fact faced with a mass of raw data, whose interpretation is half clinical, half assessment and, in any case, difficult to rebuild or check. It is like a validation by testimony.

One semiotic approach allows describing at least one analysis procedure. It comprises three stages:

(1) The observations being made in the context of a problem; it is essential to begin by making the representational map of the whole representational working field (Figures 1, 6 and 9 above) in which a search for solving can be managed by students. This does not depend on what the students have done, but on what is given to them or expected of them.

(2) This working field is a tool for breaking each student's production into segments or interpretable units according to:

- the passages he/she does or does not move to (i.e. the ways he/she finds or does not find) between the different registers of representation
- the register chosen by the student to perform a treatment

So, we obtain two kinds of significant observations.

(3) Finally, on this basis, a verifiable comparison can be carried out between the various productions collected. Naturally, this comparison can be correlated with the level of mathematical performances, without being confused with them. And, this comparison can also be extended to productions collected over long periods of time, in order to observe whether the comprehension evolves in depth or not.

Besides a diagnosis of students' particular difficulties, this procedure highlights, under the global performance of a group or classroom, the individual disparities of comprehension and preparation for further acquisitions.

Conversion as a method for discriminating meaning

The representation conversion is the most powerful tool that the semiotic approach provides. But this is beyond the scope of this overview. I will confine myself to the next question which is a recurring problem for most students and ... teachers! How do we discriminate–or make students discriminate–various kinds of statements that look alike linguistically because they use the same words? For instance, how do we discriminate a proposition from its converse, a word problem from another word problem whose solution is different–as is already the case with additive problems with only one operation–and more generally, a valid reasoning from a non-valid reasoning?

Teachers have long since noted the need for auxiliary or transitory representations, and the educational innovation possible in this area seems inexhaustible. The controversial issue is whether all supports devised provide the means for truly distinguishing the difference in meaning between two statements that look alike, and determining which statements' meanings are mathematically relevant. Most innovations often take that which precisely causes trouble for students for granted, namely, converting representations.

In reality, it is not the auxiliary representations that are important for understanding, but the conversion activity that leads to their production. The auxiliary representations must therefore meet some semiotic specifications:

- highlighting the discursive operations which underlie statements
- providing a non-iconic visualisation of their organisation
- being easy to produce and check by the students themselves

Graphs on a semantic bi-dimensional representation seem to be the best candidates and have been shown to be effective with students (Duval 1992, 1995, 2007b).

VIII. DOES THE ANALYSIS OF COMPREHENSION PROCESSES IN MATHEMATICS ENTAIL A SEMIO-COGNITIVE APPROACH TO THE DEVELOPMENT OF MIND?

This is a borderline problem with regards to the other approaches to knowledge acquisition. We cannot study the processes of comprehension and learning in mathematics without implicitly or explicitly assuming a model for the development of mind.

A consensus seems always to have existed about the need to import models from other disciplines in order to take learners and the processes and conditions of knowledge acquisition into account. Thus, the influence of Piaget's genetic epistemology has led to analysing knowledge acquisition in terms of action and concepts, regardless of any semiotic system, and considering them to be outside mental processes. Now, the priority given to the interactions between students in the classroom has led to favouring socio-cultural approaches.

The epistemological situation peculiar to mathematics, the cognitive paradox it creates and the specific problems of understanding that its learning raises compel us to widen the field of questions and observations about thinking processes. Mathematics education requires new cognitive models, more complex than the ones generally accepted. A quick evocation of two questions will be sufficient for highlighting the importance of a semio-cognitive approach, not only in mathematics education, but also for a general modelling of mind.

What is the nature of thinking (noesis)?

In mathematics, any activity requires the possibility of using various semiotic representation systems (registers), even if very often only one is explicitly chosen according to the particular contextual purpose of the activity. Hence, the crucial importance of the ability to convert the representations arises. Therein lies the first source of trouble for students. Conversion seems to be something like a magic trick that cannot be truly learned and that is not taught! The common assumption that conversion should depend on previous acquisition of the relevant mathematical concepts cannot hold for two reasons. First, it amounts to ignoring the cognitive paradox of mathematical knowledge. Next, the semiotic nature of conversion, and not its conceptual one, is evidenced by two facts that can still be checked. Students are able to convert only if the source representation and the target representation are congruent. They can have a complete mental block if the representations are not congruent (Duval, 1995a). When one reverses the direction of the conversion, it becomes quite a different task for students, and their performances may fall considerably (Duval, 2006a).

The ability to convert requires an internal **co-ordination** between the various representation systems that can be possibly selected and used. Without this internal co-ordination**, which must be constructed,** two different representations mean two different objects without any relation between them even, if they are two representations of the same object. Thinking in mathematics is a synergy between at least two mobilised registers, even when one works explicitly in only one register. From a cognitive viewpoint, thinking functions as a synergy of many different semiotic systems. No *noesis* without *semiosis*.

How can we situate semiotic, "mental" and non-semiotic representations in the whole functioning of mind?

The common assumption is the functional subordination of semiotic representations to both mental and non-semiotic representations. On the one hand, semiotic representations would be "external", whereas mental representations would be "internal"; on the other hand, non-semiotic representations would be prior to the semiotic, whereas these would only be a substitute for absent or things at a distance and non-semiotic representations. This is wrong.

Semiotic representations are neither "external" nor "internal". This opposition rests on the confusion about the production of representations. In their production, we must distinguish two aspects independent of each other: the kind of system used for producing representations and the phenomenological modes of their production. For instance, Vygotsky only paid attention to the three modes of production in the native language: mental, oral and written (visual). Thus, most of our mental representations are semiotic. The co-ordination of registers of semiotic representations provides something like an extension of our mental capacity and provides the means of the mind's development (Duval, 2007a).

CONCLUSION

Mathematics has two essential characteristics that deeply distinguish it from others forms of knowledge. Its knowledge objects are only accessible through semiotic representations and, above all, mathematical activity grows through changes and substitutions of semiotic representations. These two characteristics are at the root of problems of comprehension in mathematics learning for most students. The educational issue is about the methodology and the theoretical framework relevant to both cognitive and mathematical viewpoints, if we want to analyse these problems of comprehension.

It is essential to separate structural analysis from functional analysis. Structural analysis focuses on the variety of semiotic systems and aims at bringing out the functioning particular to each. Functional analysis focuses on the selection of one or more registers and the role given to their use. This is always done in the particular context of a problem to be solved or an activity to be organised. It can lead to opposite divergent results depending on whether one takes an expert's viewpoint or looks at actual students' productions. Functional analysis cannot be significant without a previous structural analysis. One of the major weaknesses of semiotic approaches so far is the lack of structural analysis relevant to mathematics.

It is also essential to separate the mathematical analysis of a problem from the semio-cognitive analysis of the various tasks to be performed in order to be in a position for solving it. An *a priori* complete description of the working field of a problem is a tool for distinguishing difficulties relating to conversions and those relating to managing treatment in a register. Thus, it reveals that, under the surface level where mono-register activity prevails, there is a deep level where co-ordination between registers may or may not be achieved for each student. Thinking in mathematics depends on the synergy of several registers and not on the activity of a single system. Unlike what occurs in other fields, mathematical concepts are only understandable within such a synergy.

There remains, therefore, a problem not mentioned in this overview, which is nevertheless crucial for comprehension in mathematics. This is the problem of proof. Would not the epistemological nature of proof invalidate the intrinsic role of semiotic representations that we have highlighted? To think so would amount to forgetting that a proof lies in the controlled experience of necessity. But there are three kinds of necessity, each one being completely different: the physical one, the logical or semiotic one, and the social one (Duval, 2005b). What kind of experiential necessity are mathematical proofs based on?

A real semiotic approach is ahead. It opens promising prospects for research in mathematics education.

REFERENCES/BIBLIOGRAPHY

Duval, R. (1992). Argumenter, démontrer, expliquer: continuité ou rupture cognitive. *petit x*, *31*, 37–61.

Duval, R. (1995a). *Sémiosis et pensée humaine*. Berne: Peter Lang.

Duval, R. (1995b). Geometrical pictures: Kinds of representation and specific processings. In R. Sutherland, & J. Mason (Eds.). *Exploiting mental imagery with computers in mathematics education*. Berli Heidelberg: Springer.

Duval, R. (2005a). Les conditions cognitives de l'apprentissage de la géométrie: Développement de la visualisation, différenciation des raisonnements et coordination de leurs fonctionnements. *Annales de Didactique et de Sciences Cognitives*, *10*, 5–53.

Duval, R. (2005b). Compréhension des démonstrations, développement de la personnalité et formation de la conscience individuelle. In D. Tanguay (Ed.), *Actes du colloque GDM 2005.* Montréal: UQAM.

Duval, R. (2006a). A cognitive analysis of problems of comprehension in a learning of mathematics. *Educational Studies in Mathematics, 61*, 103–131.

Duval, R. (2006b). Quelle sémiotique pour l'analyse de l'activité et des productions mathématiques. In L. Radford & B" D'Amore (Eds.), *Semiotics, culture, and mathematical thinking. Revista Latinoamericana de Investigación en Matemática Educativa (Special Issue)*, pp. 45–81.

Duval, R. (2006c). Transformations de représentations sémiotiques et démarches de pensée en mathématiques. In J.-C. Rauscher (Ed.), *Actes du XXXIIe Colloque COPIRELEM.* Strasbourg: IREM.

Duval, R. (2007a). La conversion des représentations: Un des deux processus fondamentaux de la pensée. In J. Baillé (Ed.), *Conversion* (pp. 9–45). Grenoble: PUG.

Duval, R. (2007b). Cognitive functioning and the understanding of mathematical processes od proof. In P. Boero (Ed.), *Theorems in schools: From history, epistemology and cognition to classroom.* Rotterdam: Senses Publishers.

Duval, R., & Godin, M. (2005). Les changements de regard nécessaires sur les figures. *Grand N, 76*, 7–27.

Frege, G. (1971). *Ecrits logiques et philosophiques* (tr. C. Imbert, Trans.). Paris : Seuil.

Raymond Duval
Laboratoire LCEM,
Université du Littoral Côte d'Opale

MICHAEL OTTE

METAPHOR AND CONTINGENCY

INTRODUCTION

Metaphor replaces logic as the language of change and evolution, it is said, and it aims far beyond poetry and aesthetic pleasure; "it flourishes in playful prose and high poetic art, but it is vital also at the growing edges of science and philosophy" (Quine 1978, p. 161). Metaphor represents Thirdness, "or the mental ... influence of one subject on another relatively to a third" in terms of Peirce's fundamental categories (CP 5.469), whereas logic is Secondness. Thus, for metaphor to be relevant for conceptualizing change, the latter must be understood simultaneously in terms of continuity and discontinuity and the growth of knowledge then appears neither as simply cumulative nor unpredictably disconnected.

Metaphor is paradigmatic of radical creativity; it is its very heart. Metaphor is therefore sometimes understood exclusively in terms of use, rather than content. Creativity is something completely spontaneous, or so it is believed. In general, skepticism about a complete explanation of creativity is based on the notion that a creative act gives rise to an outcome that is new in kind, was unpredictable, and has a definite character that is not reducible to the sum of its elements. Metaphorical insight is, in fact, a kind of seeing or intuition, and trying to transform this insight into an explicitly conscious process results in something like Xenon's paradox of the race of Achilles and the Tortoise. But still, creativity depends as much on continuity and redundancy as on spontaneity and chance (otherwise one would not even know how to recognize new ideas).

All our understanding is more or less metaphorical, because meaning depends on context. The apparently craziest ideas may be fertile if conceived of in the appropriate context and from the right perspective. Metaphor brings tools of inference over to new contexts and its major strength resides in this (Pinker 2007, p. 252 ff). But mathematics and logic represent, it is usually claimed, literal, decontextualized and universal knowledge. Since the 19th century, there has been in particular a "strong trend in the development of mathematics to replace seeing with understanding" (Goodman 1983, p. 63). From the point of view of conceptual mathematics, metaphor appears as useless and as aberrant speech. Mathematics is not poetry, Frege used to say.

Maybe! But those who do not appreciate poetic metaphor will most probably not esteem pure mathematics either. And mathematical discourse is even less likely to become that minutely detailed logical narration that Frege wished it to be. The search for clarity of vision does sometimes even contradict complete conceptual itemization, as the various foundational and educational crises show.

L. Radford, G. Schubring, and F. Seeger (eds.), Semiotics in Mathematics Education: Epistemology, History, Classroom, and Culture, 63–82.

The conception of mathematics as espoused by Bolzano or Frege has, from the very beginning, been contrasted with a different "model-theoretical" conception of mathematics as "diagrammatic reasoning" (Peirce), more favorable to metaphor, because conceiving of mathematics in terms of diagrammatic reasoning implies a genetic view.

All exposition of mathematical theories since Euclid's "Elements" with its "theorem-proof"-structure has essentially been metaphorical—reasoning from general ideas, rather than catalogues of particular examples—and thus belonging to an ideal world, rather than to fuzzy reality. Literal interpretation is often grotesquely beside the point. In everyday discourse, humor comes to the rescue and in science or mathematics, idealization and generalization. Metaphor implies generalization and does not contradict abstraction; it is not reductionistic, contrary to today's most prominent theories of metaphor. Colloquial language and common speech are full enough of embodied metaphors, it is true. But what is at stake is generalization and not reduction. Metaphors are generalizations establishing general ideas as in the "equation" *heat=motion* creating the abstract notion of energy. This is a paradigmatic example. How could we reach beyond the directly accessible and conceive of the general notion of energy? How did scientists arrive at a theory of electricity, for example, or at the atomic or subatomic laws? By metaphor! (see also Steiner, 1998).

Most theories of metaphor are misguided and incoherent, because they do not see this alternative and ignore the fact that generality does not necessarily mean conceptual generality. Metaphors offer quite a different view of generality: relational generality in terms of continuity and discontinuity or, as generic coherence or kinship and as something "between" the particulars, rather than "above" their distinctiveness. Generality, according to Charles Peirce's semiotic approach means primarily the objective continuity of ideas. Mathematical ideas no more exist objectively in the world than colors, it is said by reductionists. But colors and ideas are objective in as much as they have an effect on us. They are signs and signs make objective possibilities work and cause their interpretants to spread continuously, although in sometimes unpredictable directions.

The understanding of metaphor offers the possibility to be a realist about mathematical (ideal) objects without subscribing to the Platonic view that the existence of these objects is wholly independent of the symbolic means via which they emerge (Otte 2003a, 2003b).

FROM INTUITIVE CERTAINTY TO LANGUAGE AND EXPLANATION

The quest for certainty is among the most important and fundamental motivations of the modern age. It was a driving factor in the emergence of science and especially shaped the debates about the role mathematics could or should play in science and scientific knowledge. Benjamin Nelson has pointed out that the main impact of the Scientific Revolution of the 17th century came from a change in the habits of thought and, in particular, from a campaign for individual cognitive certainty, as against catholic casuistry and probabilism. This provided mathematics with a special social and epistemological role.

Aristotle had already declared that it would be "equally foolish to accept probable reasoning from a mathematician as to demand strict demonstrations from a rhetorician" (Aristotle, Nicomachean Ethics 1094b12-28). Such statements were repeated over and again up to the 16th and 17th centuries. Descartes made the problem of intuitive certainty pivotal to his philosophy. The general purpose of the *Discourse on Method* was to show how he himself had encountered the problem and dealt with it. And his *Regulae* (1628) paradigmatically stress that we are to "make it a rule to trust only what is completely evident and incapable of being doubted" and to reject merely "probable knowledge" (Rule II). "For Descartes speech was an epistemological obstacle, because it was an easy vehicle for the seductive inducements of eloquence and emotive persuasion – hence the denunciations of rhetoric that are so common in Galileo, Descartes and Locke" (Aarsleff 2001, p. xiii). Locke, for instance, laments that words "interpose themselves so much between our understanding and the truth .... that, like the medium through which the visible objects pass, their obscurity and disorder does not seldom cast a mist before our eyes and impose upon our understandings" (Locke, 1894, pp. 395-96). Book 3 of Locke's *Essay* is considered the first semiotic treatise in our history and it is in its greater parts devoted to spelling out the "imperfections" and "abuses" of words.

What Galileo, Descartes, Locke and others tried to do was to establish a new method appropriate to enlarge the mental capacities of men such that they could gain true and certain knowledge about the world from the Book of Nature itself, and these methods had to be essentially mathematical methods. However, mathematical analysis and description of Nature rarely could provide demonstrations in the traditional Aristotelian sense, that is, proofs that do not only provide "knowledge that", but give real explanations, or "knowledge why", because mathematical explorations always depend on new and, as such, personal intuitions and ideas, rather than on common words and familiar meanings. What you did not see that could not be readily communicated to you, it was believed. Right knowledge was considered a private achievement and language itself was under suspicion of being merely a box of rhetorical tricks. Communication is risky, it was believed, and could at best provide probable knowledge.

This attitude changed gradually during the 18th century and a new understanding of the nature of language and of the social constitution of man took place. Now communication was no longer risky but creative. And it was, in fact, Etienne Bonnot de Condillac (1715-1780), a major figure of Enlightenment France, famous throughout Europe during his lifetime, who first considered language as essential to mental development and to knowledge, and who was "relentlessly Anti-Cartesian" (Aarsleff, 2001, p. xii). The foundations of knowledge must be based on signs and languages, rather than intuitions. In the "Essay on the Origin of Human Knowledge" (1746), Condillac presents an empiricist account of knowledge more or less in the tradition of Locke. But Condillac's empiricism took a completely new twist in comparation to the British empiricists, because of his semiotic conception of knowledge. Whereas Locke claimed that the function of language is only secondary in the communication of ideas, which could exist independently of it, Condillac insisted that the function of language was constitutive in their formation. This claim culminated in the view that knowledge itself is a well-made language, with Algebra as the most important example.

Condillac himself commented on the insufficiencies of Locke's approach. The fact that objects or substances as parts of reality hold together complex bundles of general properties "prevented him from seeing the necessity of signs for the notions of substances (and) … prevented him from discovering the great importance of signs for the exercise of the operations of the mind" (Condillac in Aarsleff, 2001, p. 91).

Most important is the fact that Condillac realized that the role of signs is essential to an evolutionary epistemology and developmental account of knowledge, thereby establishing a genetic view of language and pointing out the indispensability of a semiotic approach to cognition and epistemology. For the Condillacians, in contrast to the Cartesians, "it is the art of speaking which leads to the art of thinking" (Joly, 1985, p. 148). This semiotic approach to epistemology went along with a radical cultural shift towards emphasis on natural sociability. Signs become signs first of all in the context of social life. In a letter to the Swiss mathematician Gabriel Cramer, Condillac writes:

> That's what my entire system comes down to in this matter. Social intercourse gives occasion (1) to change the natural cries into signs; (2) to invent other signs that we call arbitrary; and these signs (the natural and the arbitrary) are the first principles of the development and progress of the operations of the mind." (Condillac in Aarsleff, 2001, p. xxviii)

The genetic perspective on cognition and knowledge is based on the complementarity of the connotative and expressive functions of language, on the one side, and its denotative and representational properties, on the other side. "Both the representational and expressive theories of language operate at once in Condillac's texts, resulting in dual theories of rhetoric" (Hobbs-Peaden, 1993, p. 140).

Condillac considered notions, like metaphor or abstraction as "bound to be very defective" and felt that "it would probably be risky to use them", the risk being that we may fall prey to a mere play of empty words. Abstractions and metaphors are, however, indispensable, taking into account "the limitation of our mind" (Condillac in Aarsleff, 2001, p. 93) and we must reify them because "the mind cannot reflect on nothing" (Condillac in Aarsleff, 2001, p. 94). To bring concrete ideas to new and general questions requires signs and there are no signs without abstraction and generalizations. General ideas "reveal beings that are concealed from the rest of mankind" (95). Condillac endorsed a nominalist view of signs or abstractions; they do not exist really, but are functional devices essential to the development of human cognition and identity. What, in fact, is really at stake is the constitution of the human mind itself – "men and words reciprocally educate each other" (Peirce, CP 5.313). This latter view, it should be added, comes out more explicitly and fully only in the works of Herder, Schlegel and others from the Romantic movement.

And since "the mind is too limited to comprise all at once, both its being and modifications, it must distinguish them by forming abstract ideas" (Condillac in Aarsleff, 2001, pp. 93–95). These ideas, or the symbols by which they are represented, are simultaneously forms and processes of interpretation. Interpretation consists in nothing but the construction of new representations, which are to a certain degree constrained by the given one. But in course of these constructive activities it always happens that something unexpected and contingent enters, hence the risks about which Condillac speaks. One might perhaps even say that science maturated when people learned to use metaphors and analogies in more disciplined ways. Scientific "revolutions" produce in return new metaphors in abundance (the Romantic movement in early 19th century Germany provides a case in point—see Leventhal (1986, p. 250) and Otte (1989)).

Condillac recognizes the importance of the social condition for the development of human beings and he considers "signs of action", that is, the grasp of significance in some pre-linguistic way, as a first step in this social development of man. According to Condillac, the use of signs, which is essential to human development, is learned first from action and in a narrator's perspective, rather than in the role of the spectator (with respect to similar modern theories: Bruner 1990, p. 70). It begins with "natural signs", that is, items of behavior which were naturally associated with certain kinds of experience. However, from this natural basis, Condillac claims, we bootstrap ourselves into creating signs which are conventional and represent pure abstractions and these abstractions function as instruments of analysis and inference. It is thus that a language becomes "an analytical method".

Once more, one should add here in passing that philosophical idealism and language theory as it appears in the works of Herder, Schlegel and others emphasizes, in contrast to 18th century Enlightenment, historical self-consciousness (the role of the spectator) and metaphor, and questions Condillac's positing of "a level of sensate cognition or experience prior to distinct cognition" (Leventhal, 1986, p. 250).

All that counts in the novel is narration and the narrative is structured by the plot or the story. If there is no good story to tell, the most beautiful language does not help much. In poetry, things are different, reflection on structure and the poetics of language is all. And the nature of metaphor seems very different in both cases. Levi-Strauss has made the appearance of this complementarity the touchstone of the modern age. He returned many times to this issue, stating that mythical thought had been replaced during the Scientific Revolution of the 17th century not just by science; its essential aspects have been handed over to structural thought, on the one hand, and to romantic or expressive literary narration, on the other. "It seems as if music and literature have shared the heritage of myth among themselves. Music has served itself from its form and the novel has taken the non-formal rest developing, free from the constraints of symmetry, the means of free narration" (Levi-Strauss, 1971, Mythologiques IV, final chapter).

MATHEMATICS, EDUCATION AND LANGUAGE

Since the 19th century, mathematical discourse has become more and more influenced by the requirements of teaching on larger social scales and of text production for a wider readership, and mathematical institutions have come under the influence of philology. Indeed, the impact of language theory in the works of Herder or Humboldt "culminated in the plans for the construction of the University of Berlin in 1809-1810" (Leventhal, 1986, p. 243).

Mathematicians have reacted in mixed ways to these developments; on the one side, there has been the so-called movement of "arithmetical rigor", and, on the other side, the development of modern axiomatics. The program of rigorization by arithmetization endeavored to solve the foundational problems in a reductionistic manner, by defining all mathematical concepts in terms of some basic entities, ultimately natural numbers.

The axiomatic movement, in contrast, as anticipated in the work of Poncelet or Grassmann, tried to employ, so to say, a top-down strategy, solving the foundational problems of mathematics by extending and generalizing its relational structures and its rules of inference. Grassmann's dropping of the commutativity of a general product and his definition of the anti-commutative vector product furnish a pertinent example here.

The main difference between these two programs is in their views about theory and its application. Arithmetized mathematics reifies mathematical theory in its belief in a direct application, whereas, the axiomatic approach stresses that theories are realities *sui generis* that have to be grasped in their own terms, before one can think of meanings or intended applications. The reductionistic approach of the rigor movement has always been dominant in educational context, in schools and universities alike.

Along with this split in mathematics, there have been similar bifurcations in logic (Heijenoort, 1967), philosophy and the sciences. "Some philosophers have remained faithful to the Enlightenment and have continued to identify themselves with the cause of (positivistic, M.O.) science ... Other philosophers ... have concluded that science is no more than the handmaiden of technology" (Rorty, 1989, p. 3). One group thought metaphor useless or superfluous; the other emphasized its indispensable role in creative development. And, in mathematics, the same occurred.

Take for example the metaphor of the complex Gaussian plane. Members of the group of arithmetical rigor, Cauchy, for instance, did not see the geometrical representation of complex numbers as something important and conceived of these numbers in terms of pairs of real numbers. Whereas, in the hands others—think of Grassmann himself, for example—this model meant the recognition of the fact that mathematical concepts should be developed by a recursive interaction of operative and intuitive reasoning, and this acknowledgement led to new structural theories, like linear algebra or group theory, for example.

Imaginary numbers became essential algebraic tools as soon as G. Cardano (1501-1576) and R. Bombelli (1526-1572) tried to solve algebraic equations of the third degree. And things remained as they had left them until, at the turn of the 19th century, quite a number of people suddenly tried to provide new meanings and new uses for these numbers. Furthermore, by looking more closely at the works of Wessel (1745-1818), Buée (1748-1826), Argand (fl. 1806, 1813, 1814), Gauss (1777-1855), Hamilton (1805-1865), Hermann Grassmann (1809-1877) and others, one essentially ends up perceiving two things.

First, although the metaphors employed heuristically to establish the geometrical representation of imaginary numbers varied widely and drew on the greatest possible diversity of metaphysical accounts, the resulting mathematical theories were more or less all the same. Secondly, however, the geometrical representation of imaginary numbers and related developments like, for example, the discovery that there is a geometrical theory independent from measurement and distance, led to a new understanding of mathematical theories as realities in their own right, and this in turn stimulated a structural conception of algebra and axiomatical mathematics.

Arithmetized mathematics served the consolidation of pure mathematics defining a separate group of professionals and setting the training requirements for membership in such a group. Axiomatization assists cooperative efforts of developing new applications because of the way complexity is reduced by the axiomatic method (Otte, 1989).

These different attitudes correspond to what Umberto Eco called the strategies of open vs. closed texts. As a rule, mathematical text books try to pull the reader along narrowly determined paths. These texts "that obsessively aim at arousing a precise response on the part of more or less precise empirical readers ....are in fact open to any possible aberrant decoding. A text so immoderately open to every possible interpretation will be called a *closed* one" (Eco, 1979, p. 8). Closed texts are like algorithmic processes or are written in a formalized and literal manner. These texts "presuppose an average reader resulting from a merely intuitive sociological speculation". In math education, one is confronted over and again with affirmations to the effect that the "average" student will not understand this or that, in consequence supporting a reductionist teaching strategy. Such reduction tries to reduce everything to simple factual statements and to rule knowledge. Statements of fact are, however, not meaningful *per se*. And, in the world of isolated facts, everything seems equally possible and the various possible interpretations of facts are not easily transformed into some coherent whole.

"Open" texts, in contrast, project their "model readers" and develop them, that is, they create in the reader's mind a certain attitude and expectation, which transform the process of interpretation of a text into "a structural element of its generative process" (Eco, 1979, p. 9). Their meaning is not fixed either; open texts stimulate and organize their interpretations and the various interpretations of an open text form a more or less coherent network, thereby enlarging the meaning of the text.

Metaphorical statements appear to be open texts par excellence, and some mathematics teachers may not like metaphors, sometimes because of an alleged reductionism and at times because of this openness. Metaphor is not just analogy, although its cognitive value very often depends on good analogy. Metaphor makes one see something in a certain light or from a certain general perspective, which is not the direct result of one's own prior understanding. According to Aristotle "a good metaphor implies an intuitive perception of similarity in the dissimilar" (Aristotle, Poetics: 1459a).

In order to understand a metaphor as a means of demonstration, one has therefore to introduce intermediate terms breaking up the reasoning into small steps such as to make similarities or equivalencies obvious. Such reasoning occasionally takes quite unexpected directions and may in sum result in quite surprising insights and dramatic overall changes of perspective. The creativity of metaphor is thereby explained as based on chains of contiguity relations and it always involves elements of contingency. We have explained the process in great detail elsewhere using an example from geometry (Otte, 2006).

A metaphor can, according to the traditional Aristotelian view, be represented as an equation $A=B$. This symbol $A=B$ contains something equal and something different as well; therefore, it differs from $A=A$. In order that a metaphor obtain, it is essential that both be present: the difference or incompatibility as well as the equality or new compatibility. Depending on where one places one's priorities, one may see such an equation in two different ways (Otte, 1989, p. 21). One may begin with an action and a distinction, and would then have to look for a possible relationship or compatibility. Or one might, as in the Fregean interpretation of $A=B$, conceive of A and B as different representations of one and the same object, which, however, need not be a concrete existing thing.

The first view seems to be more appropriate to a genetical view of knowledge, that is, in cases where the common extension or reference is not given, but is to be gained or established by some hypostatic abstraction, such as in the case of new mathematical objects or in the case of theoretical terms, like energy, of which heat and motion are different representations. In these cases, metaphor reveals a kind of generic kinship between heterogeneous entities. Grassmann's "Theory of Extension" of 1844 seems to be the first text to explicitly mention the metaphorical character of mathematical objects or meanings, defining them exactly in terms of equivalence and distinction, and thus understanding mathematical concepts as distributive rather than collective entities (Otte 1989, p. 7 & p. 29).

In the end, it all depends on how universals (concepts, ideas, abstract objects) are conceived of. Take an example like "2+2=4". According to Frege "2+2" and "4" have the same referent, but different meanings. Such an interpretation presupposes the existence of numbers in a Platonic sense. The axiomatic approach makes no such existence claims. It explains the seemingly concrete, the fact that 2+2=4 by what is seemingly abstract and only possible, namely, by the axioms of arithmetic.

HOW TO LEARN TO DEAL WITH ABSTRACT IDEAS

It has repeatedly been argued against the Aristotelian simile theory of metaphor that the explanation of metaphor in terms of similarity is circular, because the similarities we most naturally cite in explaining what a metaphor's corresponding simile means are, as a rule, themselves figurative. The reason for this lies, exactly as in the examples of theoretical explanation presented above, in the abstract and general content of metaphor that is essential to the growth of knowledge. Indeed, Condillac had dealt with the problem already.

In a diagram, like in a theory or a work of art, synthesis of representation is realized in the construction and transformation of the representation, that is, in the process of generalization. To generalize in this way one must "see" something and to see means to construct a representation, in a way similar to the way the very essence of Monet's garden at Giverny has been realized in his paintings. Monet's achievement resists any commentary or explanation. A work of art is just a work of art; a theory is just a theory. It must be grasped as a form *sui generis,* as a second world, rather than as a pale or abstract representation of the given world, before we can inquire into its possible meanings or applications. Any creative achievement remains imperfect so long as questions about its possible literal meanings dominate its consideration. One has to open up to the world of a theory or a work of art, such that it may enlarge one's own world.

In artistic drawing, what we achieve is a line, and the line does all the work, and if it fails to do so, no philosophical commentary will rescue or repair a bad work of art. In literature or philosophy, it is the word or the sentence, in mathematics the operation or the diagram, which carries the entire weight, etc. Mastery, Paul Valery, says, presupposes that "one has the habit of thinking and combining directly from the means of imagining a work only within the limits of the means at hand, and never approaching a work from a topic or an imagined effect that is not linked to the means" (Valery, n.d., p. 40). It is the objectivity of the representation and the logic of the perspective thereby assumed, which "compels us to put some things into very close relation and others less so" (Peirce). Works of art or theories, being signs, must have a transparence and coherence in order to convey any meaning at all. "Consistency", says Peirce, "belongs to every sign as far as it is a sign; and therefore every sign, since it signifies primarily that it is a sign, signifies its own consistency" (Peirce, CP 5.313).

On a different occasion, Peirce writes:

> The work of the poet or novelist is not so utterly different from that of the scientific man. The artist introduces a fiction; but it is not an arbitrary one; .... The geometer draws a diagram, which if not exactly a fiction, is at least a creation, and by means of observation of that diagram he is able to synthesize and show relations between elements which before seemed to have no necessary connection. The realities compel us to put some things into very close relation and others less so, in a highly complicated, and in the [to?] sense itself unintelligible manner; but it is the genius of the mind, that takes up all these hints of sense, adds immensely to them, makes them precise, and shows them in intelligible form in the intuitions of space and time. (Peirce, CP 1.383)

Mathematics, like art or poetry, is dealing with metaphor. But the question is, in education at least, how can the teacher make the student grasp metaphor? The difference between metaphor and simile lies in the impact on the addressee of the representation itself and on his way of seeing things. The understanding of a mathematical proof, for example, requires more than being able to follow an argument step by step. Besides knowing what is at stake and understanding the information provided, one must also have incorporated the mathematical way of seeing the world. This becomes quite clear from Lewis Carroll's version of the race between Achilles and the Tortoise.

".... That beautiful First Proposition of Euclid!" the Tortoise murmured dreamily. "You admire Euclid? .... Well, now, let's take a little bit of the argument in that First Proposition – just *two* steps, and the conclusion drawn from them. .... let's call them *A*, *B*, and *Z*:

(*A*) Things that are equal to the same are equal to each other.

(*B*) The two sides of this Triangle are things that are equal to the same.

(*Z*) The two sides of this Triangle are equal to each other.

Readers of Euclid will grant, I suppose, that *Z* follows logically from *A* and *B*, so that any one who accepts *A* and *B* as true, *must* accept *Z* as true?"

And if not?

Achilles has, in fact, no means of compelling the Tortoise to accept this mode of inference and in particular to accept Z, if it accepts A and B, apart from writing down precisely this claim as a new rule:

"(*C*) If *A* and *B* are true, *Z* must be true. " .... etc., etc. (L. Carroll, 1939, pp. 1105–1106).

Logicians have commented on Carroll's little piece by saying that one's assumptions must be explicitly augmented by the exact mechanisms by which one is to deduce consequences from those assumptions and "what Achilles learns, to his lasting regret, is that *modus ponens* must be first granted as a rule of inference, for otherwise no conclusion can ever be reached".

But how do we learn to transfer activities from the concrete to the abstract? The answer: by means of metaphor. Lewis Carroll's version of the race between Achilles and the Tortoise shows, that one has to take one's insight strictly and act according to them. To make something an object of thought requires a perspective and as such an involvement. The acceptance of something as real and relevant seems to occur apodictically and intuitively.

The infinite regression of argumentation and explanation can only be overcome if the rule (the proposition or the concept conceived of as a scheme of action) is seen to be identical with its own application. Sign and thing become quasi identical, such that no questions about meaning arise. This is, in fact, how intuitional reasoning is commonly characterized. All insight and knowledge is ultimately based on perception (Peirce MS 316; CP 5.157) and the forming of a perceptual judgment is not a conscious logical act and thus cannot be fully analyzed and consciously decided. Trying to spell out every step explicitly would end up in an eternal race between Achilles and the Tortoise (Peirce CP 4.479; 5.148).

Metaphors are Thirds in which iconicity prevails, whereas images are Firsts and diagrams, representing relations, are Seconds. A metaphor is an icon, which mediates between the intuitive and operative elements of diagrammatic reasoning. Metaphors are meta-semiotic statements in the same way as axioms, in the sense of Hilbert or Peano, are meta-theoretical representations. Metaphorical representation, being a case of Thirdness, reaches beyond intuition (Firstness) and logical or factual truth (Secondness). Peirce's own term for this "mode of development of thought" in which metaphor plays an essential role, is "agapism":

> The tychastic development of thought will consist in slight departures from habitual ideas in different directions indifferently, quite purposeless .... The anancastic development of thought will consist of new ideas adopted without foreseeing wither they tend, but having a character determined by causes ...The agapastic development of thought is the adoption of certain mental tendencies, not altogether heedlessly, as in tychasm, nor quite blindly by the mere force of circumstances or of logic, as in anacasm, but by immediate attraction for the idea itself, ..., by the power of sympathy, that is by virtue of the continuity of mind. (Peirce CP 6.307)

DONALD DAVIDSON'S THEORY OF METAPHOR AND ITS ADVERSARIES

There is one claim that has persisted throughout the different attempts to understand metaphor since Aristotle. Metaphors are expected at least to call attention to formerly unknown similarities. The main objection against such a simile or comparison view of metaphor, writes Black,

> is that it suffers from a vagueness that borders upon vacuity. We are supposed to be puzzled as to how some expression (M), used metaphorically, can function in place of some literal expression (L) that is held to be an approximate synonym; and the answer offered is that what M stands for (in its literal use) is similar to what L stands for. But how informative is this? (Black 1962, p. 37)

As an alternative, Max Black proposes the adoption of an "interactive theory of metaphor". He sees a metaphor as composed of two subjects, a principal one and a subsidiary one. Rather than a strictly forward comparison (as in the Aristotelian view), Black argues that the work of the metaphor takes place in the interaction between these two subjects. The metaphor, Black says, making reference to the example "Man is a wolf", "selects, emphasizes, suppresses, and organizes features of the principal subject (Man) by implying statements about it that normally apply to the subsidiary subject (wolf)" (Black 1962, pp. 44–45). These implications are typically provided by received "commonplaces" about the secondary subject. Black claims that what is needed in order to understand a metaphor like "Man is a wolf" is the capacity to use the "system of associated commonplaces" (Black 1962, p. 40) connected with the word "wolf". "The effect of (metaphorically) calling a man a *wolf* is to evoke the wolf-system of related commonplaces" of a certain culture (1962, p. 41).

Now Black's "interactive theory of metaphor" seems but a socio-cultural version of the Aristotelian view, presuming that the understanding of metaphors is facilitated by cultural context. A second simple observation is that metaphors are, as a rule, symmetrical like equations, whereas Black's restriction of metaphorical phrases suggests an asymmetry between principal and subsidiary subject. However, both subjects, in fact, influence the manner in which the other is perceived. Thirdly, it should be remembered that not all metaphors have a subject-predicate form. An example accidentally at hand: "Parts of this book skate on pretty thin ice" (Rorty, 1989, preface). Metaphor has a dynamic narrative side together with a pictorial one.

The most important aspect of Black's view is that he sees metaphor as having special cognitive content, which cannot be circumscribed in literal terms. "The literal paraphrase inevitably says too much – and with the wrong emphasis" (Black, 1962, p. 46). But why is this so? With respect to the traditional Aristotelian view, one might understand this, precisely because of the openness and vagueness, which Black deplores. The metaphor leaves the similarity or supposed similarity we want to emphasize open. And the possible candidates for comparison are boundless. Everything in the world is similar to everything else in some respect. Phenomenological reality is continuous. This might transform imagination into a kind of random walk, making it impossible—for the time being—to paraphrase a metaphor in literal terms. This suggests one more possible answer, according to which metaphor might be untranslatable because it represents a kind of embodied meaning in which content and medium have become inseparable. Metaphor is then a question of language use.

The traditional and well-established convictions about metaphorical expressions have been questioned by Donald Davidson on just such an account. Davidson challenges Black's claim and the claims of many others by switching the focus of attention from the content of metaphor to what happens between a hearer and a speaker when a metaphor is used. Metaphor, according to Davidson, has no proper semantical content, but

> belongs exclusively to the domain of use. It is something brought off by the imaginative employment of words and sentences and depends entirely on the ordinary meanings of those words and hence on the ordinary meanings of the sentences they comprise. (Davidson, 1978, p. 33)

The common error, according to Davidson "is to fasten on the contents of the thoughts a metaphor provokes and to read these contents into the metaphor itself" (Davidson, 1978, p. 45).

He continues, "The central mistake against which I shall be inveighing is the idea that a metaphor has, in addition to its literal sense or meaning, another sense or meaning" (Davidson, 1978, p. 32). This idea is common to many who have written about metaphor, from Aristotle to George Lakoff and Max Black. "The idea takes many forms, from the relatively simple in Aristotle to the relatively complex in Max Black" (Davidson, 1978, p. 32). According to Davidson, "Max Black is wrong when he says: 'The rules of our language determine that some expressions must count as metaphors'" (Davidson, 1978, p. 31). Using a metaphor does not hurt the rules or structures of language, otherwise any application of a rule must be governed by rules again, *ad infinitum*. Metaphor thus belongs to pragmatics, not to semantics.

Metaphor, it is true, is intensional; its nature lies in the way of presenting something, rather than in the presented content. But the intensions and extensions of our terms or the connotative and denotative function of our language are complementary to each other and should not be completely separated from each other in order to better understand the development of meaning (Otte, 2003b, p. 212 & pp. 220–221). Davidson takes great pains, by means of an example, to point out that previous theories of metaphor did not preserve the distinction between using a word metaphorically and "learning a new use for an old word" (Davidson, 1978, p. 37). If this distinction is ignored, however, then "there is no difference between metaphor and the introduction of a new term into our vocabulary: to make a metaphor is to murder it" (Davidson, 1978, p. 34). This distinction should neither be overrun nor be made absolute—otherwise, metaphor becomes either trivial or incomprehensible.

Arthur Danto illustrates the point as follows:

> It may be one of the main offices of art less to represent the world than to represent it in such a way as to cause us to view it with a certain attitude and with a special vision. ... When Napoleon is represented as a Roman emperor, the sculptor is not just representing Napoleon in an antiquated get-up, the costumes believed to have been worn by the Roman emperors. Rather the sculptor is anxious to get the viewer to take towards the subject – Napoleon – the attitudes appropriate to the more exalted Roman emperors.... That figure so garbed is a metaphor of dignity authority, grandeur, power and political utterness. Indeed the description or depiction of *A* as *B* has always this metaphoric structure. (Danto, 1981, p. 167)

And this is so, precisely because A and B are not identical.

Danto's example also shows that both parties in the controversy, Black as well as Davidson, failed to mention one thing: that metaphor is first of all a vehicle for making important and general ideas more accessible, ideas which reach beyond everything presented explicitly in the metaphor. "Grandeur", for example, is not presented directly, no more than "energy" or "society" could be shown directly. A metaphor is a complex argument that at times cannot be broken down into atomic components and that results in implicit or intuitive convictions enabling one to act on new terrain as if it were familiar. Carroll's example in the last section shows what is meant. The "essence" of things can, as a rule, be grasped from a genetical perspective, only such that knowledge is an activity. Metaphor does not only make us see something in new ways, thereby creating new realities, but enables us also, in consequence of this, to act and reason in new ways, oriented by very general and fundamental ideas. In contrast to the reductionistic theory of metaphor prevailing in education, we claim, in particular, that metaphor is indispensable to scientific generalization.

Steven Pinker summarizes the issue as follows:

> The basic idea is that there are overarching laws of complex systems that govern diverse phenomena in the natural world. One set of laws explains why solar systems, atoms, planets with their moons fall into stable patterns of

> revolution. Another explains similarities in ecosystems, bodies and economies: in all three systems, for example energy is taken in, internal functions differentiate and resources get recycled. .... To the extent that these laws exist, scientists can discover their properties as they study the systems that are governed by them. And they are entitled to use a metaphor both as a label for that kind of system and as a means of generalizing from well-understood exemplar to a less well-understood one. (Pinker, 2007, p. 258)

Perhaps one should add, in order to further elucidate the role of metaphor, that the more abstract and general the laws are, the greater the variety of possible representations and the more powerful the metaphor. Think of mathematical concepts, like the notion of an algebraic group or a vector space, and the wide variety of applications.

THE TWOFOLD CHARACTER OF LANGUAGE

Is it reasonable to identify meaning exclusively with use, as some analytical philosophers like Wittgenstein or Rorty have suggested? Syntax represents pure instrumentality; its rules bear a nearly arbitrary relationship to the functions they fulfill. Abstract algebra is an example here. Consider the following sentence: "Colorless green ideas sleep furiously".

This is a sentence composed by Noam Chomsky in 1957 as an example of a phrase whose grammar is correct but whose meaning is nonsensical to show the relative independence of syntax and semantics. The rules that govern syntax in a language have nothing to do with the meaning that is conveyed. Chomsky's sentence may, however, be understood metaphorically and may appear as a completely acceptable phrase within a piece of poetry (in 1985 a competition was organized at Stanford University, where people were asked to compose a poem around Chomsky's phrase). Art seems as formal as mathematics—taking the term "formal" as derived from form—and mathematics abounds with metaphors exactly because of its formal character and the universal applicability of its structures.

Oliver Sacks has, in contrast to Chomsky, claimed that it becomes "increasingly clear from studying the natural acquisition of language in the child, and, equally from the persistent failure of computers to understand language, its rich ambiguity and polysemy, that syntax cannot be separated from semantics. It is precisely through the medium of meanings that natural language and natural intelligence are built up. From Boole, with his 'Laws of Thought' in the 1850s, to the pioneers of artificial intelligence at the present day, there has been a persistent notion that one may have an intelligence or a language based on pure logic, without anything so messy as meaning being involved". That this is not the case and "cannot be the case", Sacks continues, may perhaps find groundings in the new theories of the brain (Sacks, 1995, p. 116).

But the algebraic approach has existed at least since the 17th century and, ever since Descartes' introduction of the yet unknown "x" as an object of activity, it has been important to realize that we can act without knowing the meanings or essences of all the things involved in our actions. If this were not be possible, nothing new and yet unknown could ever be explored or become finally known and meaningful to some degree. But it is true that humans bring meanings to new areas of activity because they do not operate like machines. Our thinking depends on metaphors and models as much as on rules and algorithms. On the other hand, it should be remembered that we do not find our way in the world by simply consulting explicit descriptions of all things. So syntax and semantics can be grasped as relatively independent from each other and metaphorical discourse depends on exactly this (relative) independence. Davidson's analysis showed exactly this.

This relative independence can be grasped too from the study of speech aphasia. Roman Jakobson has classified, in fact, linguistic behaviour as referring to either code or context, and has accordingly described the diverse forms of aphasia in relation to disturbances of either of these references. One could call one type of aphasia a loss of predication, or, using semiotic terminology, a lack of metaphor, and the other a loss of instrumental or functional orientation (see Otte 2003b, 213-215; see Jakobson & Halle, 1956, pp. 79–86). For aphasics of the first type, context is the indispensable and decisive factor and concepts become mere *ad hoc* names. Their behavior is characterized by Jakobson as "a loss of meta-language", rendering them incapable of uttering a predication that has not been stimulated by the context at hand. In the second type of aphasia, described by Jakobson as *contiguity disorder*, the ability to construct narrative contexts is impaired and concepts cannot be put to specific purposes.

IS ALL DEVELOPMENT COMPLETELY CONTINGENT?

Rorty's project is of the same kind as Condillac's or Kant's, but it tends to the opposite direction. Whereas Condillac or Kant wanted to secure the identity of the human mind and employed general ideas or metaphor for that purpose, Rorty believes in an utmost contingency of life and wants to renounce the idea of general human nature or mind, and therefore dismisses universal ideas and hypostatic abstractions. In his most famous book, "*Philosophy and the Mirror of Nature*" (1979), he blamed the belief in universal ideas and hypostatic abstractions for the wrong emphasis our culture puts on philosophy and knowledge.

Like Condillac, Rorty believes that humanity is created through cultivated social compassion and that language and narrative are essential to this end. But unlike Condillac, he rejects the idea that language is a medium of representation between the world and us. Rorty agrees with Condillac that language and speech are essential to individual and cultural development, and he emphasizes that "changing languages and other social practices may produce human beings of a sort that had never before existed" (Rorty, 1989, p. 7), but he also believes in the utmost contingency of linguistic processes and rejects the continuity of universal ideas. Rorty, in a sense, falls back on Kant's "Copernican revolution" which had established the objectivity of the subject – the human subject is not a God-like free agent, but is part of the reality to be known – although the 18th century

Enlightenment in general did not yet grasp or comprise the idea of human evolution. But, as the example of Condillac shows, it is exactly the fact of conceiving of cognition as a semiotic process, rather than as a merely mental one, which helps to develop a genetic view on knowledge and subjectivity. Rorty, in contrast, believes that the emphasis on representation and language

> does not essentially change the Cartesian-Kantian problematic... It is the notion that human activity ... takes place within a framework which can be isolated prior to the conclusion of inquiry – a set of presuppositions discoverable a priori – which links contemporary philosophy to the Descartes-Locke-Kant tradition. For the notion that there is such a framework only makes sense if we think of this framework as imposed by the nature of the knowing subject, by the nature of his faculties or by the nature of the medium within which he works. (Rorty, 1979, pp. 8–9)

Rorty understands Davidson's theory of language and of metaphor in particular as based on a rejection of the idea that language is a medium of either representation or of expression, and therefore believes that the complementarity of these functions of language have no importance. Language and meaning are to be completely defined in terms of use, rather than referring to anything extra-linguistic. And here, Davidson's theory of metaphor comes in.

> To see the history of language and thus of the arts, the sciences, and the moral sense, as the history of metaphor is to drop the picture of the human mind, or human languages, becoming better and better suited to the purpose for which God or Nature designed them, for example, to express more and more meanings or to represent more and more facts .... Davidson puts this point by saying that one should not think of metaphorical expressions as having meanings distinct from their literal ones. To have meaning is to have a place in a language game. Metaphors, by definition, do not. ... Davidsonian philosophy of language sees language, as we now see evolution, as new forms of life constantly killing off old forms – not to accomplish a higher purpose, but blindly. (Rorty 1989, pp. 16–19)

To see cultural history and the sciences in this way is to understand them as a series of contingent redescriptions of the world or ourselves, a series which never converges. Rorty makes the conception of metaphor the cornerstone of a classification of philosophical and scientifical perspectives.

> The Platonist and the positivist share the reductionist view of metaphor. They think that metaphors are either paraphrasable or useless for the one serious purpose which language has, namely representing reality. By contrast, the Romantic has an expansionist view: He thinks metaphor is strange, mystic, wonderful. Romantics attribute metaphor to a mysterious faculty called imagination. (Rorty, 1989, p. 19)

Both sides are overstating either the objective or the subjective side of knowledge. Rorty wants to get rid of epistemology, and along with it, a whole bunch of dichotomies, beginning with the subject-object distinction. One might agree with Rorty that the modern age and its rationality type is characterized by the fact that all the classical distinctions – the theoretical and the empirical, concept vs. object, the analytic and the synthetic, a priori and a posteriori, subject and object, etc. etc. – become relative, and this seems to be expressed when we speak about the complementarity of knowledge and on the twofold character of language (see Otte, 2003b).

Rorty's interpretation of Davidson does not preserve the important distinction between using words metaphorically and giving new literal meanings to words. The common reductionistic theories of metaphor—reducing metaphor to analogy or simile—,which Davidson was criticizing, also ignored this distinction. This seems remarkable and should be attributed, we believe, to their common rejection of abstract ideas or objective universals. Both parties consider metaphor as incompatible with generality of knowledge (with respect to mathematics, see for example George Lakoff and collaborators). They endorse a reductionist view of metaphor and an anti-theoretical conception of (mathematical) knowledge. The reason for this is an exclusively predicative notion of generality – a general is an abstraction "which applies to many things" to rephrase Aristotle (see also Peirce CP 5.103). Generality comes down to us, according to Rorty or Lakoff and others, from Plato's realm of ideas and is to be rejected for that reason. This seems strange enough because everybody agrees that time and space are the sources of some of the most fundamental metaphors.

Rorty blames this Platonic origin of occidental epistemology for the wrong directions philosophy has taken in its development and for "the invention of the mind" —this being the title of the first chapter of (Rorty, 1979)— in particular. Rorty believes, like Condillac or Peirce, that minds and abstractions or universals are creating each other, in contrast to them, however, he does not believe that this evolution is more or less continuous and objectively constrained.

> There would not have been thought to be a problem about the nature of reason had our race confined itself to pointing out particular states of affairs .... But poetry speaks of man, birth and death as such, and mathematics prides itself on overlooking individual details. When poetry and mathematics had come to self-consciousness .... the time had come for something general to be said about knowledge of universals. (Rorty 1979, p. 38)

Mathematics depends, in fact, on the employment of hypostatic or reflective abstractions, otherwise it would be drowned in complexity from the very beginning. But mathematics is not just abstract conceptual reasoning and it does not make universals the causes of things. Peirce affirms that mathematical reasoning, like all necessary reasoning, is essentially "diagrammatic reasoning", and that iconicity represents the dominant character of mathematical diagrams. It is the notion of sign or diagram, as well as the use made of it, which make universals potentially effective.

A view of generality, different from Plato's, comes down to us from Aristotle's notion of essence. In a metaphor, the similarity relationship exists in between the differences, not over and above them, and the very same is true with respect to

Aristotelian essences, which thus represent a kind of objective generality different from abstract Platonic ideas or concepts. Peirce, calling himself "an Aristotelian of the scholastic wing", describes it thus:

> The old definition of a general … recognizes that the general is essentially predicative and therefore of the nature of a representamen. ... In another respect, however, the definition represents a very degenerate sort of generality. None of the scholastic logics fails to explain that *sol* is a general term; because although there happens to be but one sun yet the term sol *aptum natum est dici de multis*. …. Take any two possible objects that might be called suns and, however much alike they may be, any multitude whatsoever of intermediate suns are alternatively possible, and therefore as before these intermediate possible suns transcend all multitude. In short, the idea of a general involves the idea of possible variations [or continuity —M.O.]. (CP 5.102-103)

The twofold character of the general is expressed in the history of mathematics by two different interpretations of the continuity principle, two interpretations over which Cauchy and Poncelet quarreled (Belhoste, 1981), when the idea of pure mathematics was at stake. Metaphor is incompatible with an abstract and static view of generality, and thus incompatible with the transformation of continuity into propositional content. Metaphors are used like pointing or pictures or diagrams, to draw our attention to certain features of the world that we may or may not have been aware of previously. They want to make us see something, which in general cannot be given in propositional terms. And in teaching, they therefore sometimes have to be broken down into a quasi continuous process of non-transitive equivalences (see Otte, 2006).

Universals are signs, according to Peirce, and as such they must be applied or interpreted and, for that purpose, the subject has to create another sign, which becomes an interpretation of the first sign. This would sound similar to Rorty's account of cognitive and cultural development in terms of the permanent creation of new metaphors, were it not for his more or less tacit belief that this creation is completely autonomous and contingent. So how do meanings or habits come about? Rorty says there are no meanings beyond use. "Old metaphors are constantly dying off into literalness, and then serving as a platform and foil for new metaphors" (Rorty, 1989, p. 16).

Peirce sounds similar in places, but he believes that activity and experience create the sign as well as the subject itself – man is sign - and that signs are objective and more or less constrained. New terms are constantly being invented such as "electricity", which at first sounds awkward but then becomes familiar and infused with the meaning derived from many contexts of experience, so that, as previously stated, "men and words reciprocally educate each other" (Peirce, CP 5.313). Man's nature lies essentially in his being a part of the "Great Chain of Being". This is what Peirce means by "agapism".

Since people, according to Rorty, have no "essence"—that is, no objective history—they seem to act freely, spontaneously and at random over and over again and without feeling any constraints. But this renders sympathy or solidarity and moral values in general subject to mere chance. "Our insistence on consistency", writes Rorty, "and our consequent opposition to ideas like 'essence', 'nature', and 'foundations', makes it impossible for us to retain the notion that some actions and attitudes are naturally 'inhuman'" (Rorty, 1989, p. 189). One might, quite to the contrary, claim that the isolated individual itself is "naturally inhuman".

CONCLUSION

We started from the assumption that metaphor replaces logic as the language of change and evolution and that it is also vital to generalization and abstract thought. For conceptualizing change, the latter must be understood simultaneously in terms of continuity and discontinuity; the growth of knowledge then appears neither as simply cumulative nor unpredictably disconnected. We think that metaphor helps to understand this complementarity of continuity and contingency, because metaphor presents the unexpected and abstract as if it were concrete and familiar.

All theorizing in mathematics education and philosophy will remain futile if no sound notion of mathematical ontology is developed. We believe that the semiotic approach is very promising to cope with this problem of mathematical objects from a genetic viewpoint, thereby enabling us to make our way safely between Platonism or mentalism, on the one side, and physicalism, on the other side.

We began our argument by showing that the genetic view of knowledge and cognition came about together with the appreciation of signs and language during the 18th century, and more forcefully so in the context of higher education since the 19th century. Condillac's semiotic approach to cognition is illuminating and a first step towards an understanding of the development of human cognition and thought. His view of algebra as an analytical language becomes superseded only after people like Galois or Grassmann propose a structural conception of algebra. One might interpret this development as a transition from the instrumental to the metaphorical view. Roman Jakobson has, in fact, classified linguistic behaviour as referring to either code or context, and has accordingly described the diverse forms of aphasia in relation to disturbances of both of these references. One could call one type of aphasia a loss of metaphor and the other a loss of instrumental or functional orientation.

We then reported on a controversy between Davidson, on the one side, and Black, Lakoff and others, on the other side, regarding the notion of metaphor. We think that Davidson is essentially right, although we would certainly not share his overall view of epistemology. The one-sided and radical interpretation of his theory by Rorty indicates, even if it is not completely true to the original, that both might have missed the "twofold character of language" (Jakobson), which is, however, essential to the understanding of human development as simultaneously continuous and contingent.

REFERENCES/BIBLIOGRAPHY

Aarsleff, H. (Ed.). (2001). *Etienne Bonnot de Condillac. Essay on the origin of human knowledge*. Cambridge: Cambridge UP.
Black, M. (1962). *Models and metaphors*. Cornell UP.
Bruner, J. (1990). *Acts of meaning*. Cambridge: Harvard UP.
Carroll, L. (1939). *The complete works of Lewis Carroll*. London: Penguin Books.
Danto, A. (1981). *The transfiguration of the commonplace*. Cambridge: Harvard UP.
Davidson, D. (1978). What metaphors mean. *Critical Inquiry, 5*, 31–47.
Eco, U. (1979). *The role of the reader*. London: Hutchison.
Goodman, N. D. (1983). Reflections on Bishop's philosophy of mathematics. *The Mathematical Intelligencer, 5*, 61–68.
Heijenoort, J. v. (1967). Logic as calculus and logic as language. *Synthese, 17*, 324–330.
Hobbs-Peaden, C. (1993). Condillac and the history of rhetoric. *Rhetorica, XI11*, 135–156.
Jakobson, R., & Halle, M. (1956). *Fundamentals of language*. The Hague: Mouton.
Joly, A. (1985). Cartesian or Condillacian linguistics? *Topoi, 4*, 145–149.
Leventhal, R. S. (1986). The emergence of philological discourse in the German states, 1770–1810. *Isis, 77*, 243–260.
Levi-Strauss, C. (1971). *Mythologiques IV*. Paris: Librairie Plon.
Locke, J. (1894). *An Essay concerning human understanding*. London: George Routledge and Sons.
Otte, M. (1989). The ideas of Hermann Grassmann in the context of the math. and philosophical tradition since Leibniz,. *Historia Mathematica, 16*, 1–35.
Otte, M. (2003a). Does mathematics have objects? In what sense? *Synthese, 134*, 181–216.
Otte, M. (2003b). Complementarity, sets and numbers. *Educational Studies in Mathematics, 53*, 203–228.
Otte, M. (2006). Proof-analysis and continuity. *Foundations of Science, 11*, 121–155.
Peirce, C. P. = *Collected papers of Charles Sanders Peirce*. In Vols.umes I–VI, edited by C. Hartshorne and & P. Weiss, (Eds.), 1931–1935; Vols.umes VII–VIII, edited by A. W. Burks, (Ed.), 1958. Cambridge, Mass.: Harvard University Press (reference by volume and paragraph numbers).
Peirce, M. S., &:= Robin, R. S. (1967). *Annotated Catalogue of the Papers of Charles S. Peirce*. Cambridge: The University of Massachusetts Press.
Peirce, N. E. M., & Eisele, C. (Ed.). (1967). *The New Elements of Mathematics by Charles S. Peirce*. (Vol. I–IV)., The Hague-Paris/Atlantic Highlands, N.J.: Mouton.
Pinker, S. (2007). *The stuff of thought*. New York: Viking Press.
Quine, W. V. (1978). A postscript on metaphor. *Critical Inquiry, 5*, 161–162.
Rorty, R. (1979). *Philosophy and the mirror of nature*. Princeton, NJ: Princeton UP.
Rorty, R. (1989). *Contingency, irony, and solidarity*. Cambridge, MAass: Harvard UP.
Sacks, O. (199). The new vision of the mind. In J. Cornwell (Ed.), *Nature's Imagination* (pp. 101–121). Oxford: Oxford UP.
Steiner, M. (1998). *The applicability of mathematics as a philosophical problem*. Cambridge, MAass: Harvard UP.
Valery, P. (n. d.). *Leonardo*. Frankfurt: Suhrkamp Verlag.

Michael Otte
Institut für Didaktik der Mathematik
University of Bielefeld, Germany

WOLFF-MICHAEL ROTH

THE DAWNING OF SIGNS IN GRAPH INTERPRETATION

> But, one will counter, *what* is taken to be sign initially has had to be accessible in it and has to be captured *prior to* being made sign. Certainly, and speaking generally, it has to be available in some manner. But the question remains, *how,* in this earlier encounter, that which exists ["Seiendes"] is discovered. (Heidegger, 1977, p. 81, my translation)[1]

Semiotic studies in mathematics education are generally concerned with processes and products of semiosis, that is, the processes by means of which human beings elaborate sign-referent relations through the production of interpretants. Rarely, if at all, studied phenomena are the processes by means of which (aspects of) inscriptions come to be the signs that are the material resources for semiosis (Roth & Bowen, 2001). In this earlier work concerned with scientists' interpretations of graphs from undergraduate courses in their own discipline, we proposed two mutually constitutive processes that analysts need to attend to when studying individuals working with unfamiliar inscriptions: *structuring* and *grounding*. In the structuring process, potential signs are produced—by visual inspection and articulation of features from the inscription—that come to be related to possible references in and through the grounding process. Structuring is required, for although the graph materially exists, *how* it exists prior to the constitution of sign(s), as Martin Heidegger states in the opening quote, is a whole other matter, one that I endeavor to elucidate in this chapter. Grounding is understood as a process by means of which inscriptions, like words, are taken up and linked into some lifeworld constituting network of referring relations that we associate with the sense and experience of meaning (Roth, 2004). Much like words, I suggested in a paraphrase of Martin Heidegger, inscriptions (signs) do not *have* meanings somehow related or attached to them but *accrue to* always and already existing meanings, that is, systems of lifeworld constituting referring relations. This is so because what a sign primarily indicates is not some referent in and of itself but rather the situation *wherein* one lives, *whereby* taking-care-of occurs, and *what* is relevant (Heidegger, 1977/1996).[2] Thus, to interpret a typical but unfamiliar graph from an entry-level undergraduate biology textbook such as that presented in Figure 1, the interpreter has to define structural aspects that point to and make salient some structural aspects of a biological phenomenon *as this occurs in some familiar context*, which may be that of research or common sense. The difficulty of the task can be seen from the fact that merely 50% of the scientists involved in the study provided interpretations of Figure 1 that the professor from whose course the

L. Radford, G. Schubring, and F. Seeger (eds.), Semiotics in Mathematics Education: Epistemology, History, Classroom, and Culture, 83–102.

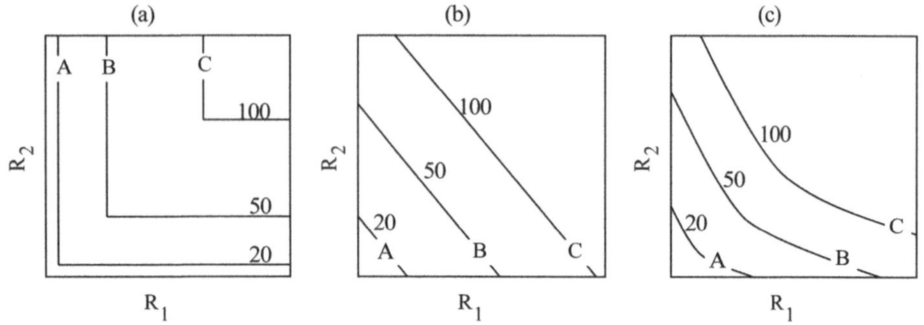

The amount a plant grows depends on a number of factors, for instance, the availability of nutrients (R). A shortage of any single nutrient can limit plant growth. Sometimes scientists study the effect of pairs of nutrients. The graphs depict three different biologically-realistic scenarios of how two nutrients (R1 and R2) might combine to affect plant growth.

Discuss the effects of different levels of the two nutrients on each amount of plant growth (20, 50, 100) in each scenario (a, b, c).

Figure 1. One of the graphs used in think-aloud protocols with scientists to understand the processes involved in graph interpretation

graph was culled accepted as correct from his second-year biology students. A standard interpretation requires seeing the plates as three depictions of the dependent variable of growth rate (specific values of which appear as “20,” “50,” and “100” in the graphs) as functions of two independent variables. The three graphs therefore are much like the lines of equal elevation (isoclines) on a geographical map, where the two resources R_1 and R_2 would correspond to the two dimensions that span a two-dimensional surface. This is a prerequisite for understanding scenario “a” as depicting the relationship of *essential* resources (for a given amount of R_2, growth rate is largely independent of R_1), *substitutable* resources (increase in R_1 can compensate for decrease in R_2 to achieve constant growth rate), and *complementary* resources (in some combinations, the total amount of resources can be minimized).

The problematic of the interpretation of graphs becomes salient when we consider the relation between the sign and its referring function, which is threefold (Heidegger, 1977/1996).[3] First, pointing (indicating) is based in the tool nature of a sign, which orients it to the *what-for* and *in-order-to* of the action (pointing).[4] Here, the structure of the terms *what-for* and *in-order-to* already indicates that a sign cannot be considered in and for itself because it is always already bound up with intentions, purposes, motives, and the objects of actions. Second, the pointing is part of an irreducible whole (“Zeugganzheit,” literally *stuff ensemble*), part of a network of references that binds together everything that is experienced as meaningful specific in situations. For example, the direction indicators of cars are part of the traffic navigation ensemble, so that in perceiving a flasher we do not think of it as such—e.g., as a filament heated by an electrical current or as a red light—but simply step out of the way of the car and steer, momentarily, off our course. Third, signs are not only as present to hand as other things and tools, but have a special status in that they explicitly make available the lifeworld as it is experienced in mundane, circumspect, everyday coping. Rather than the sign itself, it is the world or some aspect of it that becomes available in and through the sign, which itself may be completely transparent in use (Roth, 2003). When we listen to someone, we hear words rather than sounds—only while transcribing classroom videotapes, when we cannot understand what someone says, do we sometimes come face to face with the fact that words really are sounds that we hear as this or that word. To interpret a graph therefore requires the constitution of the three relations, which means, defining a sign, finding a form of reference, and articulating a part of the known world so that the three relations can be established. Each of these three dimensions is articulated in the detailed analyses that follow below.

My earlier work on the interpretation of graphs and the emergence of sense and meaning was concerned with the second process, *grounding* or semiosis, the way in which identified signs are articulated in terms of what is already familiar to (i.e. understood by) the interpreter. This work, like research in mathematics education generally, is concerned with the interpretation of already existing signs. This work says little if anything about situations where signs have not yet been identified. In this chapter, therefore, I return to the process of structuring, the often-forgotten issue in interpretation: "How do signs come into being and become what they are?" I am concerned with the dawning of structure during sessions in which scientists interpret unfamiliar graphs. Consistent with recent philosophical work concerning the phenomenon of givenness (Marion, 2002), I understand aspect dawning to be a dialectical process, involving both agency (human beings act in the world) and passivity (the world given to and disclosed by us, which is apparent each time that we encounter something that we neither anticipated nor knew). In my experience, the phenomenon of aspect dawning is generally hidden in individual interpretation sessions—though there are sometimes indications that a new aspect is being considered—but they clearly show up in sessions involving two individuals collaboratively interpreting graphs. This is not surprising because collaboration requires the participants to be aligned perceptually and conceptually and this requires them to articulate for one another what they attend to, what the objects are that they are talking about, and so forth. And the very fact that they draw on talk and nonverbal forms of communication—which, according to Heidegger, have "apophantic" function (pointing to what lets itself be seen as it is[5])—makes available to the analyst what is salient to the participants, what they do when something salient to one is not salient to the other, and how saliencies change.

Because of the limitations of a-posteriori accounts of actions generally and aspect dawning specifically—descriptions are provided through the lens of what has been achieved as a result of a task as a whole, namely, the discovery, insight, or the final interpretation—I draw on ethnomethodology and conversation analysis, methods appropriate for the research problem at hand because they are actor-centered methods. Only what actors make available to each other as resources for getting the job at hand done may figure in the "documentary method" of exhibiting the required work. I describe this work by means of which perceptual aspects come into play in collaborative graph interpretation sessions involving physicists; and I provide evidence for the pedagogies involved when one participant "teaches" another what there is to be seen in the material display toward which their attention is turned. Signs are not simply there, they emerge as situated and contingent products of the reader with an initially unstructured configuration; the end results are signs or sign complexes that have—or are discarded for not having—meaningful referents.

SIGNS AND EUROPEAN PHILOSOPHY

This study is concerned in particular with one constituent part of graph interpretation; here I take aspects of graphs as signs and graphs as sign complexes. At the most general level, "to interpret a sign means to define the portion of continuum which serves as its vehicle in its relationship with the other portions of the continuum derived from its global segmentation by the content" (Eco, 1984, p. 44). In other words, the sign is part of the material continuum pointing to another part of the continuum; and this other part of the continuum constitutes the content of the sign. Defining signs and sign relations in terms of the material continuum of which we also are a part, and which all forms of communication presuppose, is consistent with my dialectical materialist approach to cognition generally. This definition immediately raises the issue of identifying the "portion of continuum" that serves as sign vehicle: to be salient, the sign itself has to become figure, that is, a continuous material needs to be structured into figure and background so that some sign comes to stand against everything else. How to parse the material continuum generally is not problematic, unless we are unfamiliar with a phenomenon. For example, novice French speakers who have problems with the morpho-phonological fact that a normally silent letter at the end of a word is pronounced in such a way that it can be heard as the beginning of the next word—"faux ami" (false cognate) normally pronounced [fozї ami] (rather than [fo] [ami]) sometimes can be heard as [foї zami].[6] Similarly, where and how to parse an unfamiliar inscription is not self-evident. Thus, graphs are signs created for the special purpose of serving a referential function. This referential function means that some of the structure of the sign expresses the supposed structure in the world better than language. A completely arbitrary sign, such as a knot in a handkerchief, would not internally provide the resources for making inferences as to which phenomenal objects or events it is trying to point to.

Two philosophical studies constitute the backdrop against which I develop my case: Heidegger's (1977/1996) work on everyday cognition, signs, and language use developed in *Sein and Zeit*, and Ludwig Wittgenstein's (1958) comments on aspects arising from *Philosophical Investigations*. Heidegger notes that in everyday use, signs are not noted as signs. Signs are first and foremost useful things (tools), the usefulness of which lies in pointing (indicating), that is, bringing something other than themselves into (joint) awareness; pointing is a kind of referring, making a relation with something else. A sign is *not* comprehended if we *consider it as* a sign. Rather, signs allow us to encounter a relational whole at hand and the sign-in-use provides us with an orientation in and to this whole. In everyday use, the sign therefore does not stand in relation to other things—as presupposed in the interpretation of unfamiliar graphs, for example—but allows us to deal with our everyday world without making it or the sign conscious. This can be seen in the way we act when we perceive the above-mentioned direction indicators of a car. We do not perceive the flashing lights *as* signs that we have to interpret to know what they mean and what we are to do. Rather, we step out of the way or wait until the car has turned before stepping onto the pedestrian crossing. The sign does the work it is intended to do by withdrawing itself and therefore allowing the indexed phenomenon to appear. Thus, in one study of graphing at work, I show how a water technician points to a spike in a graph and announces, "This is a clogged pipe" (Roth, 2003). She does not say, "this spike *indicates* a clogged [water] pipe; she says, "this *is*." In the same study, I analyze scientists who point to one part of a graph that emerges on their computer monitor, point to it, and suggest that there is some unexplained stuff in their sample. Again, they do not say, "this graph suggests/indicates that our sample is contaminated with foreign, invisible material." They do point to the graph but talk about the phenomenon represented. In the same way, to return to the example of the direction indicators, we are aware of a turning car, not of the flasher as flasher, which makes signs so efficient. That is, the sign has an "apophantic" character in that it lets something other than itself be seen—here, a clogged pipe, some unknown stuff, a turning car. The sign thereby allows the mundane and everyday world reveal itself, whereas the sign and its relation to the everyday world disappear into the background rather than imposing their presence in and to consciousness. Signs are useful precisely because they retreat, allowing us to cope with the world rather than stop and take time out from our current pursuit to figure out what the sign means.

In ordinary pointing (deictic gesturing) and iconic gesturing, the index disappears, allowing the relevant portion of the material continuum corresponding to the content of the sign to appear. We do not see (are not aware of) someone's finger pointing at something but we see (are aware of, attend to) that something. Here I am concerned with the way in which signs emerged from embodied bodily activity, as portions of the material continuum are marked, re-marked, and remarked. But the signing portion of the material continuum does not exist independently of the content that marks out other portions of the material continuum, which, as we know from the hermeneutic literature, can only be an aspect of an already familiar world already known to the interpreting subject. Here, Ludwig Wittgenstein has contributed to an understanding of signs, which, in his treatment, have meaning only in relation to their function in mundane use.

Wittgenstein critiqued the general approach to the perception of signs and used many examples to communicate his ideas. One is the duck-rabbit (Figure 2): we do not see the mass of ink plots *as* a rabbit or *as* a duck unless we engage in a game of consciously flipping back and forth between the two ways in which the figure can appear to us. When someone speaks, we do not hear sounds that we have to interpret but we hear words—unless there is trouble, as transcribers of noisy video-

Figure 2. My personal rendering of Wittgenstein's duck-rabbit illustrates that we do not normally see signs as *signs or* as *something but as the something itself*

and audiotapes know. Thus, the normal expression would be, "See, a rabbit!" rather than "Now I see it as a rabbit." This distinction *ought* to be important to mathema-tics educators, because we find in the literature likely unjustified suggestions that students make "slope-height" confusions or have "perceptually driven misconceptions" (e.g., Leinhardt et al., 1990) when there is no clear evidence that the students *actually* confuse two aspects rather than perceiving one. There is evidence from research among scientists that participants see one rather than several aspects; and this one aspect is the material resource in and for their interpretations (Roth & Bowen, 2003).

Wittgenstein notes that with unfamiliar figures, there are moments when we come to see something in a material display for the first time, such as when we see a duck in Figure 2 after always having seen a rabbit. He called this phenomenon "the dawning of an aspect." The phenomenon comes with an essentially passive element, because we cannot intend that the duck or rabbit become an aspect unless we already know that a duck or rabbit can be seen. This passive element is readily experienced with images consisting of blotches, where we cannot provoke seeing a Dalmatian or a Holstein (Friesian) cow until "it dawns upon us," whereafter we can readily get ourselves to see one or the other figure. The existence of multiple aspects shows that the raw stimulus on the retina is not the only thing that is important to the perception of signs, and that there are dispositions that give us some structures preferentially without our being able to intend them—if we see a duck and have never seen the rabbit we cannot *intend* to see the latter unless someone tells us to look for it.

Both of these approaches are consistent with recent findings that scientific graphs bear a metonymic relationship to the context of their production: it is this relation that constitutes the meaning of meaning (Roth, 2004). Scientific graphs are portions of the material continuum referring to other portions of the material continuum in the context in which they normally appear; they exist and are used within networks of signification outside of which they mean little (if nothing). Thus, when scientists provide a reading of a graph, they tend to begin by articulating instruments, research problems, data contexts, data transformations, and so on; and graphs emerge from these complexes as constitutive moments and products. Graphs are part of the research context that points back to this context, which thereby constitutes a referential whole that undergirds any understanding of the graph.

DAWNING OF SIGNS IN JOINT GRAPH INTERPRETATION

In regular use, tools (signs) disappear; they appear and become available to consciousness precisely when they do not work as usual. But un-usability reveals circumspect use rather than the visual detection of properties. In the task of interpreting graphs, the problem is posed in the reverse. Subjects are told that there is a sign. Their task is to discover (a) a relation between the portion of the material continuum available on the paper and (b) possible portions of the material continuum that are the contents to which the former portion refers. To be at all intelligible, the task *presupposes* a relationship between one or more structures of the primary portion (ink plots on paper), the content of which relates to some other yet to be discovered portion. The task therefore requires discovering properties by structuring the display in such a way that, together with it, a sign relation can be established to some other portion of the world.

The following analyses show that signing and sign interpretation is more complex than simply looking at some display and beginning to talk about it. Relevant structure that can become a sign has itself to be established within a world that does not make a-priori distinctions between what is and what is not a sign. What it is that some speaker attends to is exhibited in and through the changes in salience with respect to what has been—although some gestures, for example, are recognized as grooming rather than as pertaining to the issue at hand, which requires the recognizability of some relationship between the changing feature and the current focus. Signs emerge from the co-variation of changes, some of which (e.g., pitch) mark others (e.g., sound, word) as important and therefore as potential signs. For example, when someone moves a pen or finger to a graph, this constitutes a change and we generally perceive it as relevant to understanding what the talk is about; when a speaker changes the pitch of her delivery, we hear it as an emphasis of the word uttered, and this emphasis mediates how we hear it.

Producing Joint Orientation

Any form of communication and collaboration requires participants to produce joint orientation, that is, sharing goals and coordinating interactions in pursuing these goals (Hoffmann, 2007). Joint orientation to a part (section) of the world is a prerequisite for an aspect to dawn (emerge). In direct collaboration, such joint attention and understanding is presupposed, being the background for collaboration as such. When joint attention to a common focal object does not exist, one or the other participant begins the work of realigning everybody to the joint task. In some instances, a current speaker might get the other's attention, or a person who has momentarily attended to another track of the inquiry requests actions that align her again with the main track. (Transcription conventions are in the appendix.)

Just prior to the following instance, Lyn had shifted her gaze from looking at the right-most graph in the task—where she also tracked her pen tip parallel to the abscissa—then Ian utters "plants in type c" (turn 82) to the last paragraph in the caption on the utterance of "survive" (turn 82). She then lifts up her head while uttering "sorry say that again," clearly shifting her gaze to Ian's face (turn 88). Her pitch rises toward the end as in a question. Lyn then changes her orientation toward the graph precisely at the moment when Ian begins to speak after a brief time lapse since Lyn finished (turn 90).

```
082 I: *clearly y can at least
       (0.25) plants in type
       ce::, (0.14) ca:n=a,
       (0.51) .hhhhh survi:ve,
       (0.36) an do extremely
       well; and thrive; (0.61)
       uh: (0.14) with less
       r=one;
083    (0.20)
084 L: <<p>point [he:re   ]
085 I:            [than with] r=two.
086    (0.33)
087    .hhhhhh
088 L: sorry say that again?*
089    (0.60)
090 I: *uh you nly need about.
       (0.58) <<dim>again i=ll
       do it independent>
       <<p>percent of axes::.>
       (0.15) you need about ten
       percent of r=one; .hhh in
       order to re (0.23)
       achieve;
091    (1.06)
092    this mea[sure of a r=one of growth]
093 L:         [a hundred percent       ]
```

Lyn then points to the fact that she has not followed (understood) what Ian has been saying while reading the caption below the graphs, and asks Ian to repeat what he was saying (turn 88). She orients herself to the plate "c" that Ian has been talking about but initially watches Ian place his pen at the right end of the diagram and places her pen at the corresponding point on the sheet in front of her. Both the right hand/pen at the bottom right of plate "c," near where the 100-line crosses, and thereby corresponding to the place where Ian's pen is placed on the corresponding display, and the left-hand index next to the 100-line constitute pointer configurations; these allow the plate "c" generally and the 100-line specifically to come to the fore and thereby become a general/specific aspect that emerges into a collective space of attention. Lyn then reorients herself toward her own sheet and, overlapping with Ian, provides a description of what can be achieved "a hundred percent" (turn 93). This utterance of "a 100," which corresponds to the line that she has been pointing to, is a possible completion of the sentence Ian had begun but not completed in turn 90. As there is a conversationally longer pause (turn 91), there is a moment for Lyn to take a turn leading to an overlap, because Ian, too, has taken the existing opportunity for a turn.

Language has an essentially indexical feature (Garfinkel & Sacks, 1986). It does not mean as such, nor does it have meaning behind every word that listeners need to find, but it allows the mundane features of the world to become available in our everyday pursuits, including more esoteric tasks in cognitive science research on the interpretation of graphs. That is, language allows us to cope with the world so that knowing a language means knowing one's way around the world generally (Davidson, 1986). It therefore stands in a dialectical, constituting-and-constituted rather than simply referential, relation to the world. This can amply be seen in the present episode, for example, when Lyn utters "a hundred percent" (turn 93). This utterance alone, independent of any situation, makes little sense because it is not bound up in the specific *what-for* and *in-order-to* structure of everyday activity. This structure therefore directly links the sign to what it does and what it is used for, so that the sign (graph) in and by itself means nothing. Thus, Lyn's utterance is intelligible because it is bound up with the situation in a network of significance, where other visible structures point back to the words. Here, the utterance "a hundred percent" is motivated by a line in plate "c" next to which we find ink traces that culturally competent members can read as 100, which is the numerical equivalent of the words uttered. At the same time, the utterance re-marks the corresponding number and the line that it seems to name, allowing it to stand out against everything else as the background. Her re-marking the number allows Ian to notice (remark) it as something currently relevant.

In this episode, the participants not only make signs visible to and for one another, as a prerequisite for noting structural features as sign, but also make visible the joint orientation toward a segment of the space they currently inhabit. In the three bodily orientations, Lyn exhibits attention to the text, attention to Ian, and attention to the display about which Ian has been talking, and about which he is requested to talk again. The visibility of their joint attention is an integral part of any thing (structure, aspect) visible within this joint space, within which the topics of talk are to be found and made visible.

Producing Visibility and Visible Features

Gestures constitute an important part of making things visible in the surrounding world. They are especially good—much better than verbal indexes ("there," "this," "that")—in marking out spaces, positions, and entities because they themselves have spatial quality. Thus, an index finger or a pen tip moving along a line in plate "c" allows the curvature of the line to become salient much better than a fixed index finger (as we observed in the previous episode). The index finger alone does not even demarcate the curve, it points to something and only the situation as a whole will show what that something is, the symmetry, the curvature, the level (i.e., 100 percent), the intercepts, the complementarity between the variables R_1 and R_2, and so on. From the perspective of communication and participative thinking, gestures are integral to the communicated content, but concretize different aspects not captured in words, which in turn concretize different aspects and contents. Both components of the communicative situation incompletely and one-sidedly concretize the said (content) of what is articulated. Here, the verb "articulate" should be understood in two ways simultaneously. On the one hand, to articulate has the sense of to utter, to express in words, to pronounce. On the other hand, to articulate, mostly used in a passive form, has the sense of to joint, to attach by a joint. Thus, speech makes apparent the joints that separate things from their surrounding world ("Umwelt"). This is the apophantic nature of speech, which allows the surrounding world to be experienced in terms of different things even without conscious reflection. Articulation therefore also *marks* what can be seen as different from the background, remarks them, as the verbal articulation presupposes perceptual articulation and vice versa, and thereby allows us to *re*-mark what we have marked before. It is evident here that I do not advocate the Sapir-Whorf hypothesis, according to which language articulates the world and shapes our perception. A study of scientists and technicians at work in different situations shows that they often make perceptual distinctions without being able to express these in words (Roth, 2005). Therefore, words, their contents, and the (semiotic) relation binding them emerge together.

The present data provide evidence for the fact that gestures orient perception, marking and remarking the structure with which it stands in a dialectical, that is, mutually constitutive, relationship. The traces on paper motivate the finger trajectory, and the finger trajectory points to the particulars of the trace. In the following episode, I underlined three parts of the transcript that corresponded to the gestures depicted in the overlaid video offprints of Figure 3. In the three instances, the gestural trace highlights the particulars of the three curves, which first descend vertically downward toward Lyn, and then move horizontally from left to right (her position). The gesture marks, because it is in and through the gestural trace that the particulars of the curve become visible; but the following of the ink trace already requires the visibility of the curve. The curves thereby come to be marked, as the finger traces out the line, and they become re-marked, marked again. Both marking and re-marking serve to remark the curves and to make them remarkable.

```
035 I: situation a we have (0.23) d=case whe:re .hhhh we have
       *two nu-* nutrients that are both *nee:ded*. (0.31) for
       *(0.86)* plant growth
036    (0.24)
037 L: needed about equally by the look of it?
038    (0.62)
039 I: and uh: yea n=fact (0.16) one wouldn=t be too surprised
       to find that (1.66) they=d done something weird with
       the axes <<dim>in order to make make em d=same; cause
       this seems to be [goin]=on.>
040 L:                  [ um ]
```

Figure 3. While Ian is talking, Lyn produces gestures that stand in a constitutive relation to the talk but are irreducible to them

By uttering "so in basically situation a we have …." Ian orients Lyn to the left-most panel, because here the letter "a" articulated in sound finds a corresponding letter "a" in the display. The sound that we hear as "a" makes salient, and thereby visible, a part of their environment marked by the ink trace "a." That there is a recognized correspondence can be seen from the way in which Lyn orients to the display in front of them, as she moves her gaze, which has been in the direction of the text below the three plates to the left-most plate. As before, Lyn not only looks toward the plate marked "a," but in changing her head position she allows Ian to see that she is orienting herself to the plate labeled "a." The utterance "situation a" therefore also has been an instruction for Lyn to orient toward a particular part, which thereby becomes the locus of attention and the topic of subsequent talk. Therefore, from everything that there is available to see, from a potentially infinite number of structures potentially available in their setting, only some are made and come to be visible in their joint attention to "the same thing."

In this episode, we can see how the possibility of a feature to serve as a sign is not only mediated by what has been and has become perceptually isolated, but also by the potential worldly affair to which the graph might be related. Thus, Ian not only talks about and makes salient the presence of two nutrients and the variable of plant growth but he *infers* that the two nutrients "are both needed." Lyn, too, makes salient both perceptual and interpretive moments. On the one hand, she articulates that the equal dependence of growth on both nutrients is available in perception, that is, "by the look of it." This expression has the sense of appearance of natural things, which is expressive of their quality or nature. She uses this expression to substantiate the inference that the current topic of talk, the two nutrients, is "needed about equally."

While Ian speaks about two nutrients, Lyn allows, in the trace of her index finger, the "two-ness" of the nutrients to become visible in the two distinct movements of which the trajectory as a whole is composed: first straight down, then precisely horizontal from left to right. She does so three times, coordinated with Ian's utterance. She articulates *for* Ian the fact that she has been attending to him and the unfolding interpretation, as she links to his talk by repeating the verb "needed," and then by verbally elaborating the symmetry already available in the earlier trajectory of her finger over and about the curve. Recent neuroscientific evidence suggests that to perceive an action, a set of mirror neurons that are also activated during the production of the action is required; even listening to action terms modulates the execution of action terms (Buccino et al., 2005). The present observations are consistent with this research, as the gesture orients the gaze that follows the finger along the lines, and the gaze thereby becomes entrained into perceiving the particulars of the trajectory enacted. Both finger and gaze engage in the marking and remarking of the ink trace as such.

The gesture is visible to everyone else oriented to Lyn. Her gesture not only serves her own orientation to a particular aspect, but also serves as a description and an instruction for others attending to her regarding what there is to be seen and how to orient one's gaze to see what she is seeing. The gesture becomes remarkable because of its iconic relation with the ink trace underneath. These gestures therefore also orient Ian and any other onlookers (the research assistant present and the analysts), though, in this particular situation, Ian engages in a first interpretation of the traces that Lyn simultaneously highlights and thereby brings to the fore. It is not that the gestures merely draw attention to something already perceived; rather, especially when, in following the curves, these occur for the first time, they orient perception, here understood not as the image that is on the retina but what is salient to the person looking at the display. The onlooker can take the gesture as relevant rather than as a grooming gesture, and the relevance has to be found in the background against which the gesture stands. The index finger points to itself to the extent that it marks the indexical nature of what is being done, "Look here, I am pointing something out."

The remarkable nature of the curves, here, lies in their symmetry, which Lyn articulates (turn 37) following Ian's comment that both nutrients are needed. Ian also marks and re-marks the symmetry, hypothesizing ("one wouldn't be too surprised") that something was done to the axes so that the curves come to be the same ("to them same"). The trajectory of the finger itself is relevant because it allows the curve to become visible in joint space, but also because it points without pointing to itself. It is not the finger that becomes a salient figure against a background but that which the finger brings to awareness. But at the same time, the index finger allows us to see the pointing of the pointing without attending to it as such. Marking is pointing, and marking is done using a number of means including sound (that we hear as words), variations in prosody (intensity, pitch, speech rate), change in body (head or more complete) orientation, and position of body parts (hands on graph or text).

Dialectic of Talk and Image

Talk is not straightforwardly related to graphical aspects and orientations. Rather, as seen in the following episode, we need to understand talk as one way in which the current communicative unit, constituted by the present material and social situation as a whole (Roth & Pozzer-Ardenghi, 2006), expresses itself. But it does so one-sidedly, therefore capturing of the communicative whole only some aspect in the same way that the "wave nature" captures only part of the phenomenon of

Figure 4. This overlay of three moments shows how Lyn (right) and Ian produce two different gestures as perceptual features alongside the same stretch of talk

light or that the "beaver" (tail) captures only a part and incompletely denotes a Canadian 5-cent coin ("a nickel"). In any one situation and at any one moment, very different but still equivalent perceptual features may accompany the same talk. This is especially noticeable in the current database of joint graph interpretations, where one person talks while both individuals gesture iconically and deictically (pointing) (Figure 4).

```
043 I    [anyway] but=u thats tha:ts: how many realistic
        (0.76) whereas in •`b we have a very 'clear and strong
        correlation
044     (2.56)
045     which loo:ks very much li:ke you need (0.30) one unit
        of * r=one about for ev- (0.23) fer every one * unit of
        r=two *
```

In this episode, Ian produces a trace perpendicular to the three lines on the graph. It is a line that constitutes a diagonal, which can be seen as constitutive of a curve in which every unit of increase in R_1 corresponds to an increase of a unit in R_2, just as described in turn 45. But for hearing/seeing this reading and the perceptual trace as equivalent, the producer and recipient of the communication have to be familiar with functions that relate changes in one variable to changes in another. Lyn, on the other hand, places her index and small finger onto the axes R_1 and R_2. She does reach the second axes just ahead of the word "unit," and therefore clearly before the utterance of "r=two." She is attending to Ian, whose utterances therefore become an a-posteriori description of what Lyn has been pointing out. This shows how Lyn, currently listening, is aligned with the development of ideas in the discourse. We therefore need to understand the communicative situation as a whole, as one of *participative thinking* (Bakhtin, 1993), where aspects of participation and thinking made available never capture the communication in its entirety. Communication and participative thinking are irreducible to any one of the concretizations in which they are made available to other participants and, therefore, to the analyst(s) vicariously participating in the session.

The gestures make salient different aspects. On the one hand, Lyn points out the two axes, pointing first to R_1 and then to R_2; and she does so at the points where the 20% curve intersects with each, respectively. On the other hand, Ian produces a trajectory and therefore an ephemeral trace that does not have an equivalent ink trace on the paper. And yet, this trace, produced contemporaneously with the utterance, is part of the same communicative unit; and the two, speech and gesture, are irreducible moments of this same unit (Vygotsky, 1986). Because the present unit of analysis is the situation as a whole (Vygotsky, as Heidegger, emphasizes the fact that thinking and consciousness always reflect the situation as a whole, including all referring relations), the stretch of talk stands in a mutually constitutive relation to different visual features. Hyperbolically formulated, we must not replace the gestures and images (photographs, drawings) thereof by text, which would falsify and misrepresent essential features available only perceptually but not verbally.

The visibility of different graphical features is an important but frequently misunderstood or misrepresented aspect of graph interpretation. Thus, children and students frequently are said to *confuse* the height and slope of a graph or to make iconic errors, that is, interpret what is visibly at hand. In the present episode, we see two highly trained individuals express different perceptual features to accompany the same text, the same shared communicative context. As we generally do not question the objectivity of what appears to us in our perception—unless there is trouble or unless we have reason to believe that there is an optical illusion—analysts need to focus on what participants do with those features that they *actually* perceive and attend to in order to understand the forms of reasoning enacted. Because of the essentially passive element in perception—we are *given* phenomena rather than actively producing them (Marion, 2002)—the production of interpretations based on what we objectively see cannot be theorized appropriately in terms of misconception, iconic error, or, where applicable, by slope-height confusions. In my research, scientists tend to engage in the same kind of thinking as students when slopes rather than heights are salient in their perceptions, which they make available in their gestures (Roth & Bowen, 2001).

Prosodic Pointing

In the episode of the previous subsection, we can also clearly see how prosody is used to mark aspects to be attended to. The "whereas" marks that a contrast is coming, which occurs immediately thereafter as Ian utters "in b." Here, it is not only the letter "b," associated as it is with a second panel next to the one previously noted as "situation a," but the utterance of "b" stands out from the utterances surrounding it in the sudden shift in pitch level and speech intensity (Figure 5). That is, the pitch jumps from what it was for "in" and what it becomes in the "we," and the speech intensity jumps equally with respect to the syllables before and after. These speech parameters call attention to the "b," but do so in the background in the same way that the tracing index finger calls attention to the trace beneath it, marking the trajectory as an (iconic) sign without drawing attention to the sign or the sign of a sign. The "b" is not just any letter but it is made to stand out, and it stands out in both its intensity and pitch level.

The sound that we hear as "be" (transcribed as [bi:] in the conventions of the International Phonetics Association) is salient and significant in a network of relations that subtends it as salient. There is an ink trace "b" on the sheets of paper that Ian and Lyn have marked out in and through their orientation. This ink trace motivates the utterance of "be" ([bi:]), which in turn orients the participants and us to the central of the three plates. The sound that we hear as "be" is marked as standing out by further markers (signs), increased pitch and intensity levels. Thus, the salience of the plate marked "b" is the result of a series of mutual and mutually constitutive references. The plate and the three curves it features become available and significant as part of a network of significations.

RETHINKING GRAPH INTERPRETATION

In this chapter, I exhibit the work that allows the dawning of graphical aspects said to be signs of phenomena. This work consists in and is the product of sequentially ordered communicative contributions consisting not only of talk, as per conversation analytic tradition, but also of other material productions that come to constitute signs and sign relations and, in this, bring to the fore (allow us to see) a network of significance. This work involved (a) establishing joint orientation and specification of an aspect of the setting that becomes the topic of talk, (b) gesturing to articulate (structuring the field, expressing the structure) features that possibly serve as signs for something else, (c) articulating the situations differently and by different means, and (d) producing sets of features that subtend the salience of some against others as background and support. As a result of this work, certain graphical features come to be articulated against a background that also would allow other features to become visible. The actually visible features are the materials available for building the relation on which the nature of the sign as sign is built.

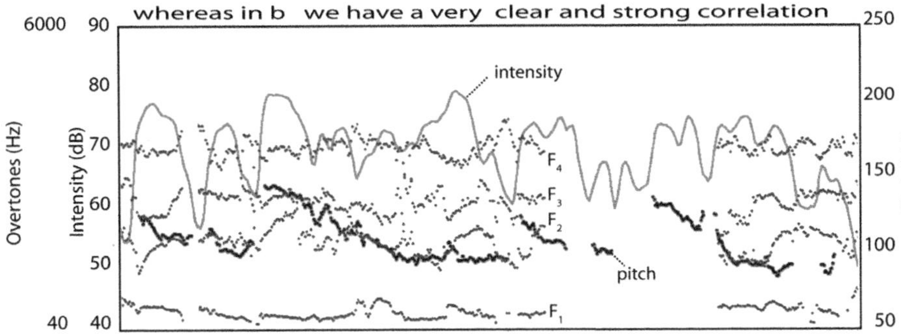

Figure 5. Various prosodic means, including speech intensity, pitch, and speech rate are used to mark (indicate) some aspects of the sound stream, raising their salience. Pitch is the base frequency (F_0) of a voice; the frequencies F_n are the higher order frequencies

Here, I highlight the way in which the visibility of features comes to be grounded. Whether some feature is in fact a viable sign depends on the possible worlds in which it might have a useful function. Any decision that there is no corresponding world to which a signing relation can be established, therefore, would lead the interpreters back to continue the work of seeking possible structures in the display. Thus, once Ian had noted the symmetry of the curves in plate "a," he supposed that the producers of the graph might have done something to the axes not immediately available on the task sheet (Figure 1) itself. Here, the feature that had become salient and therefore visible appeared strange, and, though it did not lead to questioning perception, it led to the hypothesis that it was an artifact of the way in which the axes were produced and scaled.

Graph interpretation turns around the problematic of understanding signs in everyday use where its particulars are revealed in use rather than in contemplation. Use makes apparent the *what-for* and *in-order-to* dimensions of mundane activity. It is this in-order-to orientation that the interpretation of a sign attempts to reconstruct. This is achieved when subjects find in their own actions the relevant possible uses of the sign as tool. As circumspect use of a tool presupposes a disclosed world, the interpretation of a sign that has anything to do with potential use is possible only if there is a contemporaneous world (significant relations) in which the sign might have a place. This study shows how graph features are constituted as signs. But these features emerge from a network of mutually constitutive forms of pointing (signing), including the general spatial area of joint attention, features of the sound stream to be attended to, attention to the general parts of the display and to specific parts of the display, and the relation of features of trajectories (hand, pen) and features in the display. It is only in this network of relations enacted against an invariant ground that anything can become salient structure with signing function. It is in this network of relations that signs and their tool function (reference) can emerge.

IMPLICATIONS FOR RESEARCH AND PRAXIS

This chapter has implication for the way we think about and theorize graph interpretation specifically and interpretation of inscriptions—means of communication other than text, including photographs, drawings, maps, graphs, diagrams, and mathematical formulas and notations—generally. Based on the present research, we must not assume that what a researcher takes to be available as a sign actually is available to research participants in their perception. In more technical terms (cognitive psychology, artificial intelligence), we must not make a-priori assumptions about the participants' *domain ontology*, the collection of wholes and parts currently present in and to consciousness. This ontology not only differs between individuals but also changes within individuals across time. A research method such as the one employed here allows researchers to tease out just what is salient to participants, especially when they work jointly on tasks, as they continually orient one another and request others to be oriented toward currently relevant detail. Participants make salient relevant detail through a variety of devices including words, gestures, body movements, body orientations, body positions, and prosody. Together, these devices constitute and are constituted by a network of signifying relations as the result of which some signs and referring processes come to stand out against and are subtended by all the other referring relations.

The chapter has important implications for rethinking the claims about mathematics education research that explain certain think-aloud protocols as evidence for slope-height "confusions" and iconic misconceptions. For there to be a confusion, at least two things—here height and slope—need to be present simultaneously but fused into one. Whether this is the case is an empirical issue—there is evidence in the present database that even experienced physicists attend to slope when the "correct" answer requires attending to height; and there is evidence that these physicists interpret iconic (surface) features rather than the "deep structure" cognitive psychologists are interested in (Roth & Middleton, 2006). That is, the physicists in this study interpret what is present in their perception, and thereby, to use Heidegger's expression, "ready-to-hand" (ready to use without conscious reflection) *in-order-to* achieve some goal that it is useful for (i.e., the *what-for*). My research therefore contributes to a revalorization and re-appreciation of the lived worlds that we bring to situations and engender, the lifeworlds that are given to us in our perceptions, and the always-real structures which we use in our (discursive, material) actions.

NOTES

1 Heidegger uses the present participle *das Seiende* in noun form, often translated as "beings," though perhaps better translated as "that which exists" because of the parallel construction in the singular tense. Unlike in German and French, the noun form of to be, *being* [Ger. *das Sein*, Fr. *l'être*], falls together with the present participle, being [Ger. *das Seiende*, Fr. *l'étant*], which would lend to confusion. In English, the singular present participle in German and in French translations is thereofore always rendered in the plural form. I sometimes produce my own translations to construe English texts that are less "arcane" to the eyes and ears of normal readers/speakers of English than the official translations. In making reference to Heidegger, I do not claim to extend the philosopher's work or to give an interpretation that is done in his spirit. I am not and do not pretend to be one of Heidegger's exegetes. I am a cognitive scientist to whom Heidegger's work has given, as it has given to others in the field of cognitive science and artificial intelligence (e.g., Agre & Horswill, 1997), many starting points for developing my own thinking concerning a number of dimensions of everyday cognition. Heidegger, for present-day ethnomethodologists, is useful because he assists us in attending to the world as it appears to us in everyday coping, which is structured very differently from the rationalistic ways most cognitive scientists, psychologists, and educators present it to us.

2 Heidegger discusses the notions of sign and reference in ¶17 of *Seit und Zeit* entitled "Reference and Signs" (p. 76–83 [77–77]), where the square brackets indicate the page numbers in the English version, which also notes the page numbers in the original German version. This ¶17, however, does not discuss signs and references in themselves. Rather, Heidegger is concerned with the nature of things in everyday coping, which, as he states, are bound into systems of referring relations. In ¶16, Heidegger discusses how the worldliness of the "surrounding world" (some semioticians, such as Thure von Uexküll, use the term *Umwelt*, other scholars prefer the term *lifeworld*) exhibits itself in beings or things we encounter "within this world" (*innerweltlich*). The discussion of the referring

relations continues into ¶18, "Relevance and Significance: The Worldliness of the World" (p. 83 [77]). The primary function of signs is discussed on p. 80 [74].

[3] Heidegger (1977/1996) discusses the triple character of the sign on p. 82 [77].

[4] Because I have previously used the term *grounding* to denote the process by means of which a sign comes to be linked to a person's familiar world (lifeworld), I here use the verb "to base" for translating or translating Heidegger's German verb "fundieren" (to ground, to base, to found). Heidegger also uses the noun *Fundament* (foundation), which shows his concern for the relations on which a thing or phenomenon is build.

[5] This formulation, "letting itself be seen as it is," points to a fundamental experience of passivity in human experience. For example, the experience of finding out about something by touching it requires both agency, moving the hand so that the fingertips come to touch the object, and passivity, allowing the object to affect our senses and, therefore, our sense-making effort (e.g., Roth, 2007). Jean-Luc Marion's (2002) *Being Given: Toward a Phenomenology of Givenness* conjugates the different dimensions of this fundamental human experience.

[6] I make use here of the standard transcription convention that the International Phonetics Association adheres to. The technical linguistic concepts describing the phenomenon are "liaison" and "Sandhi." In spoken French, such liaisons are everyday features, and especially confusing when the speaker appears to pause before pronouncing the next word, which then begins with the plural "s" of the previous word, giving, to the uninitiated, a word that cannot be found in the dictionary. The issue of parsing is rendered more difficult because, in cases of words beginning with "s," the sound is not due to the liaison, such as in [fo] [semblant] (*faux semblant*, false resemblance), which, in rapid speech, becomes [fo ï semblant].

REFERENCES

Agre, P., & Horswill, I. (1997). Lifeworld analysis. *Journal of Artificial Intelligence Research, 6*, 111–145.

Bakhtin, M. M. (1993). *Toward a philosophy of the act*. Austin, TX: University of Texas Press.

Buccino, G., Riggio, L., Melli, G., Binkofski, F., Gallese, V., & Rizzolatti, G. (2005). Listening to action-related sentences modulates the activity of the motor system: A combined TMS and behavior study. *Cognitive Brain Research, 24*, 355–363.

Davidson, D. (1986). A nice derangement of epitaphs. In E. Lepore (Ed.), *Truth and interpretation* (pp. 433–446). Oxford: Blackwell.

Eco, U. (1984). *Semiotics and the philosophy of language*. Bloomington, IN: Indiana University Press.

Garfinkel, H., & Sacks, H. (1986). On formal structures of practical action. In H. Garfinkel (Ed.), *Ethnomethodological studies of work* (pp. 160–193). London: Routledge & Kegan Paul.

Heidegger, M. (1977). *Sein und Zeit*. Tübingen: Niemeyer. (J. Stambaugh translated the English version *Being and time* [1996], published by State University of New York Press, Albany.)

Hoffmann, M. H. G. (2007). Learning from people, things, and signs. *Studies in Philosophy and Education, 26*, 185–204.

Leinhardt, G., Zaslavsky, O., & Stein, M. K. (1990). Functions, graphs, and graphing: Tasks, learning, and teaching. *Review of Educational Research, 60*, 1–64.

Marion, J.-L. (2002). *Being given: Toward a phenomenology of giveness*. Stanford, CA: Stanford University Press.

Roth, W.-M. (2003). Competent workplace mathematics: How signs become transparent in use. *International Journal of Computers for Mathematical Learning, 8*, 161–189.

Roth, W.-M. (2004). What is the meaning of meaning? A case study from graphing. *Journal of Mathematical Behavior, 23*, 75–92.

Roth, W.-M. (2005). Making classifications (at) work: Ordering practices in science. *Social Studies of Science, 35*, 581–621.

Roth, W.-M. (2007). Theorizing passivity. *Cultural Studies of Science Education, 2*, 1–8.

Roth, W.-M., & Bowen, G. M. (2001). Professionals read graphs: A semiotic analysis. *Journal for Research in Mathematics Education, 32*, 159–194.
Roth, W.-M., & Bowen, G. M. (2003). When are graphs ten thousand words worth? An expert/expert study. *Cognition and Instruction, 21*, 429–473.
Roth, W.-M., & Middleton, D. (2006). The making of asymmetries of knowing, identity, and accountability in the sequential organization of graph interpretation. *Cultural Studies of Science Education, 1*, 11–81.
Roth, W.-M., & Pozzer-Ardenghi, L. (2006). Tracking situated, distributed, and embodied communication in real time. In M. A. Vanchevsky (Ed.), *Focus on cognitive psychology research* (pp. 237–261). Hauppauge, NY: Nova Science.
Selting, M., Auer, P., Barden, B., Bergmann, J., Couper-Kuhlen, E., Günthner, S., Meier, C., Quasthoff, U., Schlobinski, P., & Uhmann, S. et al. (1998). Gesprächsanalytisches Transkriptionssystem. *Linguistische Berichte, 173*, 91–122.
Vygotsky, L. S. (1986). *Thought and language*. Cambridge, MA: MIT Press.
Wittgenstein, L. (1958). *Philosophical investigations* (3rd ed.). New York: Macmillan.

Wolff-Michael Roth
Lansdowne Professor, Applied Cognitive Science
University of Victoria
Canada

APPENDIX: TRANSCRIPTION CONVENTIONS

In this chapter, I draw on transcription conventions common to conversation analysis enhanced by transcription features specific for researchers interested in marking prosody (Selting et al., 1998). I add specific features for transcriptions that include offprints. The transcription is neither grammatical—see punctuation—nor consistent with spelling rules but constructed to exhibit the sounds. The turns have been numbered for the entire session.

Feature in context	Explication
(0.25)	Time in hundreds of seconds
.hhhhh	Inbreath, each "h" corresponding to 0.1 seconds
Hh	Outbreath, each "h" corresponding to 0.1 seconds
survi:ve	Colon indicates lengthening of phoneme, each colon corresponding to 0.1 seconds
r=one	Equal sign means "run-in" of the phonemes or "latching" of different speakers, meaning no pause between phonemes
084 L: <<p>point [he:re] 085 I: [than with]	Square brackets in consecutive turns indicate extent of overlapping speech, features
;.,?	Punctuation marks indicate movement of pitch toward end of utterance segment, down, strongly down, up, and strongly up, respectively
<<p>point>	Triangular brackets mark prosodic features, here "piano," that is, lower than normal intensity
<<pp>point>	"Pianissimo," much lower than normal speech intensity, next to inaudible intensity
<<dim>point>	"Diminuendo," decreasing speech intensity
*	The asterisk marks correspondence between talk and offprint
two nu-	Two asterisks and underline are used to mark the extent of a gesture made visible through an overlay of two or three offprints
•`b, 'clear	Arrow and diacritics indicate movement of pitch: upward jump, downward and upward contour of phonemes that follow
ev-	Phoneme stops with level pitch movement, as in mid-word or mid-sentence

NORMA PRESMEG

TRIGONOMETRIC CONNECTIONS THROUGH A SEMIOTIC LENS

In learning the definitions of sine, cosine, and tangent of an angle in trigonometry, it is necessary for students to coordinate and be able to move fluently amongst four sign vehicles that represent these objects. These sign vehicles belong to the registers of the triangle definitions, the coordinate plane, the unit circle, and the graphs of these trigonometric functions. Data from research projects are used to illustrate issues that arise in the course of students' learning. These issues include compartmentalization and use of prototypes, generalization, compression, use of metaphors, use of gestures, use of visual sign vehicles, idiosyncratic notation and sliding interpretants, the power of falsification, and use of a tool as a sign. After presenting data instantiating these categories, some implications for the teaching of trigonometric definitions in these four registers are drawn.

OVERVIEW

In each of two research projects, the central question being investigated was as follows: *How may teaching facilitate students' construction of connections amongst registers in learning the basic concepts of trigonometry?* In the first investigation, with Susan Brown in her "Advanced Enriched Algebra and Trigonometry" class in a Chicago high school, in addition to observation of her teaching, and collection of students' work, four students were interviewed six times each over a period of three months. In the second investigation, with Jeffrey Barrett and Sharon McCrone in Jeff's university course mainly for prospective teachers, "Geometric Reasoning: Geometry as Earth Measures," three students were interviewed twice each to interrogate their knowledge of the trigonometry that Jeff was including in the course. Analysis of more than 300 pages of transcript data from both sets of student interviews provided the categories of issues arising in students' learning, but the transcripts from the Chicago study are the main focus of this chapter. The specific question addressed here is as follows: *What aspects enable or constrain the making of connections amongst signs in learning the basic concepts of trigonometry?* Finally, some suggestions are made regarding the teaching of trigonometry. A strong implication of the investigation is that by highlighting structures and patterns across domains, teaching that encourages students to make connections may foster generalizations and help to combat the phenomenon of compartmentalization.

L. Radford, G. Schubring, and F. Seeger (eds.), Semiotics in Mathematics Education: Epistemology, History, Classroom, and Culture, 103–119.

THEORETICAL LENSES

Three theoretical lenses were used in interpreting the data in this study, namely, Peircean semiotics (Peirce, 1992, 1998), Duval's (1999) conversions amongst registers, and Hitt's (1998) levels of understanding of systems of representation.

In analyzing the transcripts for the purpose of investigating the ways that students construct, or fail to construct, connections amongst the triangle definitions of sine, cosine and tangent of an angle, the coordinate plane and unit circle, the sinusoid and tangent graphs, and trigonometric identities, two of Peirce's (1992, 1998) ten triads were used. These triads are as follows:

- object, representamen, and interpretant;
- signs that are iconic, indexical, or symbolic according to students' interpretations.

With regard to terminology, because Peirce himself used the word "sign" in different ways, sometimes referring to the representamen, but more often referring to the whole triad of object, representamen, and interpretant, in this paper I shall use the term "sign vehicle" when referring to the representamen, and "sign" when referring to the whole triad.

According to Peirce (1992), *trichotomic* is the art of making three-fold divisions. By his own admission, he showed a proclivity for the number three in his philosophical thinking. "But it will be asked, why stop at three?" he wrote (Peirce, 1992, p. 251), and his reply to the question is as follows:

> [W]hile it is impossible to form a genuine three by any modification of the pair, without introducing something of a different nature from the unit and the pair, four, five, and every higher number can be formed by mere complications of threes. (ibid.)

Accordingly, he used triads not only in his semiotic model including object, sign vehicle (or representamen) which stands for the object in some way, and interpretant based on the relationship between sign vehicle and object, but also in the types of each of these components. These types are not inherent in the signs themselves, but depend on the interpretations of their constituent relationships between sign vehicles and objects. In a letter to Lady Welby on December 23, 1908, he wrote as follows.

> I define a Sign as anything which is so determined by something else, called its Object, and so determines an effect upon a person, which effect I call its Interpretant, that the latter is thereby mediately determined by the former. My insertion of "upon a person" is a sop to Cerberus, because I despair of making my own broader conception understood. I recognize three Universes, which are distinguished by three Modalities of Being. (Peirce, 1998, p. 478)

For the purpose of using some of Peirce's triads in examining data on the teaching and learning of trigonometry, I shall embrace his "sop to Cerberus" and acknowledge that different individuals may construct different interpretants from the same sign vehicle, thus effectively creating different signs for the same object.

To illustrate the differences among iconic, indexical, and symbolic signs by using some of Peirce's examples, in an iconic sign, the sign vehicle and the object share a physical resemblance, e.g., a photograph of a person representing the actual person. Signs are indexical if there is some physical connection between sign vehicle and object, e.g., smoke invoking the interpretation that there is fire, or a sign-post pointing to a road. The nature of symbolic signs is that there is an element of convention in relating a particular sign vehicle to its object (e.g., algebraic symbolism). These distinctions in mathematical signs are complicated by the fact that three different people may categorize the "same" relationship between a sign vehicle and its object in such a way that it is iconic, indexical, or symbolic respectively, according to their interpretations. In practice the distinctions are subtle because they depend on the interpretations of the learner—and therefore, viewed in this way, the distinctions may be useful to a researcher or teacher for the purpose of identifying the subtlety of a learner's mathematical conceptions if differences in interpretation are taken into account.

As an example, let us examine the quadratic formula in terms of this triad. The roots of the equation ax2+bx+c = 0 are given by the well known formula

$$x = \frac{-b \pm \sqrt{b^2 - 4ac}}{2a}$$

Because symbols are used, the interpreted relationship of this inscription with its mathematical object may be characterized as symbolic, involving convention. However, depending on the way the inscription is interpreted, the sign could also be characterized as iconic or indexical. The formula involves spatial shape. In my original research study of visualization in high school mathematics (Presmeg, 1985), many of the students interviewed reported spontaneously that they remembered this formula by an image of its shape, an iconic property. However, the formula is also commonly interpreted as a pointer (cf. a direction sign on a road): it is a directive to perform the action of substituting values for the variables a, b, and c in order to solve the equation. In this sense the formula is indexical. Thus whether the sign vehicle of the formula is classified as iconic, indexical, or symbolic depends on the interpretant of the sign.

Peirce's model includes the need for expression or communication: "Expression is a kind of representation or signification. A sign is a third mediating between the mind addressed and the object represented" (Peirce, 1992, p. 281). In an act of communication, then—as in teaching—there are three kinds of interpretant, as follows:

- the "*Intensional* Interpretant, which is a determination of the mind of the utterer";
- the "*Effectual* Interpretant, which is a determination of the mind of the interpreter"; and
- the "*Communicational* Interpretant, or say the *Cominterpretant*, which is a determination of that mind into which the minds of utterer and interpreter have

to be fused in order that any communication should take place" (Peirce, 1998, p. 478, his emphasis).

It is the latter fused mind that Peirce designated the *commens*. The commens proved to be an illuminating lens in examining the history of geometry (Presmeg, 2003), and this third triad of interpretants is a useful one to bear in mind when considering implications of the results of the trigonometric investigation for teaching.

These lenses are augmented by Duval's (1999) notion of *registers*, taken in this paper to mean *types of sign vehicles*. There are four distinct registers in the trigonometry investigated in this research, namely, triangle definitions, sign vehicles of trigonometry in the coordinate plane, the unit circle, and in the graphs of trigonometric functions.

Using the theoretical lens of registers rather than Peirce's signs, in an enlightening analysis that is relevant to the conversions among registers of the trigonometry students, Hitt (1998) identified five levels of understanding of the concept of function, as follows.

> Level 1: Imprecise ideas about the concept (incoherent mixture of different representations of the concept).
>
> Level 2: Identification of different representations of the concept. Identification of systems of representation.
>
> Level 3: Translation with preservation of meaning from one system of representation to another.
>
> Level 4: Coherent articulation between two systems of representation.
>
> Level 5: Coherent articulation of different systems of representation in the solution of a problem.

These levels of understanding cast further light in interpreting students' conversions amongst the four registers of trigonometry.

EMPIRICAL SETTING AND DATA COLLECTION

In spring semester of 2006, informed by Sue Brown's knowledge of the students in her year-long class in a Chicago high school, "Advanced Enriched Algebra and Trigonometry", four students were chosen from the 30 in the class to provide a range of cognitive styles and abilities. These students, Laura, Raj, Jim, and Brian (pseudonyms), were interviewed six times each over the period from January to April, addressing the time in which they were learning trigonometry. The audio-recorded task-based interviews were of 15-35 minutes' duration. The tasks were chosen by Sue and the author in collaboration, and included some tasks from Sue's dissertation study (Brown, 2005) for the purpose of later comparison. An example of an elementary task from the first interview is presented in figure 1.

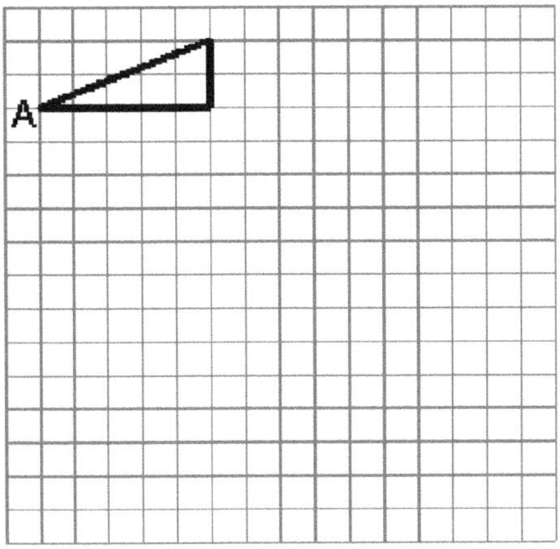

The diagram shows a right triangle for which A is an angle. On the grid, draw a triangle with an $\angle Q$ for which $\tan \angle Q$ is twice as large as $\tan \angle A$.

Figure 1. A task in the first interview (Brown, 2005).

The discrepancy between the students' interpretants in response to the iconic sign vehicle in the diagram of figure 1, in Sue's class and in the investigation with prospective elementary school teachers (in collaboration with Jeff Barrett and Sharon McCrone), is illustrated starkly by the fact that most of the prospective teachers doubled both the opposite side and the adjacent side, in attempting to double the tangent of angle A. Of the four students' interviewed in Sue's class, only Raj made this error. He saw his own mistake when questioned, and modified his interpretant by considering the special case in which $\tan \angle A = \frac{2}{5}$ would need to become $\frac{4}{5}$ in order to double the tangent. Most of the students in Sue's class provided a correct new sign vehicle spontaneously when they encountered this item in a test. The prospective teachers' project provides interesting comparisons but will not be addressed in this chapter.

In addition to the interviews, Chicago data sources included notes from seven observations in Sue's class, documents consisting of students' work on tests and quizzes, and Sue's reflections on ways that she attempted to facilitate students' construction of connections between registers in trigonometry. The methodology of this part of the research could be regarded as a teaching experiment, which includes cycles of joint reflection based on interviews with students, followed by further teaching, in some ways comparable to developmental research (Gravemeijer, 1994). It is beyond the scope of this chapter to describe the teaching experiment in detail. However, some aspects of Sue's pedagogy are described in drawing implications for teaching in the final section.

Some of the transcribed interview data from the Chicago project and that with prospective teachers—consisting of a total of more than 300 pages of transcriptions from the 30 interviews—will provide a source for the results reported in this chapter. Analysis of the transcripts consisted of coding the data according to themes that appeared to be relevant to aspects that enable or constrain the making of connections amongst registers in trigonometry. These themes and relevant examples are provided in the next section.

RESULTS AND ANALYSIS

Themes identified in the data analysis were as follows. (Letters in parentheses refer to the students, Laura, Raj, Jim, and Brian, in whose interviews, 1-6, these themes were identified.)

- Need of students for specificity in a sign vehicle (L4, R4).
- Compartmentalization and prototypes (L2&4, R2, J2, B2).
- Generalization and logic (R2, J4&5, B2): a "mathematical cast of mind" (Krutetskii, 1976) (B1,4,5&6).
- The power of falsification (R4&5, J4, B4&6).
- Compression, reification, encapsulation (B1,2,4,5&6).
- Spontaneous metaphors (R2, J2,3,4,5&6, B1,3&4) and taught metaphors (L6, J1, B1&6).
- Notation and sliding interpretants (L1&5, R5).
- Gestures as sign vehicles (R5, J3).
- Visual sign vehicles (L6, J2,4,5&6, B4&6).
- Use of a tool as sign (R3).

Examples of the limiting effects of compartmentalization (lack of the perception of possible conversions amongst registers) and certain prototypes in trigonometry were illustrated in a PME-30 plenary presentation (Presmeg, 2006) and will not be repeated in detail here. However, it should be noted that prototypes may affect conversions amongst registers in both negative and positive ways (Doerfler, 1991): the mnemonic advantages of memorable qualities of a prototype such as a right triangle with "horizontal" and "vertical" legs may be offset by the negative aspect that the prototype may prevent students from recognizing properties when the triangle is in a different orientation—and thus hinder conversions (Presmeg, 1992). In this chapter I shall illustrate examples of the following categories: *Need for specificity, Generalization and logic: a "mathematical cast of mind"* (Krutetskii, 1976), *The power of falsification, Visual sign vehicles, Notation and sliding interpretants,* and *Use of a tool as a sign.*

Need of students for specificity in a sign vehicle (Laura 4, Raj 4)

Tasks for interview 4 were as follows: “Check true or false. Illustrate with an example on the unit circle, sinusoid graph, or triangle.

(1) $\cos\theta = \cos(-\theta)$?

(2) $\sin(2\theta) = 2\sin\theta$?

(3) $\sin\theta = \sin(\theta+180^{\circ})$?”

Laura originally said that the first statement is false. Then she drew a diagram of an angle labeled 25° in the first quadrant: “Say it’s 25° for the angle. Well, if it was *negative* 25 ... it would be 25 off of zero, as well ... so the cosine would be the same.” Thus her interpretant for this sign was that the statement is true. She had difficulty connecting this interpretant with the cosine graph, in which she did not give the angle a specific value: “Well, it’s kinda hard to explain.”

Raj also worked with specific sign vehicles, namely, 30° and 330° on the unit circle: “It shares the same side so it has to be true” was his correct interpretant.

There is a sense in which the iconic signs Laura and Raj constructed for specific values of θ in the unit circle were interpreted generally. However, this generality is more apparent in the interpretants of Jim and Brian, as follows.

Generalization and logic (Jim 4, Brian 4, Laura 4) and falsification (Raj 4)

Jim preferred visual sign vehicles in all cases. However, these iconic signs were much more overtly general in his interpretants than in those of Laura or Raj. For the first task he, too, used $\theta=30^{\circ}$ and hence concluded that $-\theta$ is the same as 330° in a circle: “The cosine will stay the same, still cover the same distance on the x.” His *general* interpretant was one of *symmetry*. “The special case doesn’t matter. And then it’ll be the same thing on the other side of the second quadrant.” He also connected this symmetry with points reflected vertically on the cosine graph, which he drew.

For the second and third tasks in the fourth interview, **Raj** worked with particular sign vehicles again, which quickly led to the correct interpretant, by the **falsification principle**, because both of these statements are false. The power of falsification was evident in many of the transcripts: a specific example in this case leads to the general result that the statement is false.

Laura, by contrast, constructed a general sign vehicle for $\sin(2\theta)$: “it refers to the period—two cycles in 360°”; and for $2\sin\theta$, making a strong connection between the radius of a circle, and the amplitude of the corresponding graph. The discrepancy between the two graphs, which she drew separately, resulted in her interpretant that $\sin\theta$ does not equal $\sin(2\theta)$, and thus the statement is false. For $\sin\theta = \sin(\theta+180^{\circ})$ she again used the particular value, $\theta = 25^{\circ}$, and drew a circle diagram to conclude from positive and negative values of y that the statement is false. Although she used falsification, she did not refer to this principle.

The most general task in all six interviews came in the form of an "extra credit" item on a test the class had done, which was discussed in interview 4 when time permitted (figure 2). Of the four students interviewed, only **Brian** addressed this task. His thinking during aspects of trigonometric problem solving in all six interviews could be characterized by what Krutetskii (1976) called **a mathematical cast of mind.**

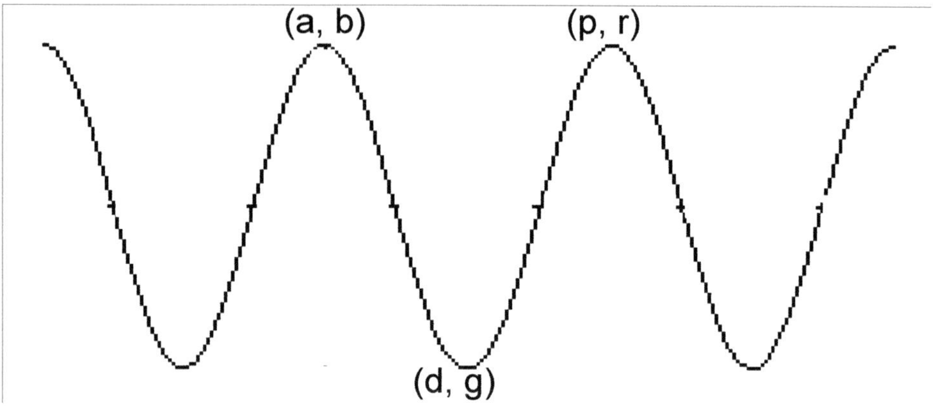

Give an equation of this sinusoid.

Figure 2. The most general task of all, from an "extra credit" test question

Brian placed a y axis through the point (a, b), and drew a midline horizontally through the sinusoid. "You're gonna have to find four things," he declared, namely, the *midline*, *amplitude*, *period*, and *phase shift*. He recognized that with this placement of axes he should work with the cosine. His final equation was as follows:

$$y = \frac{(b+g)}{2} + b - \frac{(b+g)}{2} \cos \frac{360^\circ}{p-a} (\theta - a)$$

Each part of the equation was an interpretant for a different general sign vehicle in the graph. The transcript of this part of the interview (over several pages) illustrates not only the generality but also the facility of Brian's mathematical connecting as he converted fluently between graphical and algebraic registers. The part of the equation that gave him most pause was the sign (positive or negative) of the second term, b. He originally gave this sign as negative. The average of y-coordinates b and g yields the y coordinate of the midline. Then the amplitude of the sinusoid is that value subtracted from b. After negotiating the meanings of the signs, he recognized that a negative sign for the second term would have the effect of flipping the graph—reflecting it over the horizontal axis. Hence the sign before the second term should be positive. He should have placed parentheses around the amplitude, ($b - \frac{b+g}{2}$), but otherwise his final equation is correct.

Generalization and the power of falsification (Brian 4)

Brian was specific, in the fourth interview, about the power of falsification:

> You can prove something false by having one example false! Because if you have a generality you assume that it would work for all cases and if you can find one case where it doesn't work, then, you have to change your generality. You know, it'll be false. On the other hand, proving it true, you'd either have to prove it not false [by contrapositive] … or you'd have to prove it so many times, that it will just, all special cases, then you'd have to prove them true. (Brian, Feb. 27, 2006)

This statement of the principle of falsification may be regarded as a higher-order sign vehicle that reifies a mode of functioning that goes beyond the context of trigonometry, although it was expressed in this trigonometric context.

Visual sign vehicles (Jim 4 & 6)

The next themes to be illustrated in this chapter are the power and pitfalls of iconic signs. Most of **Jim**'s sign vehicles were interpreted as visual icons. For instance, in response to "$\sin \theta = \sin (\theta + 180^{\circ})$?" he superimposed a sine graph with a phase shift of 180 degrees on the original sinusoid, seeing that the values of angles were opposites on the two graphs, although their absolute values were the same. In this case the interpretant was effective, but for Jim this was not always the case.

In the sixth interview, students were asked to describe connections between what they saw as the "big ideas" in the trigonometry that they had learned. In connecting the graph of $y = \cos \theta$ with the unit circle, **Jim** drew the diagram in figure 3 and tried to connect it iconically with his graph.

The quadrants and positive and negative signs do synchronize with the cosine sinusoid graph, as Jim indicated in the interview by referring to the crucial points of intersection and zero values on the y axis (1, 0, -1, 0). However, the interpretant breaks down—and hence the connection between registers breaks down—when cosine is defined as the x coordinate of a point on the unit circle. Interestingly, Jim's diagram conforms to the standard way that angles are conceived by surveyors, a symbolic sign measuring angles clockwise from due north. When he realized the discrepancy in his two interpretants, Jim was able to re-draw his circle diagram to conform to the conventional symbolic sign in which angles are measured counter-clockwise from the positive x-axis.

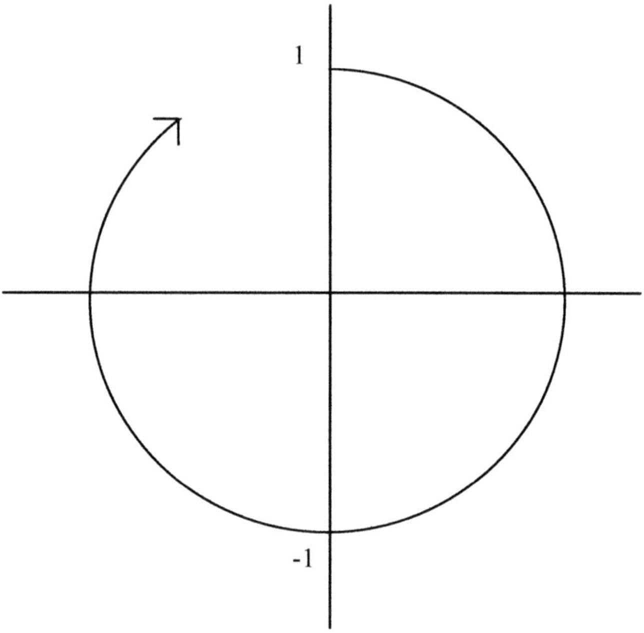

Figure 3. Jim's iconic sign connecting a unit circle with the graph of y = cos θ

An unhelpful sign that may persist (Jim 6)

Trying to link the circle definition of tan θ as y/x with its graph, Jim made *four* attempts at drawing the graph of y = tan θ, each time ending up with the last part as shown. In this example and the icon shown in figure 3, analysis of the definition was necessary to correct the mistake (symbolic sign). Jim seemed to have an image (iconic sign vehicle) of the idea (object) that the graph had to come down from the asymptote in some way. In figure 4 he has already partially corrected his conception in the case of the asymptote at 90°. In an earlier version he drew only one asymptote as in the last part of figure 4.

Notation and sliding interpretants (Laura 1&5)

It happened in interviews with Laura and with Raj that a double meaning for the notation that they chose for a particular problem gave them difficulty. For instance, in the first interview, Laura was explaining to the interviewer how to find the sine of an angle in the second quadrant. She had drawn a right triangle in the second quadrant and assigned variables to the lengths of the sides: x was the vertical leg and y the horizontal. That notation caused her to write the coordinates of a point in the second quadrant as (-y, x). The sliding meanings for x and y caused confusion, because in the unit circle, the x coordinate is defined as the cosine of the angle, and y is the sine. Again in the fifth interview, Laura had been asked to simplify a trigonometric expression in which tan x occurred. (The complete expression was (sin x + tan x)/(1 + sec x).) When I asked her to explain *why* tan x = (sin x)/(cos x) the sliding interpretant of x as a variable angle, and the x coordinate in the plane, again caused difficulties. For Laura, the definition of the tangent of an angle in the register of the coordinate plane was y/x, which came down to the sine of the angle divided by the cosine in the unit circle. The variable angle x in yet a different register (simplification of a trigonometric expression) proved to present too many unrelated interpretants for Laura at first.

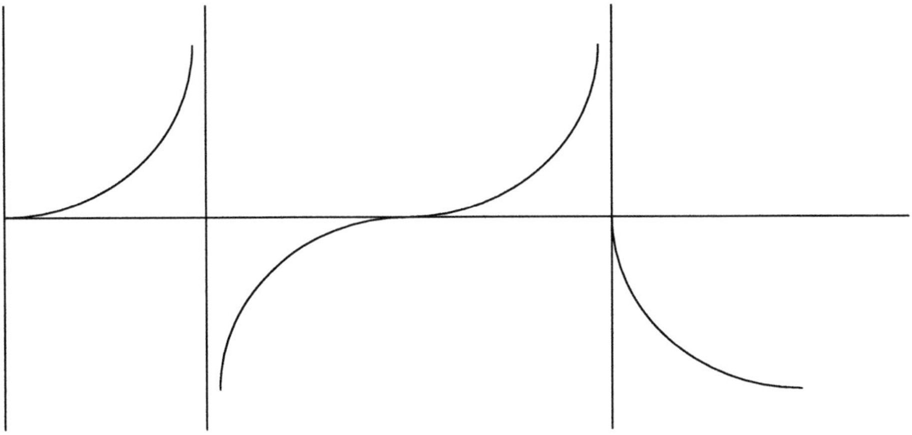

Figure 4. Jim's persistent sign vehicle for the graph of $y = \tan \theta$

Use of a tool as a sign (Raj 3)

As a final example of aspects that either constrain or facilitate conversions amongst register of trigonometry, I present the facilitative use of a tool by Raj in the third interview. He had been asked to do the following task.

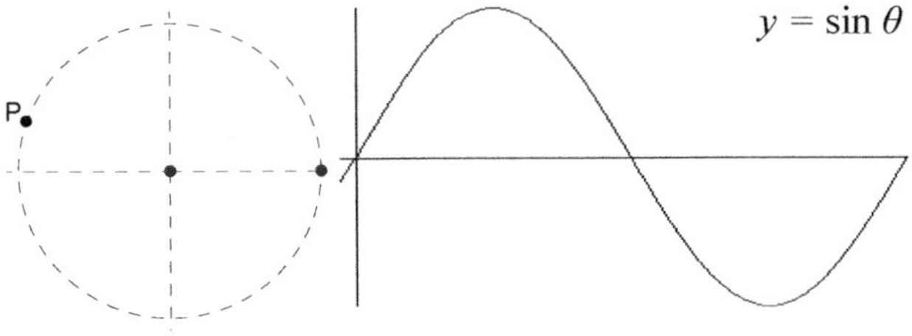

Figure 5. A task in the third interview

On the unit circle, P is the terminal point of a rotation. Mark the point that corresponds to P on the graph of $y = \sin \theta$.

Recognizing that he needed the y coordinate of the point P, Raj tentatively indicated with his finger where the corresponding point should go on the sine graph. Then, after a silence of 15 seconds, he asked, "Can I tear a piece off this?" indicating the question sheet. After my affirmative reply, he carefully tore a rectangular piece from the bottom right corned of the sheet, and placing it against the y coordinate of P, he measured the distance from P to the x axis, making marks on his tool. Then he transferred this distance to the sine graph and drew a vertical line to the sinusoid to indicate where P should be placed. It appeared that *distance* was the factor that linked these two registers, and Raj's use of a home-made tool to measure this distance enabled the precise placement that he was looking for. In semiotic terms, the sign vehicle of the unit circle had to be reconciled with the sign vehicle of the sinusoid, and the interpretant, "measure the y distance" provided a means of completing the task, using his constructed tool.

Synopsis

Returning to the categories summarized at the beginning of this section, it is useful to take a wider view and question whether or not these elements facilitated conversions amongst the trigonometric registers. Based on the data analysed, a judgment of whether each element was useful in allowing students to convert fluently amongst registers is provided in parentheses.

- Need of students for specificity in a sign vehicle. (Occasionally, students who required specific numbers could interpret these in a general way, allowing for conversions. More often, specific numbers tied thought to the original, given, sign vehicle.)
- Compartmentalization and prototypes. (Compartments in thinking, by definition, hindered conversions amongst register. Prototypes, on the other hand, could limit thinking but also had mnemonic advantages that sometimes allowed for conversions.)
- Generalization and logic: a "mathematical cast of mind". (This aspect appeared to be the key to fluent and dependable conversions amongst registers.)
- The power of falsification. (Falsification usefully combined the specificity of a particular case with the general principle that it takes only one counterexample to negate a mathematical statement. Conversions were seldom required in this case, but the generality of the principle facilitated conversions.)
- Compression, reification, encapsulation. (These elements were useful in facilitating the bird's eye view that allowed for conversions. Generalization was often implicit in these processes.)
- Spontaneous metaphors and taught metaphors. (Both kinds of metaphors were powerful in the interpretants constructed by students. Sometimes they were specific to one register, but in other cases they facilitated conversions.)
- Notation and sliding interpretants. (Students' idiosyncratic notation often took on unintended meanings, such as labelling A as the side of a triangle and then confusing it with the vertex or the angle of the triangle. Sliding interpretants hindered conversions.)

- Gestures as sign vehicles. (Analysis of these was beyond the scope of this paper, but they showed potential as means to support conversions.)
- Visual sign vehicles. (These manifested both strengths and impediments, depending on the generality of their interpretants.)
- Use of a tool as sign. (In the one case in which this element appeared, it was powerful in facilitating conversion from the unit circle register to that of trigonometric graphs.)

DISCUSSION AND SOME CONCLUSIONS

From the evidence illustrated in the foregoing, some preliminary conclusions may be drawn regarding aspects that enable or constrain the making of connections amongst trigonometric signs.

It is noteworthy that conversions amongst registers, in themselves, do not constitute a robust understanding of trigonometry. In an interesting study of college students' understanding of exponential relationships, Rajeev Nenduradu in his doctoral research described the case of Mike, who could move amongst graphical, tabular, and algebraic registers, and yet whose understanding of the connections was tenuous (Presmeg & Nenduradu, 2005). In the current study of trigonometric connections the same phenomenon is evident. All four of the Chicago students interviewed could make connections at some level amongst the main registers of trigonometry, namely the triangle definitions of trigonometric ratios, definitions of those ratios on the coordinate plane, coordinates of points on a unit circle, and the graphs of sine, cosine, and tangent. The making of connections amongst these big ideas was the explicit topic of the sixth interview, and all four of the students made at least some explicit connections. As Sue commented, it is a huge step when students can see that the sinusoid graphs take two of the three variables in the unit circle (x, y, and the angle θ), and rearrange these variables on the coordinate axes in a new way: that now θ is on the horizontal axis, and the vertical axis now calibrates the distance that previously belonged to x (cosine graph), or to y (sine graph).

Using the theoretical lens of registers rather than Peirce's signs, in an enlightening analysis that is relevant to the conversions among registers of the trigonometry students, Hitt's (1998) five levels of understanding of the concept of function are presented again:

> Level 1: Imprecise ideas about the concept (incoherent mixture of different representations of the concept).
>
> Level 2: Identification of different representations of the concept. Identification of systems of representation.
>
> Level 3: Translation with preservation of meaning from one system of representation to another.

Level 4: Coherent articulation between two systems of representation.

Level 5: Coherent articulation of different systems of representation in the solution of a problem.

It is clear that Nenduradu's Mike had not attained the higher levels in this taxonomy, although he appeared to move fluently amongst registers for exponential functions (Presmeg & Nenduradu, 2005). Because there was no preservation of meaning, it is likely that his conversions could not even be classified at the third level. Of the four students in the Chicago trigonometry study, it is only Brian who could be judged to have reached level 5 in his mathematical processing. The striking level of generality he attained in the interviews suggests that it is this bird's eye view that sees general principles when others see particulars, which could be said to characterize the highest levels of coordination. This conclusion resonates with Krutetskii's (1976) identification of "The ability for rapid and broad generalization of mathematical objects, relations, and operations" (p. 350), in the processing component of his structure of mathematical abilities of "competent" mathematics learners, which was confirmed in my early research (Presmeg, 1985).

Compartmentalization, spontaneous and taught metaphors, and, gestures, are all aspects that were striking components of the students' construction of trigonometric signs and their connections in the interviews. These aspects have not been illustrated in this chapter, but they are important. Compartmentalization was described in a previous paper (Presmeg, 2006), and spontaneous and taught metaphors in another (Presmeg, 2007). There is enough evidence from the examples presented in this chapter to conclude that generalization and associated aspects such as compression and reification play an essential role in robust interpretation of trigonometric connections. As noted, the difference between Brian's interpretants and those of the other three interviewees lay in the level of generality he attained in interpreting sign vehicles. It is not that he moved beyond the use of iconic signs. It is rather that these iconic signs became nested in signs at a higher level of generality, so that he worked with principles rather than unconnected specific cases. When he did work with specific cases, they were sign vehicles for more general trigonometric objects: in this way, they became indexical because they pointed to a principle. At the same time, the principles themselves were symbolic signs that conformed to the general conventions of trigonometry, as accepted in Popper's world number three—objective knowledge that is the product of human intellectual activity (Popper, 1974).

A necessary prelude to interrogation of the teaching question—how teaching may foster the connections amongst trigonometric registers—was the identification of salient aspects in the students' signs. Unlike Jeff's class of prospective elementary school teachers, in which geometry was the focus although some trigonometry would be introduced, the main goal in Sue's trigonometry class was to foster skill in converting amongst signs as students build up comprehensive knowledge of trigonometry concepts. Early in our collaboration, Sue listed ways in which she tries to facilitate connected knowledge in her class—actions that were confirmed in the researcher's observations of her lessons, and in documents such as tests and quizzes. Some of Sue's facilitative principles that have the intent of helping students to move freely and flexibly amongst trigonometric registers are summarized as follows:

- connecting old knowledge with new, starting with the "big ideas", providing contexts that demand the use of trigonometry, allowing ample time, and moving into complexity slowly;
- connecting visual and nonvisual registers, e.g., numerical, algebraic, and graphical signs, and requiring or encouraging students to make these connections in their classwork, homework, tests and quizzes;
- supplementing problems with templates that make it easy for students to draw and use a sketch, or asking students to interpret diagrams that are given;
- providing contextual ("real world") signs that have an iconic relationship with trigonometric principles, e.g., a model of a boom crane that rotates through an angle θ degrees, $0<\theta<180$, on a half plane;
- providing memorable summaries in diagram form, which have the potential of becoming for the students prototypical images of trigonometric objects, because these inscriptions are sign vehicles for these objects;
- providing or requiring students to construct static or dynamic computer simulations of trigonometric principles and their connections, in many cases giving a sense of physical motion;
- using metaphors that are sometimes based on the students' contextual experiences, e.g., a bow tie and the boom crane.

Sue's facilitative teaching actions for fostering connections in trigonometry have the status of intensional interpretants, because in the communication arena of her classroom they serve as "determinations of the mind of the utterer" (Peirce, 1992, p. 281). It would require a finer grain to determine from the transcript data to what extent they comprise effectual interpretants (determinations of the mind of the interpreter), and whether the "fused minds of utterer and interpreter" are in evidence, thus pointing to the commens required for successful communication. However, from the students' own words and actions it is possible to determine the influence, and efficacy, of some aspects of Sue's pedagogy. The students were all accustomed to the requirement, as in the task in figure 5, that they move between unit circle and graphs of the trigonometric functions, with templates provided in some cases, in all their written work. Further, the metaphor of the "bow tie" that Sue introduced for work in the four quadrants (Presmeg, 2006, 2007) was mentioned spontaneously by three of the four students interviewed. After the introduction of this iconic sign, not one of the four students erroneously resorted to drawing triangles to the y axis, although both Laura and Jim had done so previously.

The overriding aspect brought to the fore through an analysis of the interview data is the importance of constructing interpretants that are of a general nature when working in the different trigonometric registers. Seeing "the big picture" appears to enable students fluently to convert amongst the registers, as was evidenced particularly in Brian's interview data. A number of Sue's pedagogical principles work in this direction: encouraging and expecting connections and conversions, providing memorable summaries in diagram form, using computer simulations and real world examples, all require a bigger picture, However, it is still an individual matter how students construct their interpretants from the sign vehicles provided. There is still room for individual preferences, such as Jim's proclivity to work with iconic signs. Because of Sue's espousal of principles that aim to help students move flexibly and fluently amongst registers, all of her students had been exposed to these conversions in trigonometry. In the final interview, devoted to this topic, all of the four students interviewed achieved some success in this regard.

REFERENCES

Brown, S. A. (2005). *The trigonometric connection: Students' understanding of sine and cosine.* Unpublished Ph.D. dissertation, Illinois State University.

Dörfler, W. (1991). Meaning: Image schemata and protocols. In F. Furinghetti (Ed.), *Proceedings of the 15th Annual Meeting of the International Group for the Psychology of Mathematics Education* (Vol. 1, pp. 17–32).

Duval, R. (1999). Representations, vision and visualization: Cognitive functions in mathematical thinking. Basic issues for learning. In F. Hitt & M. Santos (Eds.), *Proc. 21st Conf. of the North American Chapter of the Int. Group for the Psychology of Mathematics Education* (Vol. 1, pp. 3–26). Columbus, Ohio: ERIC Clearinghouse for Science, Mathematics, & Environmental Education.

Gravemeijer, K. (2004). Educational development and developmental research in mathematics education. *Journal for Research in Mathematics Education, 25*(5), 443–471.

Hitt, F. (1998). Difficulties in the articulation of different representations linked to the concept of function. *The Journal of Mathematical Behavior, 17*(1), 123–134.

Krutetskii, V. A. (1976). *The psychology of mathematical abilities in schoolchildren.* Chicago, Illinois: University of Chicago Press.

Peirce, C. S. (1992). *The essential Peirce.* (Volume 1, edited by N. Houser & C. Kloesel, Eds.). Bloomington, IN: Indiana University Press.

Peirce, C. S. (1998). *The essential Peirce.* (Volume 2, edited by the Peirce Edition Project, Ed.). Bloomington, INndiana: Indiana University Press.

Popper, K. (1974). The philosophy of Karl Popper. In P. A. Schilpp (Ed.), *The Library of Living Philosophers.*, edited by Paul Arthur Schilpp (Ed.). La Salle, Illinois: Open Court.

Presmeg, N. C. (1985). *Visually mediated processes in high school mathematics: A classroom investigation.* Unpublished Ph.D. dissertation, University of Cambridge.

Presmeg, N. C. (1992). Prototypes, metaphors, metonymies, and imaginative rationality in high school mathematics. *Educational Studies in Mathematics*, 23, 595–610.

Presmeg, N. C. (2003). Ancient areas: A retrospective analysis of early history of geometry in light of Peirce's "commens". Paper presented in Discussion Group 7, *Semiotic and socio-cultural evolution of mathematical concepts*, 27th Aannual Mmeeting of the International Group for the Psychology of Mathematics Education, Honolulu, Hawaii, July 13–18, 2003. Subsequently published in the *Journal of the Svensk Förening för MatematikDidaktisk Forskning (MaDiF)* (Vol. 8, pp. 24–34), December, 2003.

Presmeg, N. C. (2006). A semiotic view of the role of imagery and inscriptions in mathematics teaching and learning. In J. Novotna, H. Moraova, M. Kratka, & N. Stehlikova (Eds.), *Proceedings of the 30th Annual Meeting of the International Group for the Psychology of Mathematics Education,* (Vol. 1, pp. 19–34). Prague, July 16–21, 2006.

Presmeg, N. C. (2007). The power and perils of metaphor in making internal connections in trigonometry and geometry. *Proceedings of the 5th Conference of the European Society for Research in Mathematics Education*. Larnaca, Cyprus, February 22–26, 2007.

Presmeg, N. C., & Nenduradu, R. (2005). An investigation of a preservice teacher's use of representations in solving algebraic problems involving exponential relationships. In H. Chick & J. Vincent (Eds.), *Proceedings of the 29th Annual Meeting of the International Group for the Psychology of Mathematics Education*, (Vol. 4, pp. 105–112). Melbourne, July 10–15, 2005.

Norma Presmeg
Mathematics Department,
Illinois State University

MICHAEL N. FRIED

BETWEEN PUBLIC AND PRIVATE: WHERE STUDENTS' MATHEMATICAL SELVES RESIDE

> "It is doubtless true that although learning mathematics may ultimately be done in private, in school mathematical performance is necessarily a public event." (Tobias & Weissbrod,1980, p. 68).

> "Proof is an essentially public activity" (Bell, 1976, p. 24).

> "...I also found a remarkable thirst [among other mathematicians] for conversation and discussion about our private experiences and inner beliefs" (Davis and Hersh, 1981, p. 4)

> "It is clear that one cannot *prevent* students and teachers alike from using these ubiquitous forms of signification in mathematics. In this regard, both metaphors and metonymies may be constructed as very private, personal, and ripe with meaning for an individual. But the processes of classroom discourse will also Involve metaphors and metonymies that are public and open to negotiation" (Presmeg, 1997, p. 277).

INTRODUCTION

The quotations above were garnered from a variety of sources, three from papers in mathematics education and one from a well-known popular book on mathematics. What they have in common is the use of the words "public" and "private" to describe one aspect or another of mathematical life. The frequency of this choice of words is, perhaps, not surprising since the public/private dichotomy is important in so many other aspects of our lives—and it may be that their importance in mathematics education derives from that wider range. With that in mind, this paper will look at notions of public and private and their specific relevance to mathematics education.

Although the theme relates to mathematics education in general, my own thoughts about it began, in part, with what I saw as a problematic marriage between mathematics education and history of mathematics (Fried, 2001). The problem was finding common ground among the presuppositions and commitments of these two disciplines—a problem because these presuppositions and commitments are in certain key respects inconsistent. Without going into detail, what was suggested was that mathematics education could accommodate history if mathematics education be understood as promoting "our self-knowledge as mathematical

L. Radford, G. Schubring, and F. Seeger (eds.), Semiotics in Mathematics Education: Epistemology, History, Classroom, and Culture, 121–137.

beings," a knowledge of ourselves as a kind creature that does mathematics (Fried, 2007).[1]

But placing "self-knowledge" at the center of mathematics education brings us to ask what kind of self *is* the mathematical self. Luis Radford took up the question of self directly at the 2006 Semiotics meeting at Haus Ohrbeck, Germany (also Radford (2008)). My own talk at that meeting (Fried, 2006) also took up this question, albeit implicitly. Referring the Greek mathematical world, the position maintained was that our mathematical selves are defined through the system of signs with which we communicate and think about mathematical ideas. Since this system exists, in Saussure's terminology, as a synchrony for a given community at a given time, it has an essentially historical character. To illustrate this, the case of the constellation of signs "similar," "equal," and "similar and equal" from Greek mathematical discourse was taken up; the argument then was that by coming to terms with the Greek mathematical discourse being genuinely Greek, we come to terms with our own mathematical identity formed within our own community of "mathematical speakers." The mathematical self, in this regard, seems to be wholly *public* in nature since it is constituted in terms of our collective rather than individual character.

However does not the very notion of "self" embrace a sense of the individual, the idiosyncratic, the *private*? History too points as much to the private as it does to the collective in its focus on particular, ungeneralizable thinkers and contexts. Yet, one might still ask whether a private mathematical self is not merely a comforting illusion. I will want to claim that, on the contrary, a private self, along with the public self, is an essential aspect of our mathematical selves.

This paper, then, considers these two domains in mathematics education—the public and the private—and where in them our mathematical selves reside. The discussion will be divided into three parts. In the first part, I shall discuss the public source of a mathematical self. This section will be an expansion of the position alluded to above, namely, that a historico-semiotic view of mathematics, stressing as it does our identity within a community of discourse, leads to a view of ourselves as public creatures. The second, and main, section will then raise the question, somewhat eclectically, of the status of a private domain in the midst of the seemingly overwhelming public existence implied by the historical and semiotic considerations of the first section. The aim here is not to repudiate a public self, but to show the untenability of an utterly public or an utterly private self. Finally, in the third concluding section, I shall try and argue that mathematics education needs to see itself as mediating between students' public and the private mathematical selves.

I. MATHEMATICAL SELF AS A PUBLIC ENTITY

Mathematics as public activity is evident in many ways. In an academic context, of course, the consummation of mathematical activity, as with every other academic discipline, is precisely publication. True, publication in learned societies, journals, and conferences is a relatively modern phenomena, beginning more or less in the

17th century, picking up speed in the 19th century, and reaching its present scale only in the 20th (for a review, see Bartle, 1995); still, communication of mathematical ideas in one form or another, widely or narrowly, openly or secretively, freely or selectively, has always been part of mathematical life. The qualifications just alluded to are not unimportant; nevertheless, what is essential is that the mere association between mathematical activity and communication places mathematical life within the range of community life and, to that extent, also public life. And although this is particularly true for the academy, it is also how we present mathematics to our pupils, even at an early age. Thus the NCTM *Principles and Standards* (2000) urges that communication as "...an essential part of mathematics and mathematics education" (p. 59) be made clear to all students, beginning with pre-kindergarten children!

Communication in mathematics and a mathematical community, not necessarily the academic community, are inseparably linked and mutually define each another: emphasizing communication in mathematics and mathematics education is only a way of emphasizing that mathematics rests on collective activity, and, conversely, recognizing the existence of a mathematical community, even the restricted one of a mathematics classroom, is only a way of recognizing that in mathematics people speak to one another. There is no surprise here since communication and community themselves are inseparably linked. Thus, Aristotle says that it is speech, *logos*, that makes human beings political, creatures whose life is community life.[2] But behind *logos*, behind dialogue, is a commerce of signs within a common shared system of signs, making speech and dialogue possible in the first place. Thus, a semiotic framework—though semiotics, in principle, takes in more than *logos*, even the *phōnai* of non-human living beings, and not only their *phōnai* (e.g., Sebeok, 1968)—is, in fact, at the very heart of the link between community and communication and, simultaneously, their fullest expression. Indeed, this social orientation of semiotics is almost built into the very idea of semiotics. This is particularly pointed in Saussure's (1974) approach to semiotics with its emphasis on language, which Saussure refers to as a *social institution.*

Whether mathematics should be thought of as itself a language, popular though that comparison may be, is, perhaps, unimportant. In his essay on the semiotics of mathematical discourse, Ard (1989) considers how symbols and words, for example, come together to produce mathematical *texts*: as texts, mathematical discourse can be treated as belonging alternatively to language or to a recognizable class of concrete, visual, objects. In either case, mathematical texts are semiotic objects for which the insights and methods of Bakhtin are appropriate in the first case and those of Eisenstein in the second (Ard, 1989, p. 261).

The mention of the filmmaker-semiotician Sergei Eisenstein is telling. Ard has in mind specifically Eisenstein's montage theories, but by placing mathematical texts in the context of cinema drives home the fact that a mathematical text, like a film in a theater, is an entity existing for an audience; its semiotic character—the signs that make it available for the understanding of the reader or viewer, regardless whether the signs are elements of a language or a montage of sensible images and symbols—puts it, as in a theater, in a public realm.

So, the public side of mathematics, starting with its institutional setting where it translates into the drive to publish, derives at a fundamental level from mathematics' semiotic character. I wish to underline this point. Referring to the semiotic nature of mathematics does not mean speaking abstractly about its symbols, as if they were detached from the users of the symbols and the kinds of discourse engendered by them; it means seeing mathematics as visible, communal, *public* activity, or of the system of signs allowing such activity. As beings engaged in this kind of activity, our mathematical selves are, at this level, also public selves: from this perspective, we are what we are by how we are with and through others. For this reason, Radford (2006a) in working out this position (see also Radford, 2006b; this volume), uses the illuminating, though slightly paradoxical, phrase, *individu communautaire* or *je communautaire* [*communal individual* or *communal "I"*].

What has been said so far gives reason to believe that the mathematical self is a public self. But this line of thinking can be taken even further, namely, that a mathematical self does not just happen to be a public self, but that a non-public mathematical self is not even tenable. Among the most famous, and profound, arguments implying the latter conclusion is, surely, Wittgenstein's argument against the possibility of a private language in his *Philosophical Investigations* (Wittgenstein, 1958).

Much can be said about this argument (and much *has* been said!), but let the following remark suffice. While the private language argument ostensibly refers to the problem of describing inner sensations, Kripke (1982) has pointed out that the crux of the argument lies in the nature of rules and is rooted in Wittgenstein's paradox that no action can be determined by a rule, since some rule can be attached to any action (Wittgenstein, 1958, part I. §201). For us, the emphasis on rules is important, for it makes the application of the private language argument to an rule-driven activity such as mathematics incontrovertible. But besides bringing mathematics and other rule-driven activities, "language-games" to use Wittgenstein's famous phrase,[3] squarely into the picture, seeing the private language argument in terms of rules highlights one of its main conclusions, namely, that rules only have meaning within a community: it is only within a community of rule-users that thinking one was obeying a rule would be *not* be the same thing as obeying it. Wittgenstein's private language argument, then, suggests that private mathematical activity simply does not make sense, and with that, the idea of a private mathematical self ceases to be coherent.

To summarize, the argument for public character of the mathematical self begins with the centrality of communication in mathematical activity. This reflects the semiotic nature of mathematical activity. As such, it is activity directed towards a community, that is, an activity demanding the participation of others, and thus governed by a system of signs visible and open to others—in short, public activity. As creatures that engage in mathematics, then, we are public beings. Wittgenstein's arguments further strengthen this line of thinking by suggesting that it is not even *tenable* to define the mathematical self as a private self.

II. PRIVATE AND PUBLIC

The conclusions above are extreme and concern extremes. First, Wittgenstein. The private language argument, taken positively, asserts that forms of life governed by rules, by language, by systems of signs, exist within the context of a community, that is, that they have an irremovable social component; the argument then proceeds by asking whether any of these forms of life can be even defined without any reference at all to a community—in other words, taken as totally private matters. But this does not mean that there is no private life within mathematical life or that such life has no significance whatsoever; it only means that such a mathematical life alone is inadequate. On the other side, in saying that the mathematical self is a public self, does this mean that it is *totally* a public self? The latter, is how I think one must understand Peirce's somewhat strange statement that we are ourselves signs. In a way, that statement (appearing, incidentally, in a paper, which, like Wittgenstein's private language argument, is an attack on Cartesianism) epitomizes the arguments in the last section leading to the conclusion that our mathematical selves are public selves:

> ...the word or sign which man uses *is* the man himself. For, as the fact that every thought is a sign, taken in conjunction with the fact that life is a train of thought, proves that man is a sign; so, that every thought is an *external* sign, proves that man is an external sign (CP 5.314) [all emphases in the original].

In this section, I want to make a call for moderation. In particular, I want to show that a private self cannot be dismissed when considering a mathematical self, or any other kind of self related to a life semiotic in nature. The intention is not to reject the view developed in the last section—indeed, private life is itself, as we shall see, a social construction: it is only to claim that the obvious associations of the word "self" with notions of privacy ought to have weight in our thoughts about mathematical self.

To start, the recognition of private life has a clear and basic place in politics and in law. In a certain sense of course private life is as antithetical to politics as it is to public life, since the latter can nearly be identified with political life. Private life, however, is not for this reason ignored in political thinking, but becomes, especially in early modern political thought, a central question. Thus, in Locke (*The Second Treatise on Government*, §123), for example, the main motive for political societies is the preservation of property, which, for him, is always private property:[4] one can put a fence around one's estate in the state of nature, but only the political state can prevent trespass, that is, sanction the fence. In effect, governments protect one's private domain and, by implication, one's private life.
Through this political context, the division between public and private also becomes an issue for law. In fact, the public-private dichotomy is, perhaps, most conspicuous in legal settings. One of the earliest cases in which the question of privacy arose concerned the poet Alexander Pope.[5] The case centered on a book of personal letters from and to Pope, Swift, and other literary figures—all published without the authors' permission by a bookseller named Curl. Pope subsequently sued Curl in 1741, and won on the basis of a very Lockean argument that letters are

owned by the writer that produced them. So, Pope's right to keep the letters unpublished was established by interpreting them as property—and the judgment was put explicitly in terms of property, not privacy, even though that was at the heart of the affair. In other cases too in the course of the 19th century the term "privacy" was not used explicitly even though it was becoming increasingly clear that something like privacy, rather than just property or just breach of confidence, was actually in question. For example property rights were again cited in *Prince Albert vs. Strange and Others* (1849) where the objection was not the publication of works (etchings) by Prince Albert but a mere catalogue of them; breach of confidence was cited in *Abernethy vs. Hutchinson* (1825) where the suit involved the publication of a distinguished surgeon's lectures for students.

Ernst & Schwartz (1962) argue that "the catalyst" for the discussion of privacy as a legal concept, finally, was sensationalist journalism and, especially, the advent of instantaneous photography, where a photograph could easily be taken without a person's consent. Could the use of a photograph taken in this way be viewed as breach of contract, or breach of trust, or violation of the rights of property? Spurred by such questions, Samuel Warren and Louis Brandeis wrote in 1890 what has become a landmark paper, "The Right to Privacy" (Warren & Brandeis, 1890). In that paper, they demonstrated the failure of existing legal concepts to show what was truly at stake in cases like *Abernethy vs. Hutchinson* and *Prince Albert vs. Strange*. In do so, they more or less created privacy as a legal notion and made clear that its protection was no less than the protection of a person's own self. Thus, referring specifically to the inadequacy of treating questions of privacy in terms of property they write:

> The principle which protects personal writings and all other personal productions, not against theft and physical appropriation, but against publication in any form, is in reality not the principle of private property, *but that of an inviolate personality*. (Warren & Brandeis, 1890, p. 205; emphasis added).[6]

Now, in this legal setting what is emphasized is the *right* to privacy. This has obvious relevance to mathematics students and school classrooms, not to speak of ethical questions in mathematics education research, but it is not the point I wish to emphasize here. Rather, for the purposes of the present paper, the attention to privacy in the law shows, first, that a person's private domain is something real and significant enough to demand protection, and, second, that the private domain is related, in this setting, fundamentally to the idea of self.

A third point hinted by the history of legal cases concerning privacy is that, by its mere identity as a legal concept, privacy is to that extent a socially constructed notion, just as property is. But the legal setting only brings this out; the considerations of jurist are only a reflection of privacy's broader cultural character with a cultural history quite apart from that of law. As the historian Yvon Thébert writes, "Private life is a product of social relations and a defining feature of every social formation. Hence private life is subject, from time to time, to radical redefinition; to attempt to trace its history as a matter of continuous evolution, independent of sharp discontinuities in other areas of social life, would be misleading" (Thébert, 1987, p.320). Thébert goes on then to show, for example, the expression of private and public life in Roman architecture, where private and public space are literal spaces, and where those spaces in turn defined the bounds of public and private life.

The history of reading provides further examples of how privacy is a cultural entity with a history. This is most clear in connection to silent reading. Although there is some debate as to when exactly silent reading became common (see Knox, 1968), it is clear in the 4th century silent reading was unusual enough that Augustine should dwell on his teacher Ambrose's silent reading with some sense of wonder (*Confessions*, VI. 3). Toulmin (1979) even remarks that onlookers thought that Ambrose was a kind of magician in being able to grasp the meaning of words on a page without moving his lips. The contrast between reading aloud and reading silently is precisely the contrast between public and private, and, it can be argued that the development of silent reading was also a development of a private space—as much as the private rooms in Roman architecture—and of a private self. Silent reading is solitary, exclusive of other people; when Ambrose read silently to himself, Augustine relates, he was to be left alone and not to be disturbed. Silent reading is private reading and involves a very different relationship to a text, one that involves appropriation to oneself—"...the text itself, protected from outsiders by its covers, became the reader's own possession, the reader's intimate knowledge, whether in the busy scriptorium, the market-place or the home," writes Manguel (1996, p. 50) in his history of reading. The physical spaces where one reads—in rooms, in gardens, in armchairs, and, most of all, in bed—themselves should not be ignored, and Manguel describes them in a chapter entitled, appropriately, "private reading" (Manguel, 1996, pp. 149–161).

Private life, therefore, is evident in contexts of politics, law, architecture, reading—all contexts connected to social or cultural life and, therefore, to public life or life in the community of others. These contexts are also deeply semiotic, relying as they do on communication of meaning through signs. And, as I hope is clear from the last section, the semiotic nature of these human enterprises is tied to their public nature inseparably. But the expressions of private life, which I have referred to in this section—setting a fence around a piece of property and calling it one's own, making for oneself etchings and refusing to publish their existence in a catalogue, enjoying a world created by a book read in bed in a room behind a shut door—are no less expressions of this semiotic character and are themselves acts of semiosis. So, we have a paradoxical situation in which public life engenders private life, with a semiotic force at the center of both. It appears that the life of signs, which, as argued above, points towards the community, towards the kind of self captured in Luis Radford's (2006a, 2008) phrase *je communautaire* also, at the same time, generates a place apart from the community, pointing to a self who is a *je solitaire* [*solitary "I"*] .

One might think that relationship between public and private described above can be adequately understood in terms of Vygotskian internalization: what we call "private" is only the internal version of external socially constituted forms; in this light, there is nothing paradoxical in public life defining private life. Yet, put this way, one's private world, and, accordingly, one's private self, seems only a sort of illusion. This, I think, was not Vygotsky's own view. If it were, it would be difficult to grasp why Vygotsky opened the last chapter of *Thought and Language* with the poet Osip Mandelstam's words:

> The word I forgot
> Which once I wished to say
> And voiceless thought
> Returns to shadow's chamber
> (quoted in Vygotsky, 1986, p. 210)

In fact, that last chapter in Vygotsky's book is, beyond the peculiarities of inner speech, very much concerned with poetry, and how thoughtful speech, poetry included, arises out of that "shadowed chamber" of one's interior world. For internalization is only half the story: after external speech has become internal speech, it must eventually reemerge in discourse and in art.[7] The difficulty of that reemergence is illustrated by Vygotsky in Uspensky's poor peasant, who "must address an official with some life-important issue, cannot put his thoughts into words. Embarrassed by his failure, he retreats and prays, asking the Lord 'to give him a concept'" (Vygotsky, 1986, p. 249). Nor is that difficulty a difficulty of the peasant alone: Vygotsky goes on to say that the peasant's problem is just that of thinkers and writers, "How to put thought into words" (Vygotsky, 1986, p. 250). Thinkers and writers do not necessarily pray when they cannot get their thoughts out or crystallize them into an expressible form; however, they do retreat into the kind of private realm where another might ask the Lord "to give him a concept."[8] For thought in the form of inner speech, Vygotsky tells us over and over in that chapter, cannot be simply made vocal. It must be worked out—one might say, worked outwards. Vygotsky recognized that the way in which thoughts reemerge as understandable words is a "complex, dynamic process involving the transformation of the predicative, idiomatic structure of inner speech into syntactically articulated speech intelligible to others" (Vygotsky, 1986, pp. 248–249). It is because the process does involve the idiomatic structure of inner speech, I would argue that it must at a certain stage involve activity in a private unsharable space.

This space where thoughts are worked out—sometimes a physical space, but, in any case, a "zone of immunity" (Duby, 1987, p. viii) where one is left alone—is the true private world that, as discussed above, derives from the public world. Far from illusory, then, a private space seems to be a necessary part of public semiotic activity. What happens in that private space where thought is prepared for "publication," is following hunches, sometimes going up blind alleys, scrawling remarks diagrams barely distinguishable from doodles, sensing (but not proving) the rightness and wrongness of claims in images, metaphors, analogies and associations; it is the space where thought can be unsystematic and not called to account for the directions it takes. One can see concrete traces of this kind of activity in the scraps of paper, sketches,[9] scribbled over manuscripts, and notebooks left by people on their way to bringing thoughtful works to light. These traces are often lost to the garbage bin,[10] but often enough remain in worksheets for poems, manuscripts for musical compositions, and mathematical notebooks

(Ramanujan's notebooks, for example, can be seen in Andrews, 1979).[11] The insights such provide into the selves of the thinkers (not only great ones, I might add, but also our young students) justifies the great interest collections of manuscripts hold for researchers into literature and the history of mathematics, among others. This does not mean, of course, that the private aspect of such collections is lost—not only in the sense of legalities and the inviolability of self at the foundation of such legalities, but also as a part of creative work closed to outside observers. Thus, Charles Abbott, who organized a large collection of poetic sketches and worksheets in the 1940's, wrote about the project that, "The objection has been occasionally raised by certain poets that our aims constitute an express invasion not only of personal privacy *but also of the essential secrets of the art*" (from Arnheim, *et al*., 1948, p. 29; emphasis added).

Another, perhaps more radical, set of ideas pointing to the necessity of private world, and which are very closely connected to a sense of self in such activities as mathematics, are those of the scientist and philosopher of science, Michael Polanyi. Polanyi's philosophical career began, as he tells it, as a reaction to the view (maintained by J. D. Bernal), "...that the pursuit of science should be directed by the public authorities to serve the welfare of the society" (Polanyi, 1958, p. ix). What began as a defense of intellectual freedom ended up as a view of knowledge where knowledge is understood as never quite explicit and transparent. In other words, knowledge can never be subject to public authorities because its development, acquisition, and mode of operation in scientific thought can never be explicated in purely public terms; knowledge always contains a tacit and personal component. Even when one is working in a completely conscious and focused manner on some matter, there is a tacit, "subsidiary," awareness guiding and motivating one's thoughts.[12] Although Polanyi uses the word "personal" rather than private, he clearly has in mind what one *usually* means by referring to one's "personal" self, namely, a private self. Of course his point is not that everything in the scientists' or mathematician's world is personal in this way, private, unavailable to the public: it is that a private or personal component is always present, so that the scientists' world can never be completely open to inspection, and, subsequently, to control by a public authority.

So, our selves, considered as meaning-making beings, mathematical selves, scientific selves, have irreducible private and public aspects: neither an utterly public self nor an utterly private self is an adequate categorization of our mathematical selves. This inadequacy of the exclusively public or exclusively private is, I think, marvelously illustrated by Umberto Eco's recent novel, *The Mysterious Flame of Queen Loana* (Eco, 2005). The hero of the novel is Giambattista Bodoni, a seller, collector, and connoisseur of antiquarian books—"Yambo" to all who know him. The story begins with Yambo in hospital bed having suffered a stroke. Yambo suffers a second stroke later in the book. Each stroke brings with it a different kind of loss of self. With the first, Yambo cannot remember his name, who his wife and children are, where he is from, what his political sentiments are, or whether he ever loved the beautiful Sabilla, who works in his book studio: but he remembers texts, characters, and images in painful detail.

His doctor, Dr. Gratarolo, explains that he has lost episodic memory but has retained semantic memory, which is also called "public memory" (p.12). So, in the first two parts of the book, Yambo tries to find himself amid this world of collective, public, signs. What is his name?

> "My name is Arthur Gordon Pym."
> "That isn't your name."
> Of course, Pym was someone else. He did not come back again. I triad to come to terms with the doctor.
> "Call me...Ishmael?"
> "Your name is not Ishmael. Try harder."
> A word. Like running into a wall. Saying *Euclid* or *Ishmael* was easy, like saying, *Jack and Jill went up a hill*. Saying who I was, on the other hand, was like turning around and finding that wall (pp. 6–7).[13]

The events of his own life are revealed to him as if memories of someone else's life (e.g., p. 286). They are not in his own possession. At one point, he refers poignantly to his own autobiography, "Written by someone else" (p. 40). Even in the attic of his childhood home, among all the signs that formed the world of his youth, his self lacks completeness and coherence, except for a few passing moments in which he feels the "mysterious flame," a frisson of vague familiarity. Signs alone do not constitute a self without a passionate attachment to them, to use Polanyi's language. A utterly public self is a self that is ill.

Yambo's second stroke brings him to a second extreme. Before the second stroke, Yambo's ability to communicate with others is completely unimpaired, but all that passes between him and others lacks personal commitment. With second stroke, almost everything returns that was lost before;[14] he is again author of his own autobiography. However, this last stroke has left him comatose. He has returned to himself, but he has lost all possible contact with others. The second stroke has, therefore, reduced his self to an entirely private self. Here, he has no control of his memories: they must "come of their own accord, following their own logic" (p. 325). Yambo's life in this world is a little like the enactment of Wittgenstein's private language argument: with no interaction, no check, the meaning of Yambo's own thoughts becomes dubious. Again, they are not in his own possession. He says, "Perhaps I am, indeed, in a coma, but am dreaming within it, not remembering" (p. 417) and remarks about the unsure reality of dreams that "Indeed, I have often thought that in our dream we take over other people's memories" (p. 417), so that, as before, his memories are not his own. So, poor Yambo, relieved only by death, or so it seems, at the end, loses his self at both extremes, purely public and purely private. An integral self must somehow have one foot in the one world and the other foot in the other world.

III. MATHEMATICS EDUCATION—A BRIDGE BETWEEN PUBLIC AND PRIVATE

At this point, we ought to ask whom all this makes any difference to. It probably makes little difference to working mathematicians, or poets, whether their work is fundamentally public, private, or a blend of the two: they, for the most part, simply do what they have to do. However, I would like to claim that the awareness of public and private domains and questions of how public and private selves are coordinated and how they develop are crucial for education. This is particularly true for mathematics education, where public forms, in the past, tended to receive almost exclusive emphasis.[15]

To see the relevance of our discussion until now to mathematics education, to start, some attempt should be made to illustrate the kinds of classroom mathematical practices and activities that can be may be placed under the respective headings of public and private. I say "some attempt" because as applied to real student actions, public and private are a little like Weberian "ideal types"; the distinction between public and private is not hard and fast, and boundaries between the two spheres can shift or be drawn in different ways. Indeed, under different pedagogical practices or assessment regimes the same mathematical activity may be termed private or public. If students' homework, for example, is discussed, collected, and marked, it is likely to take on a public character; in a different pedagogical setting, homework might be given only to reinforce classroom material and never be seen by anyone except the students themselves. Some aspects of mathematical activity, on the other hand, seem to resist relocation from sphere to sphere; for example, mathematical papers or projects as public and individual preliminary reflections as private would be hard to reclassify as private and public without stretching the matter. The following list of public and private mathematical activities fairly clearly belong of the latter group (Fried & Amit, 2003):

Table 1. Public and Private Mathematical Activities

Public	*Private*
• Precise use of standard notations & representations	• Employment of individualized notations and representations
• Justification by and presentation of formal proofs	• Calculations & Heuristic sketches
• School Examinations	• Reflections, deliberations, and false starts
• Mathematics papers	• Visualization, metaphors and analogies

In recent years, mathematics education has done much to show the centrality of all the aspects of mathematical activity listed; research in mathematics education has brought out specifically those aspects which we have placed in the private domain (e.g., English, 1997; Kelly & Lesh, 2000). Yet, when one looks closely at practices in mathematics classrooms, one sees a blurring of the two domains.

In Fried & Amit (2003, 2006), for example, the character of mathematics classroom notebooks and practices connected with their use were examined.[16] What prompted that particular investigation was the observation in one of the classes studied that at the beginning of each class period, when the teacher, whom we call Danit, presented new material, gave examples, and went over exercises from the previous lesson, the students' notebooks almost always lay closed on the students' desks. This raised obvious questions for us as researchers. Did the students not find Danit's examples and lessons worth recording? Did Danit not think that the students ought to write down her remarks and examples so that they might be able to think about and refer to later? In several of the interviews, we asked the students about this. They replied that Danit lets them to choose whether they use their notebooks or not. On the face of it, then, it seemed that students' notebooks were considered by Danit as the students' own property, and students could use their notebooks as they saw fit. In fact, however, the use of notebooks among the students bore no individual stamp; it was determined by Danit. Indeed, we observed several cases in which she *explicitly* told the students what to write in their notebooks. Moreover, she sometimes asked to inspect the students' notebooks. Knowing that Danit might eventually inspect their work, students would have bear in mind Danit's reaction if nothing were written, or if what was written down has been written badly, or wrongly, even if she did not inspect the notebooks in fact.

In a second classroom, taught by a man whom we call Sasha, lessons were evenly divided between those completely directed by the teacher in front of the class and those in which the students worked collaboratively on problems at their seats. The former were carefully planned and presented. Sasha's students, unlike Danit's, did open their notebooks at the start of the class and copied Sasha's lessons as they appeared on the blackboard; of course, they also used their notebooks when they worked on problems. Like Danit, however, Sasha also determined how the students' notebook should be used and how they should appear. In one lesson, for example, he said explicitly, "I want your notebooks to reflect what is here [on the board]." Also like Danit, Sasha's students know that their notebooks are always open to inspection. In a dramatic incident it was made clear to us that students' notebooks in Sasha's class could be inspected not only by Sasha at any time, but by the rest of the class as well: dissatisfied with how an exercise in one student's notebook appeared, Sasha picked up the notebook and held it open for the whole class to see, explaining all the while how exercises *should* appear in the students' notebooks.

That students' notebooks may at any time be inspected means that they are a public matter—and to this extent they cannot be understood as a place to record students' private thoughts about what they are learning: desultory reflections; false starts; mistaken conclusions and their, perhaps embarrassing, corrections. Where such things as preliminary work for exercises and calculations are found is not in students' notebooks, but, to their teachers' continual annoyance, on students' desks! Students learn that notebooks must contain finished work that can be confidently held open to view. Writing in notebooks often becomes a kind of rehearsal for that part of school life in which the students are under the greatest scrutiny, namely, the examination. This was evident in students interviews: several students remarked that they record in their notebooks the sort of exercises likely to appear on examinations rather than the explanations behind them. Here too, the teacher seemed to be the source of that view. Sasha, for example, said in one of our interviews with him: "A [good] exam paper goes with the notebook."

Yet, both Sasha and Danit seemed to want to recognize something like the students' private domain, a space in which they make their own decisions and their own mistakes and reflect on what they have learned and how they learn. Thus, when Danit *says* that the students can choose what and when they write in their notebooks, she, in effect, appeals to this private domain. Sasha, too, while his actions tend to make the notebook a public object, does not aim that his students be adept only at public performance: in the interviews with him, he referred many times to students' thinking and, particularly, to their 'seeing' geometry. He wanted geometry to belong to the students' *own* world of thinking and seeing. This too seems to point to the students' private domain. So, in both these classrooms, there is a certain tension between the desire that mathematics become part of students' own worlds, their private domains, and the treatment of the notebook – which, because it is truly the students' own, is a natural locus for the private domain – as an utterly public object.

The case of students' notebooks supplies us with a concrete example of where private and public worlds intersect in the mathematics classroom. It also shows how unclear the division between these two worlds can be in actual classroom practice. That lack of clarity must translate in some way into students' ways of relating to mathematics and how they see themselves with respect to mathematical activity, their mathematical selves. Are their mathematical selves to be summed up by a finished product open to inspection—the kind of activity that can be judged and discussed by a common system of meanings—or by the process of working out their mathematical thoughts, allowing what they have learned and internalized to emerge according to their own idiosyncratic images and, perhaps, according to their own peculiar interests and passions? What this paper has shown, I hope, is that neither alternative is adequate by itself: the mathematical self is realized in both a public and a private domain and is, in that sense, both a public and private entity.

If this is so, it remains an important task for mathematics education to bridge those two domains and develop the two corresponding aspects of students' mathematical selves. History of mathematics, with which I began this paper, can play a part in fulfilling that task. Of course history has no monopoly on this role: one can be made aware of the public and private aspects of mathematical self otherwise; after all, the study of notebooks was not historical. History, however, does pose the question of the public and private in natural way. For in pursuing a historical understanding of mathematics the historian must contend with these same two worlds: one tries to understand particular individual thinkers both from the point of view of their own public world as well as their own idiosyncrasies. The historian is interested both in the common language of thinkers at a given point in time, what the signs of mathematical discourse meant to a past mathematical community, and also in the individual mathematical thinkers working through their ideas, their intellectual struggles, their leaps of intuition and their missteps. Historians of mathematics, like mathematics educators observing their students, act both as readers of published works and as *voyeurs* trying to peak at the unpublished mathematical lives of their subjects.

In sum, then, what I have tried to argue in this paper, albeit indirectly, is that mathematics education, in the broad sense of developing mathematical self knowledge, must coordinate two worlds in which we live: a public world and a private world. Living purely in a public mathematical world—and thus this is a purely hypothetical situation—one has the means to formulate responses to questions, frame proposition, or react to given contexts, but conviction, appreciation, interest, and perhaps even understanding become problematic. In this hypothetical situation, one is a little like Searle's "Chinese room" with a human face. The purely interior world, the private mathematical world, however, is just as inadequate. One has no true command of one's thoughts, precisely because one lacks the means of communicating them. Mathematically, one is like Yambo when he has sunk into his coma-like state, where no one outside knows whether his mind is alive or not. Mathematics education must learn to recognize the necessity of both these worlds by becoming deeply aware of their nature and their possibilities.

NOTES

1. This view of mathematics education and the place of history in it owes it origin to Collingwood's view of history itself as, "...knowledge of the past in the present, the self-knowledge of the historian's own mind as the present revival and reliving of past experiences" (Collingwood, 1993, p.175).
2. See *Politics*, 1253a10: "...more than any bee or any herding animal, it is clear that a human being is a political animal, and, because nature does nothing in vain, it is the human being alone among living beings that possesses speech", my translation. In this connection, it is suggestive to recall that, conversely, a person who is outside the doings of the *polis*, a non-public, or *private* person, would have been called, in Greek, an *idiōtēs*, from which we derive of course the pejorative, *idiot.* It is an etymology perfectly consistent with Aristotle's statement at the start of the *Politics* that "...he that is without a city on account of nature and not on account of fortune is either lower or better than man" (1253a).
3. A case could easily be made for drawing an analogy, at least, between sign systems and "language games."
4. This is clear from Locke's entire theory of property, set out in the *Second Treatise*, §§25-51.
5. Most of the material in this discussion comes from Ernst & Schwartz (1962).
6. In the footnote to this passage, however, Warren and Brandeis do make the connection between idea of property and self: "The very meaning of the word 'property' in its legal sense is 'that which is peculiar or proper to any person; that which belongs exclusively to one'. The first meaning of the word from which it is derived – *proprius* – is 'one's own'."
7. This is particularly the concern of the last two sections of Chapter 7 of *Thought and Language*. The same down and up movement is also discussed fully in Radford (2006b) where, for example, Radford speaks of the reflexive nature of thinking, which is indeed "...a *reflection*, that is, a dialectical movement between a historically and culturally constituted reality and an individual who refracts it (as well as modifies it) according to his/her own subjective interpretations and feelings" (p.7; emphasis in the original). My own interest here is in the realm, which I identify with the private realm, where that refraction takes place. I might also add that besides the parts of *Thought and Language* cited here, one can see hints of that private realm also in Vygotsky's discussions of *play*, where play is not always collective play, but often solitary play such as a girl with a doll pretending she is a mother (see Vygotsky, 1978, "The Role of Play in Development").

[8] This is illustrated beautifully in an account of M. C. Escher's working habits, as described by his son: "A new concept could take months, sometimes years of incubation before it led to a print…[h]is moods changed between irritated abstraction and relaxed discussion of some small problem, between restless pacing behind his closed door and sudden announcement that he had found some satisfying solution. During this period of gestation father demanded complete quiet and privacy. The studio door was closed to all visitors, including his family, and locked at night…" (quoted in Roberts (2006), p. 217)

[9] To distinguish such sketches, or "working drawings," from completed works, the art critic John Berger refers explicitly to the private and public: "A [working] drawing is essentially a private work, related only to the artist's own needs; a 'finished' statue or canvas is essentially a public, presented work—related far more directly to the demands of communication" (Berger, 1960, p. 24).

[10] In our work in the *Learners' Perspective Study* (see Fried & Amit, 2003), the most interesting student productions were often found either in the garbage bin or written on the desks—in both cases, those productions would otherwise have conveniently vanished by the next morning. These kinds of productions are not pursued to be shown.

[11] It is worth remarking that, as visual objects, such manuscripts are quite similar in their general appearance: arrows, cross-outs, highlighting, doodles, marginalia in all of them.

[12] These ideas clearly have analogues in the work of Heidegger (as Polanyi himself points out) and Merleau-Ponty. But what makes Polanyi particularly apt in the present discussion is his principal interest in science and mathematics and the intrinsically non-public personal element in specifically scientific and mathematical knowledge.

[13] "My name is Arthur Gordon Pym" are the opening words to a story by Edgar Allen Poe, "The Narrative of Arthur Gordon Pym of Nantucket," while "Call me Ishmael" is of course from Melville's *Moby Dick.*

[14] The name of this part of the book, accordingly, is "hoi nostoi," which are the Greek words connoting the return home, as in Odysseus' return to Ithaka.

[15] In literature education, by contrast, students' private worlds are generally given much more weight than in the sciences. Emphasis on journal writing in mathematics education represents a change in this regard (e.g., Borasi & Rose, 1989; Clarke, et al., 1993).

[16] This was study was carried out within the framework of the *Learners' Perspective Study* (LPS) (see Clarke et al., 2006)

REFERENCES

Andrews, G. E. (1979). An introduction to Ramanujan's "lost" notebook. *The American Mathematical Monthly, 86*(2), 89–108.

Ard, J. (1989). From Sign to Text in Mathematical Discourse. In Y. Tobin (Ed.), *From Sign to Text: A Semiotic View of Communication,* (pp.255-273). Amsterdam: John Benjamins Publishing Company.

Arnheim, R., Auden, W. H., Shapiro, K., & Stauffer, D. A. (1948). *Poets at work* (Introduction by Charles D. Abbott). New York: Harcourt, Brace and Company.

Bartle, R. G. (1995). A brief history of the mathematical literature. *Publishing Research Quarterly, 11,* 3–9.

Bell, A. W. (1976). A study of pupil's proof-explanations in mathematical situations. *Educational Studies in Mathematics, 7,* 23–40.

Berger, J. (1960). *Permanent red: Essays on vision.* London: Methuen & Co. Ltd.

Borasi, R., & Rose, B. J. (1989). Journal writing and mathematics instruction. *Educational Studies in Mathematics, 20,* 347–365.

Clarke, D. J., Waywood, A., & Stephens, M. (1993). Probing the structure of mathematical writing. *Educational Studies in Mathematics, 25,* 235–250.

Clarke, D., Keitel, C., & Shimizu, Y. (Eds.). (2006). *Mathematics classrooms in 12 classrooms: The insiders' perspective.* Rotterdam, The Netherlands: Sense Publishers.

Collingwood, R. G. (1993). *The idea of history.* Oxford: Oxford University Press.

CP=Peirce, C. S. (1931–1958). *Collected papers* (8 Vols.). Cambridge, MA: Harvard University Press.
Davis, P. J., & Hersh, R. (1981). *The mathematical experience*. Boston: Birkhäuser.
Duby, G. (1987). Foreword to a history of private life. In P. Veyne (Ed.), *A history of private life, Vol. 1 From Pagan Rome to Byzantium* (A. Goldhammer, Trans., pp. vii–ix). Cambridge, MA: Harvard University Press.
Eco, U. (2005). *The mysteious flame of Queen Loana* (G. Brock, Trans.). Orlando, FL.: Hartcourt, Inc.
English, L. D. (Ed.). (1997). *Mathematical reasoning: Analogies, metaphors, and images*. Mahwah, NJ: Lawrence Erlbaum Associates, Publishers.
Ernst, M. L., & Schwartz, A. U. (1962). *Privacy: The right to be let alone*. New York: The Macmillan Company.
Fried, M. N., & Amit, M. (2006). The Israeli classroom: A meeting place for dichotomies. In K. Clark & Shimitzu (Eds.). pp. 209–220.
Fried, M. N. (2001). Can mathematics education and history of mathematics coexist? *Science & Education, 10*(4), 391–408.
Fried, M. N. (2006). Equality and similarity in Greek mathematics: Semiotics, history of mathematics and mathematics education. Talk given at the conference, *The promises and problems of a semiotic approach to mathematics, the history of mathematics and mathematics education*. Bielefeld, Germany, July 13–15, 2006.
Fried, M. N. (2007). Didactics and history of mathematics: Knowledge and self-knowledge. *Educational Studies in Mathematics, 66*(2), 203–223.
Fried, M. N., & Amit, M. (2003). Some reflections on mathematics classroom notebooks and their relationship to the public and private nature of student practices. *Educational Studies in Mathematics, 53*, 91–112.
Knox, B. M. W. (1968). Silent reading in antiquity. *Greek, Roman and Byzantine Studies, 9*, 421–435.
Kripke, S. A. (1982). *Wittgenstein: On rules and private language*. Cambridge, MA: Harvard University Press.
Manguel, A. (1996). *A history of reading*. New York: Penguin Books.
National Council of Teachers of Mathematics (NCTM). (2000). *Principles and standards for school mathematics*. Reston, VA.
Polanyi, M. (1958). *Personal knowledge*. New York: Harper & Row, Publishers, Inc.
Presmeg, N. C. (1997). Reasoning with metaphors and metonymies in mathematics learning. In L. D. English (Ed.), *Mathematical reasoning: Analogies, metaphors, and images* (pp. 267–279). Mahwah, NJ: Lawrence Erlbaum Associates, Publishers.
Radford, L. (2006a). Communication, apprentissage et formation du je communautaire. In B. D'Amore & S. Sbaragli (Eds.), *Proceedings of the 20th National Italian Conference, "Incontri con la Matematica"* (pp. 65–72). Bologna, November.
Radford, L. (2006b). Elements of a cultural theory of objectification. *Revista Latinoamericana de Investigación en Matemática Educativa* (Special Issue on Semiotics, Culture and Mathematical Thinking) pp. 103–129.
Radford, L. (2008). Di sé e degli altri: Riflessioni su un problema fondamentale dell'educazione [The self and the other: Reflections on a fundamental problem in education]. *La Matematica e la sua didattica*, 22(2), 185–205.
Roberts, S. (2006). *King of infinite space: Donald Coxeter, The man who saved geometry*. New York: Walker & Company.
Saussure, F. (1974). *Cours de linguistique générale*. In C. Bally & A. Sechehaye (Eds.), with the collaboration of A. Riedlinger. Paris: Payot.
Sebeok, T. (1968). *Animal communication*. Bloomington, IN: Indiana University Press.
Thébert, Y. (1987). Private life and domestic architecture in Roman Africa. In P. Veyne (Ed.), *A history of private life, Vol. 1 From Pagan Rome to Byzantium* (A. Goldhammer, Trans., pp. 313–409). Cambridge, MA: Harvard University Press.
Tobias, S., & Weissbrod, C., (1980). Anxiety and mathematics: An update. *Harvard Educational Review, 50*(1), 63–70.

Toulmin, S. (1979). The inwardness of mental life. *Critical Inquiry*, *6*(1), 1–16.
Vygotsky, L. (1986). *Thought and language*. (A. Kozulin, Trans. and Ed.). Cambridge, MA: MIT Press.
Vygotsky, L. S. (1978). *Mind in society: The development of higher psychological processes* (M. Cole, V. John-Steiner, S. Scribner, & E. Souberman, Eds.). Cambridge, MA: Harvard University Press.
Warren, S. D., & Brandeis, L. D. (1890). The right to privacy. *Harvard Law Review*, *4*(5), 193–220. Retrieved April 7, 2007, from http://www.law.louisville.edu/library/collections/brandeis/
Wittgenstein, L. (1958). *Philosophical investigations* (3rd ed., G. E. M. Anscombe, Trans.). New York: Macmillan Publishing Co., Inc.

Michael N. Fried
Program for Science and Technology Education
Ben Gurion University of the Negev, Israel

GERT SCHUBRING

PROCESSES OF ALGEBRAIZATION IN THE HISTORY OF MATHEMATICS: THE IMPACT OF SIGNS

Wenn wir die Zeichen zerstören, verlieren wir die Orientierung [If we destroy the signs, we will loose the orientation]
From the movie: Die große Stille [The Profound Silence]
La Grande Chartreuse

INTRODUCTION

Although my starting point is to state that the processes of algebraization are among the most marked characteristics of the historical evolution of mathematics, I have to admit that this point is not uncontroversial. In fact, there has been an extended and sharp debate implying that no such evolutionary process did exist – since all essential steps had already been achieved by the Greeks. Since this debate unravelled methodological dimensions important for my topic I will begin by exposing key issues of this debate.

GREEK ALGEBRA WITHOUT SYMBOLS?

This debate had been provoked by Sabetai Unguru in 1975 and been taken up in particular by Bartelt van der Waerden, Hans Freudenthal and André Weil.[1] The key issue of this debate had been how to interprete Book II of Euclid's Elements. For van der Waerden, this Book constitutes the kernel of what he understands to be the Greek geometric algebra. He went even so far to qualify this Book II as "the start of an algebra textbook, dressed up in geometrical form" (van der Waerden 1963, p. 118). According to him, "the line of thought is always algebraic, the formulation geometric" (ibid., p. 119).

Actually, this interpretation was not restricted to a part of Euclid, but it comprised several other Greek mathematicians. Van der Waerden, for instance, affirmed:

> Theaetetus and Apollonius were at bottom algebraists, they thought algebraically even though they put their reasoning in a geometric dress (ibid., p. 265).

L. Radford, G. Schubring, and F. Seeger (eds.), Semiotics in Mathematics Education: Epistemology, History, Classroom, and Culture, 139–155.

Other historians of mathematics have adopted this reinterpretation of Greek geometry, since the end of the 19th century: Paul Tannery, Thomas Heath, Hieronymus Zeuthen, and Otto Neugebauer. Zeuthen, an important Danish historiographer affirmed since 1896 that 'the Ancients' knew to treat all forms of equations of the second degree (Zeuthen 1896, p. 50).

I am calling this school of thought, which supposes the existence of a Greek geometric algebra, a "revisionist" school – since all the centuries before, mathematicians had agreed that Greek geometry canonized the synthetic method, emphasizing each particular case and renouncing decidedly any algebraizing attempt at generality.

This revisionism raises several essential methodological issues. A first one has been evoked by Otto Neugebauer's justifying his transcription of Apollonius's geometry into algebraic language. According to Neugebauer, this transcription had in no ways affected the content of Apollonius's mathematics:

> I did not change Apollonius's text except in its exterior form (Neugebauer 1932, p. 250).

Neugebauer pretended hence – and with him the entire revisionist school of a geometric algebra – that the mathematical content is independent of its form, language and the symbols used to express it. While intended to facilitate understanding old texts, this approach is clearly anhistoric.[2]

Another methodological issue is addressed by the evident question: Allowing the assumption that Greek mathematical thinking was indeed algebraic – how can one explain that the Greeks' disguising this thinking by a geometric language? Hans Freudenthal gave three answers:

> First, historical: it just happened that the Greek end of the torturous path through foundations of mathematics, Eudoxos' theory, was so excellent that the Greeks did not aspire a better one.
>
> Second, philosophical: though in daily use by laymen as well as mathematicians, fractions were taboo in highbrow mathematics, because mathematics forbade the division of the unit.
>
> Third, traditional: Once canonized, the Elements were sacrosanct, liable to additions, but not to change. The mathematical community was small. To be understood within it, you had to quote Euclid and to speak his language (Freudenthal 1977, p. 191).

Regarding the first answer, Unguru followed Wilbur Knorr in refuting that there had taken place the famous, but only pretended crisis of foundations in the pre-Eudoxian period. The second answer contained no pertinent point, according to Unguru. And to the third answer, Unguru put as question:

> Fine, but why did Euclid then adopt the very same language? (Unguru 1979, pp. 558–559).

The third controversial point of the debate is of genuinely methodological nature and concerns our key issue in particular, the role of signs and of symbols. Unguru had emphasized in his discussion of the propositions of Euclid's Book II that neither the propositions nor the proofs do show any equation. Furthermore, there is no use of unknowns and likewise not of symbols and, consequently, no operations on symbols. Unguru had therefore sharply criticized Heath who – although correctly translating Euclid's text of propositions and their proofs – presented these propositions in his extended commentaries as entirely algebraic texts.

In their replies to Unguru's critique, Freudenthal and van der Waerden claimed that a statement made in words is completely identical to a statement made in symbols. The most radical comment in this regard was published by André Weil, however. Weil, well known for his strong reactions, not only in mathematics but also in history of mathematics, declared any reflection about signs and symbols to be superfluous:

> As everyone knows, words, too, are symbols. The content of a theorem does not change greatly, whether it is expressed in words or in formulas: the choice is, as we all know, mostly a matter of taste and of style (Weil 1978, p. 92).

The necessity of a deeper reflection on the relation between statements in words and corresponding statements in symbols, however, was already clear from considerations by van der Waerden on Book II of Euclid's Elements. His algebraic approach to this book led to a contradictory situation concerning propositions II.5 and II.6, which he could explain away. In both propositions, a segment of a straight line is cut into two unequal parts. While II.5 deals with the difference of these two sub-segments, II.6 deals with their excess:

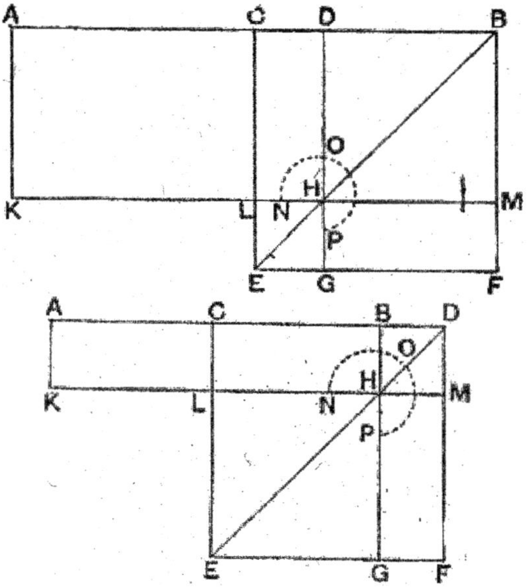

Euclid II.5 Euclid II.6

Due to his zeal to find algebraic formula in Euclid, van der Waerden had introduced designations *a* and *b* on the one hand and *x* and *y* on the other hand and labelled with them the sides in the two figures as shown in Heath, but applied according to his own interpretation:

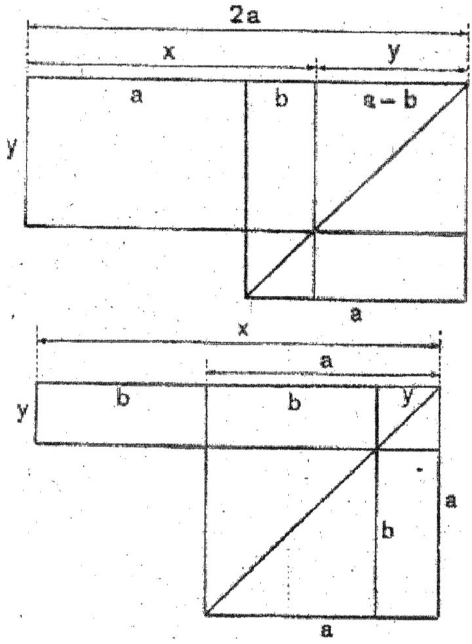

van der Waerden 1954, 120.

In fact, he had succeeded in labelling the sides in both figures in such a manner that he was able to deduce the same formula from the labelled figures:

$$(a+b)(a-b)=a^2-b^2$$

Perplexed, van der Waerden observed this "strange double form": "Why should two propositions be given for *one* formula?" (van der Waerden 1954, 120).

Actually, his perplexity reveals the profound difference between the classical synthetic method and the modern analytic one. Since modern times, one had been conscious of the marked contrast between these two methodologies. While the mathematics of the 'Ancients' treated each case separately and independently, without searching for related cases, which might be regarded as following a common pattern, the analytic method strove for generality. What were for the Moderns just variations of the same property were, for the Greeks, new cases, due to a different position of some lines within a figure.

Van der Waerden's perplexity implies, hence, the failure of the algebraizing approach. Greek geometry was geometry and not algebra. It occurred within another era and another culture that Greek geometric approaches were transposed into algebraic ones.

In fact, it was within the Islamic civilization that a basic principle of the Greek method of comparing geometric entities was bypassed, namely, the principle of homogeneity of the magnitudes to be operated with. Thanks to this rupture with the geometric origins, as initiated by Abū-Kāmil (ca. +900), Arab mathematicians were able to develop algebra as a new discipline (Djebbar 2001, p. 7).

THE EMERGENCE OF THE USE OF SYMBOLS IN ALGEBRA. AN OVERVIEW

In order to better analyze the process of emergence of algebra, Nesselmann's three-stages-model of the evolution of algebraization should be briefly recalled. This model, proposed in 1842, is surely no generally valid model, but it is useful as a heuristic device:

- the first and lowest stage is called *rhetorical* algebra: the entire mathematical procedure, with all its operations is expressed in words. Since there are no symbols introduced and used, words of the respective language serve to express the mathematical meaning.
- In the second stage, the *syncopated* algebra, the presentation of the mathematical issue is basically rhetorical, too, but one introduces here for frequently used concepts and operations always the same abbreviations – instead of the full words. One would be able, however, to return in any moment from the abbreviation to the original full word.
- The third stage, that of *symbolic* algebra, represents all forms and expressions by a language of symbols, which is constituted independently of normal language. There is practically no way back from the operations on the symbols to a rhetorical text (Nesselmann 1842, p. 302).

Quant au procédé <de calcul> de la racine approchée, tout nombre dont on veut <chercher> la racine et qui n'est pas un carré parfait, est situé entre deux carrés successifs dont l'un est plus petit que lui et le second plus grand.

Entre les racines de deux carrés successifs, il y a toujours un, et la différence entre les deux carrés est égal à deux fois la racine du plus petit des deux, plus un. Il est aussi égal à deux fois la racine du plus grand moins un. Il est aussi égal à la somme des deux racines. Ceci est évident d'après le produit de la racine du plus grand par elle-même après sa partition en la racine du plus petit, plus un. Sache le.

Soit a le nombre dont on cherche à approcher la racine carrée.
Il existe deux entiers n et m tels que :

$$n < a < m$$

avec $n = k^2$ et $m = (k+1)^2$.

On a : $\sqrt{m} - \sqrt{n} = 1$

$m - n = 2\sqrt{m} - 1 = 2\sqrt{n} + 1$ car

$m = \sqrt{m}.\sqrt{m} = (\sqrt{n} + 1)^2 = n + 2\sqrt{m} + 1.$

Bensmina, Boufioua1993

Almost right from the beginning of mathematical activity in the Islamic civilization, Greek geometry had been transposed to deal with equations. Al-Khāwrizmi and Abū-Kāmil, for instance, dealt with solving equations of the second degree. Characteristic of their mathematical texts is their practice of a rhetorical algebra. This shall be illustrated by a text of a mathematician of the Maghreb, Ibn al-Bannā (1256-1320), on the approximation of square roots, translated by M. Aballagh literally into French and then transcribed into symbols of present mathematics:

A next illustration comes from European Middle Ages, from Italy, where one finds the highest level of contemporary European mathematics. The *Trattato di Fioretti* of the abacus master Antonio de Mazzinghi, of ca. 1360, developed an impressive number of algebraic problems:

> del chongunto loro, adunque el loro chongunto è la radice di 43. Ora che abbiamo trovato che, fra amendunj, le dette quantità sono la radice di 43, e noj diremo facciasj, di radice di 43, 2 parti che ll'una multiplichata per l'altra faccino 8. Dove dimeçeraj la radice di 43, che sia la metà la radice di 10 3/4, e questo multiplicha in se medesimo, fanno 10 3/4, e di questo trai 8, rimanghono 2 3/4 del quale piglia la radice che è radice di 2 3/4 et trala di radice di 10 3/4, rimanghono radice di 10 3/4 meno la radice di 2 3/4; e tanto è la prima parte. E la seconda sia radice di 10 3/4 più radice di 2 3/4. E diremo che l'una quantità sia la radice di 10 3/4 meno la radice di 2 3/4, cioè la minore parte, e la maggiore sia radice di 10 3/4 più radice di 2 3/4. E chosì è trovato 2 quantità che i loro quadrati sono 27 e la multiplichatione dell'uno nell'altro è 8; chome volavamo.
>
> Ma per aguagliamenti dell'algibra anchora possiamo fare: ...

Mazzinghi [1967], 28

Yet, only the numbers are given by symbols; the remaining text is rhetorical, however.

In modern times in Europe, Petrus Ramus (1515-1572) was one of the influential textbook writers. The invention of the printing press in the second half of the 15th century and the growing number of printed mathematical books promoted at least the use of abbreviations. In Ramus's *Algebra* – here its second edition of 1586 – one finds, however, a use of several symbols in an operational manner, such as the signs for plus and minus. The text is not easy to read for us, however, since he does not use the signs for powers of the unknown x, but understands algebra as dealing with figurate numbers, so that $7x$, being a linear quantity is written as $7l$, and $8x^2$, being a quadratic quantity, hence $8q$, etc.

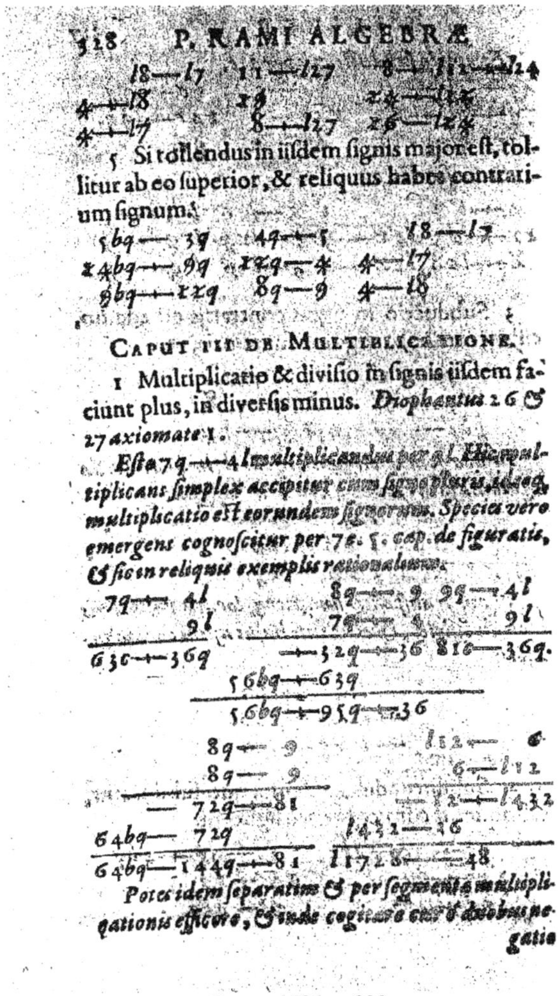
328 P. RAMI ALGEBRÆ

5 Si tollendus in iisdem signis major est, tollitur ab eo superior, & reliquus habet contrarium signum.

CAPUT III DE MULTIPLICATIONE.

1 Multiplicatio & divisio in signis iisdem faciunt plus, in diversis minus. *Diophantus 26 & 27 axiomate 1.*

Esto 7q + 4l multiplicandus per 9l. Hic multiplicans simplex accipitur cum signo plus, idque multiplicatio est eorundem signorum. Species vero emergens cognoscitur per 7e. 5. cap. de figuratis, & sic in reliquis exemplis rationalibus.

Potes idem separatim & per segmenta multiplicationis efficere, & inde cogitare cur è duobus negatis

Ramus 1586, p. 328

Another example of the difficulties presented by the symbolic practices of early modern texts is illustrated by Bombelli's explanation of his solution for $x^3 = 15x + 4$. His use of symbols for operating with roots is difficult for us particularly because of a missing designation for the unknown:

Among the first modern textbooks applying Descartes's notational innovations for designating constant and variable quantities, was Antoine Arnauld's *Nouveaux Élémens de Géométrie* of 1667. His notational conventions for symbols are quite near to ours so that his text appears readable without efforts regarding the signs used:

$\overset{3}{\smile}$
1. Eguale à 15. $\overset{1}{\smile}$ p. 4.

5.		2.
5.	.	2.
25.		4.
5.		125.
125.	R.q.p.di m.121.	

2   2

Somma R.q.p.di m.121. Resta R.q.p.di m. 121.
R.c.L 2.p.di m. 11. R.c.L. 2.m.di m. 11.
Lato 2. p.di m. 1. 2.m.di m. 1.
Sommati fanno 4. che è la ualuta del Tanto.

Andersen 1990, p. 180

CINQUIEME THEOREME.

XLV. LORS que 4 termes sont proportionnels ils le seront encore.

1. En transposant les termes de chaque raison, ce qui s'appelle *Permutando*.

2. En les prenant alternativement, c'est à dire en comparant les antecedens ensemble & les consequens ensemble : le 1er terme au 3e, & le 2e au 4e, ce qui s'appelle *Alternando*.

3. En comparant chaque antecedent plus son consequent avec son consequent, ce qui s'appelle *Componendo*

4. En comparant chaque antecedent moins son consequent avec son consequent, ce qui s'appelle *Dividendo*.

Exemples. Si	*b*.	*c*	::	*f*.	*g*.
	5*x*.	3*x*	::	5*y*.	3*y*.
Permutando	*c*.	*b*	::	*g*.	*f*.
	3*x*.	5*x*	::	3*y*.	5*y*.

Arnauld 1667, p. 34

THE EMERGENCE OF A SYMBOLIC ALGEBRA IN MEDIEVAL MAGHREB

According not only to Nesselmann, but also to the standard lore in the historiography of mathematics, it was in Europe in modern times, that a symbolic algebra became established. The overview just given seems to confirm this assessment. Recent research on the history of Arab mathematics in the Maghreb, in particular on the Djerba manuscript, shows much earlier developments, however. Since these are revealing also for the transitions from the syncopated to the symbolic stage, an account of these researches should be given here.

In standard accounts of the history of Arab mathematics, the focus is on persons who were active in the eastern part of the immensely extended Islamic civilization, on Egypt, Iraq and Iran. In Grattan-Guinness's long list of Arab mathematicians, there are just two exceptions, named as being from Spain (Grattan-Guinness 1997, p. 115). There are, however, at least two indications hinting towards another direction. Firstly, the numbers known to us as the Hindu-Arab ciphers existed, after their transmission form India, in two forms: as the Eastern Arab ones and as the Western Arab ones, or the Maghrebinian ciphers. Actually, those transmitted to Europe and hence those today universally used are the Maghreb ones.

And, secondly, the first appearance of a text which used mathematical symbolism was a text produced in al-Andalus. Strangely enough, this text is known – up to now – only in a Latin translation of about 1180, published by Boncompagni in 1851. This translation had for a long time been attributed to Gerard of Cremona, but erroneously, as known today. The Arab original had been an adaptation of one of Al-Khw~rizmi's texts. It is interesting to see the symbolism used there: for fractions, on the one hand, and for the unknown and its power, on the other hand:

2	3	4
3	4	5
c	r	d

In modern notation, this means: $\frac{2}{3}x^2 + \frac{3}{4}x + \frac{4}{5}$. The symbol $\underline{c}$ abbreviates *census* = square of the unknown, $\underline{r}$ abbreviates *radix* = root and $\underline{d}$ for *dragme* = drachme/money (Boncompagni 1851, p. 421; cf. Moyon 2007). One is therefore lead to the supposition that the origin of signs and the use of symbolism has been in al-Andalus and in the Maghreb during the Middle Ages.

A first confirmation for this supposition can be found in a remark by Ibn al-H~'im (1352-1412), himself a scholar from Egypt, who was aware – without practising it himself - that there were mathematicians who used special symbols for designating the powers of an unknown in a polynomial:

> De même, en écriture indienne ou ghūbar, ils [= les spécialistes de la terminologie] attribuent à chaque espèce un signe; comme le Shīn[3] pour les Shay [les choses], le Mīm[4] pour le Māl [le carré], le Kāf[5] pour les cubes et ainsi de suite, et ils n'attribuent aucun signe existential au nombre [la

> constante], il s'ensuit que l'absence de signe est un signe (quoted from Abdeljaouad 2002, p. 24).

His remark that the absence of a symbol constitutes itself a symbol shows a revealing consciousness for the impact of signs.

The remark by Ibn al-Hā'im was made in a commentary of the year 789 H/1380 CE on an algebra manuscript written itself in 1191 in Sevilla, in al-Andalus. The original text was written by Ibn al-Yāsamīn, in the then frequent form of a poem:

Al-Urjūza fil jabr wal muqābala (1191, Sevilla)

Ibn al-Ha'im's commentary had the title:

Sharh al-urjūza al-yāsminiya fil jabr wal muqābala (1380, Egypt).

The important point here is, however, that Ibn al-Hā'im's text became itself edited later on, in a highly innovative manner, by one such "specialist of the terminology", by a yet unknown mathematician in the Maghreb who transposed the original text into a symbolic text – by transcribing the rhetorical text in the margin of the manuscript sheets into formulas (see the reproduction of a characteristic page). This edited copy has been detected by the Tunisian historian of mathematics Mahdi Abdeljaouad who has named it the "Djerba manuscript", since it is preserved on the isle of Djerba, in private possession. The extant copy dates from 1157 H/1747 CE, but one can reasonably assume that the original transposition occurred quite earlier on. In fact, there are numerous mathematicians in the Maghreb between the 13th and 15th centuries who not only used a rather coherent symbolism, but who also explicitly commented and explained it (Abdeljaouad 2002, 20 sqq.).

Abdeljaouad has presented his findings in 2002, at the 7th *Colloque Maghrébin d'Histoire des Mathématiques Arabes*. Some of his main results shall be reported here.

In order to make the analysis better understood I will give here firstly the Arab-Maghreb ciphers:

١, ٢, ٣, ۴, ۵, ۶, ٧, ٨, ٩, ٠

And then the symbols as introduced and used in the Maghreb:

Equality ل equal يعدل ya'ōdil

Operators

	of action		of result
add	الى	plus	و
subtract	من	minus	الا
(multiply) by	في	times	في
(divided) by	على	divided	مقسوم

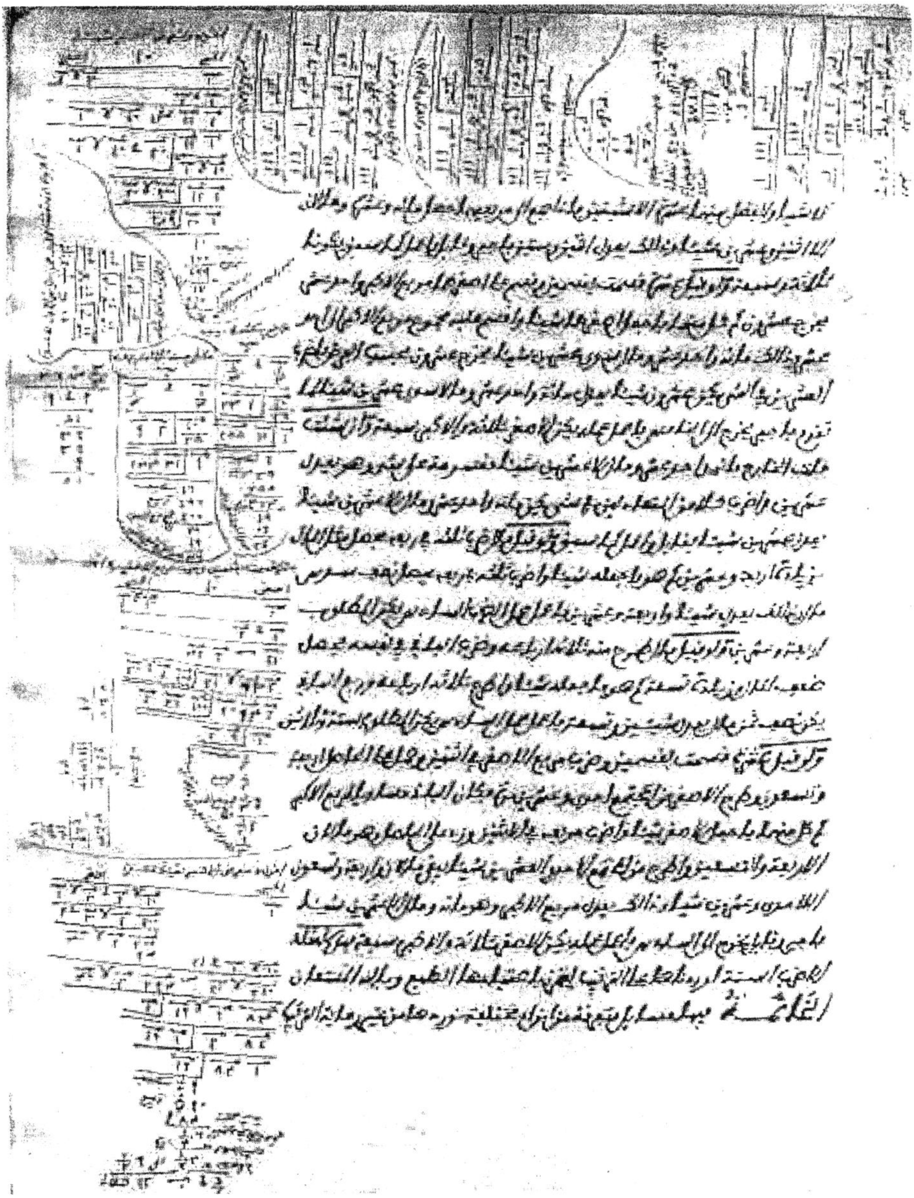

One of the sheets of the Djerba manuscript

Variables, etc.

Number	عدد	عـ
x	شيء	شـ
x^2	مال	مـ
x^3	كعب	كـ
$\sqrt{}$	جذر	جـ

Whereas the ordinary unabbreviated words are used for the operators (first table above), and the last letter of the word in the case of *ya'ŏdil* ('equal'), variables and its powers are represented by the first letter of each word. One should note, in the right column, that a horizontal bar is added to the left of the respective letter in order to make it a proper sign.

Moreover, it will be useful to note what *al-jabr* and *al-muq~bala* mean originally. Thus, *al-jabr* demands to suppress the negative terms within an equation:

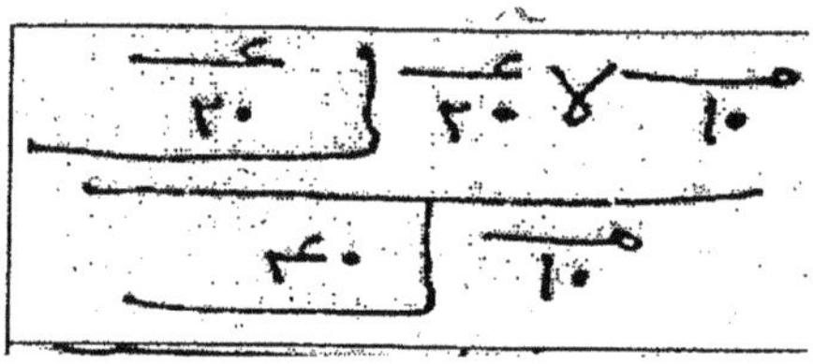

Thus $10x^2 - 20 = 20$ becomes transformed into $10x^2 = 40$.[6]

And *al-muqabala* demands that terms of the same power are put together on one side of the equation:

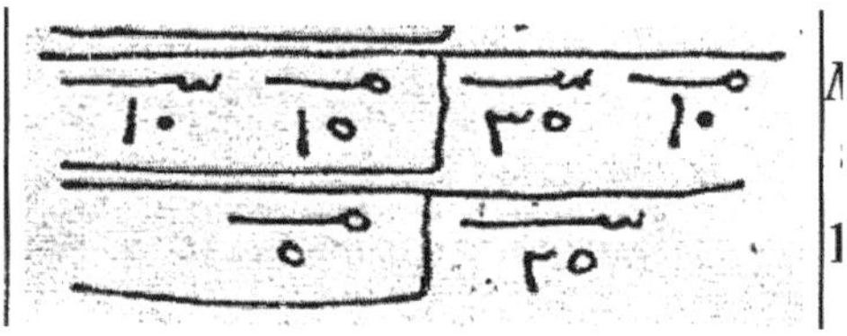

The equation $10x^2 + 35x = 15x^2 + 10x$ gets reduced to $25x = 5x^2$. Simple juxtaposition of two terms means that they have to be added (Abdeljaouad 2002, p. 48).

Given these preliminaries, we are now able to illustrate a particular feature of this symbolism. It is distinguished between operators of action and operators of state. The first operators express actions to be performed – say an addition, a subtraction, etc., the second ones state the result of the actions taken. The operators of action are indicated by the stereotypically used words in the left row (to be added to; to be subtracted from; to be multiplied by; to be divided by). The operators of state use the words shown in the right column; these correspond to the expressions plus, minus, multiplied by, divided by also used nowadays, but in our times indistinctly also for the foregoing action.

Here follows an example for the case of subtraction:

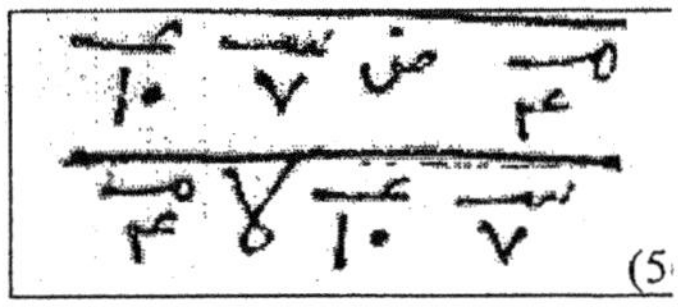

This stands for: Subtract $4x^2$ from (7*x*+10). [You find] $(7x+10)-4x^2$ (ibid., p. 59).

I am showing here a task for multiplication and I am leaving it to the attentive reader to transcribe it into our notation.[7] One has to note only that higher powers than the third are indicated by composition of the respective symbols; hence the fourth by *mīn mīn*, the fifth by *mīn kāf* and the sixth by *kāf kāf*, and that the trema over the *shīn* is often omitted:

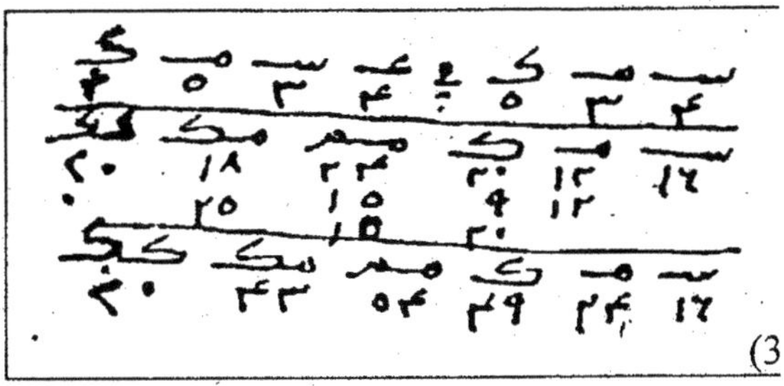

It becomes clear that the text in the margin has no longer a merely illustrative function with respect to a still dominantly rhetorical text, it constitutes the transition to a purely symbolical text with an operative character. To close this section, an example will demonstrate these operational uses:

One discerns here enormous progress towards an operational symbolism; in particular, the degree of powers of the unknown is generalizable – although in a finite manner. On the other hand, one will observe a certain shortcoming within this approach: it works for one unknown, but not for several unknowns (see Radford 1997).

Folio 43a Divisions de monômes	$2x^2 : 10x$ ——— $\frac{1}{5}x$	مـ سـ ٢ على ١٠ ——— سـ ١/٥
Folio 45a Divisions de polynômes I	$10x + 10x^2 : x + 2$ ——— $\frac{10x + 10x^2}{x + 2}$	سـ مـ سـ عـ ١٠ ١٠ على ١ ٢ ——— سـ مـ ١٠ ١٠ مــــق سـ عـ ١ ٢
Folio 45a Divisions de polynômes II	$20x^2\ 10x : 5x$ ——— $4x-2$	مـ سـ سـ ٢٠ لا ١٠ على ٥ ——— سـ عـ ٤ لا ٢

(Abdeljaouad 2002, p. 77).

AN ANALOGOUS CASE: THE NOTION OF LIMIT

The case presented above is not the only one in the history of mathematics where originally geometric concepts became algebraized. Another case, from the history of the calculus, arises in connection with the notion of limit.[8] For the purposes of founding the new infinitesimal calculus, it had been introduced by Newton, but in an entirely "rhetorical" manner, without any sign or operational theory. Until the last third of 18th century, mathematicians who used the notion of a limit continued to refer to it in this rhetorical manner, while introducing first but superficial definitions of the concept. From the years 1770 on, however, one can observe first syncopated approaches, with some symbolisations using abbreviations such as "lim.", adopted by L'Huilier and Cousin, or L, adopted by Carnot. The subsequent developments showed that there need not be uniform processes in the different mathematical communities. While at the "periphery", in Portugal, there was established in 1794 a theory operating symbolically on limits, the practice of dealing syncopatedly with abbreviations like "lim." – thus for instance Cauchy – was continued in France until the first half of the 19th century. A decisive breakthrough was achieved within the German algebraizing school, by Enno H. Dirksen (1788-1850).

Whereas Cauchy had applied for series of functions the same symbols as for numerical series, without indicating the variables, - s, s_n and r_n -, Dirksen had developed proper symbols for series of functions, taking several variables into consideration:

$$S_n = f(x,n)$$

and thanks to his novel symbol for the limit, taking into account the respective variable index:[9]

$$\overset{n=\infty}{Gr} \text{ and } \overset{\Delta x=0}{Gr}$$

he was able to distinguish and investigate multiple limit processes and was therefore able to detect the problem in Cauchy's proof of the continuity of the sum of a convergent series of continuous functions (Schubring 2005, 467 ff.).

CONCLUDING REMARKS

One would misunderstand the historical process of algebraization if one disregards the connection between mathematical form and content, as did Neugebauer and others in their interpretations of Euclid and Apollonius. In fact, this inseparability of form and content constitutes a basic pattern of any semiotic approach.

It could be shown the deep epistemological differences between the synthetic, non-symbolic approaches and the analytic ones. Whereas, for the first, different geometrical configurations present separate and independent cases so that one would not search for related cases or common structures, the algebraic method strove for generality.

A break with this static methodology meant the transformation of formerly geometrical problems by Islamic mathematicians; they did not only use symbols, but were also cognizant of the symbolic character of their mathematics as is nicely shown by their awareness that even "… the absence of a symbol constitutes itself a symbol". The case of the Djerba manuscript reinforces that a truly symbolic mathematics was developed in the Islamic civilization, that symbols were used *operationally*.

Algebraization in the sense of a growing introduction and use of symbolism shows the decisive impact of signs on the evolution of mathematics. It contributes to making hitherto implicit assumptions explicit, making them amenable to critical investigation, and leading, thus, to the refutation or improvement of propositions resting on them. At the same time, algebraization suggests generalizations of mathematical concepts and, hence, allows progress towards new areas of mathematical research.

NOTES

1 An earlier, but still lonely critic had been Jacob Klein in 1936. See its English translation 1968.

2 A modern, reliable edition of Apollonius is Fried/Unguru (2001), which is a refutation of Zeuthen's and, derivately, Neugebauer's approach to editing Greek geometric texts.

3 i.e., the Arab letter شـ.

4 i.e., the Arab letter مـ.

5 i.e., the Arab letter كـ.

6 One has to be aware that – while words are read from the right to the left in Arab language and while formulae therefore are read from the right to the left, too, the numbers are read – like the Indians did – from the left to the right!

7 The resolution of this task is given at the very end of the paper!

[8] A general, important reference fort he history of notations is Florian Cajori (1928/29).
[9] As a German romantic, Dirksen translated the Latin/French, limit' into the German word, Grenze' and abbreviated it hence as, Gr'.

REFERENCES

Abdeljaouad, M. (2002). Le manuscript mathématique de Jerba: Une pratique de symboles algébriques maghrébins en pleine maturité. In *Actes du 7e Colloque Maghrébin sur l'Histoire des Mathématiques Arabes*, Juin 2002. Marrakech: ENS Marrakech.

Abdeljaouad, M. (Ed.). (2003). *Ibn al-Hā'im: Sharh al-urjūza al-yāsminiya fil jabr wal muqābala.* Tunis: Association Tunisienne des Sciences Mathématiques.

Abdeljaouad, M. (2005). The eight hundreth anniversary of the death of Ibn al-Yāsamīn. Bilaterality as part of his thinking and practice. In *Actes du Huitième Colloque Maghrébin sur l'Histoire des Mathématiques Arabes* (pp. 1–30), Décembre 2004. Tunis: ASTM.

Andersen, K. (1990). Algebraische Lösung der Gleichungen dritten und vierten Grades in der Renaissanc. In E. Scholz (Ed.), *Geschichte der Algebra: eine Einführung*. Mannheim: BI-Wissenschafts-Verlag, 157–181.

Arnauld, A. (1667). *Nouveaux élémens de géométrie*. Paris.

Bensmina, Y., & Boufioua, A. (1993). *Quelques aspects des mathématiques d'Ibn al Banna de Marrakech (1254–1321).* Première Université d'Été d'Histoire et d'Épistémologie des Mathématiques. Montpellier.

Boncompagni, B. (1851). Della vita e delle opere di Gherardo cremonese. *Atti dell'Accademia Pontifia de' Nuovi Lincei.* Session of 27 June 1851.

Cajori, F.. (1928/1929). *A history of mathematical notations* (2 Vols.). London: Open Court.

Djebbar, A. (2001). *Une histoire de la science arabe: Introduction à la connaissance du patrimoine scientifique des pays d'Islam. Entretiens avec Jean Rosmorduc.* Paris: Éd. du Seuil.

Freudenthal, H. (1977). What is algebra and what has it been in history? *Archive for History of Exact Sciences, 16*, 189–200.

Fried, M., & Unguru, S. (2001). *Apollonius of Perga's Conica: text, context, subtext.* Leiden: Brill.

Grattan-Guinness, I. (1997). *The Fontana History of the Mathematical Sciences: The Rainbow of mathematics.* London: Fontana.

Klein, J. (1968). *Greek mathematical thought and the origin of algebra* (E. Brann, Trans.). *With an app. containing Vieta's 'Introduction to the analytical art'.* Cambridge, Mass: M.I.T. Press.

Mazzinghi, A. De' (1967). *Trattato di fioretti: nella trascelta a cura di Mo. Benedetto; secondo la lezione del Codice L. IV. 21 (sec. XV) della Biblioteca degl'Intronati di Siena.* A cura e con introduzione di Gino Arrighi. Pisa: Domus Galilaeana.

Moyon, M. (2007). La tradition algébrique arabe du traité d'Al-Khwârizmî au Moyen Âge latin et la place de la géométrie. In E. Barbin & D. Bénard (Eds.), *Histoire et enseignement des mathématiques* (pp. 289–318). Paris: INRP.

Nesselmann, G. H. F. (1842). *Versuch einer kritischen Geschichte der Algebra; nach den Quellen bearbeitet. Teil 1. Die Algebra der Griechen.* Berlin: Reimer.

Neugebauer, O. (1932). Studien zur Geschichte der antiken Algebra. *Quellen und Studien zur Geschichte der Mathematik, Astronomie und Physik, 2*, 1–27.

Radford, L. (1997). L'invention d'une idée mathématique: La deuxième inconnue en algèbre, *Repères* (Revue des Instituts de Recherche sur l'Enseignement des Mathématiques), juillet 1997, *28*, 81–96.

Ramus, P. (1586). *Arithmetices libri duo, et algebrae totidem*; à Lazaro Schonero emendati et explcati. Francofurdi, apud heredes Andreae Wecheli.

Schubring, G. (2002). *Die Kontroverse über die 'geometrische Algebra' der Griechen.* Conference Fakultät für Mathematik, Universität Bielefeld, December 2002.

Schubring, G. (2005). *Conflicts between Generalization, Rigor and Intuition: Number Concepts underlying the Development of Analysis in 17– 19th century France and Germany.* New York: Springer.

Unguru, S. (1975). On the need to rewrite the history of Greek mathematics, *Archive for History of Exact Sciences, 15*, 67–114.

Unguru, S. (1979). History of ancient mathematics. *Some reflections on the state of the art, ISIS, 70*, 555–565.

Waerden, B. L. van der (1954). *Science awakening. Vol. 1 Egyptian, Babylonian and Greek Mathematics*. Groningen: Noordhoff.
Waerden, B. L. van der (1976). Defence of a shocking point of view. *Archive for History of Exact Sciences, 15*, 199–210.
Weil, A. (1978). Who betrayed Euclid? *Archive for History of Exact Sciences, 19*, 91–93.
Zeuthen, H. G. (1896). *Geschichte der Mathematik im Altertum und Mittelalter*. Copenhagen: Höst.

RESOLUTION OF THE MULTIPLICATION TASK

$$
\begin{array}{l}
(4x + 3x^2 + 5x^3)(4 + 3x + 5x^2 + 4x^3) \\
\hline
16x + 12x^2 + 20x^3 + 24x^4 + 18x^5 + 20x^6 \\
\qquad\quad 12 \qquad 9 \qquad 15 \\
\qquad\qquad\quad\;\; 20 \qquad 15 \qquad 25 \\
\hline
16x + 24x^2 + 49x^3 + 54x^4 + 43x^5 + 20x^6
\end{array}
$$

Gert Schubring
Institut für Didaktik der Mathematik
Fakultät für Mathematik
Universität Bielefeld

VICENÇ FONT, JUAN D. GODINO AND ANGEL CONTRERAS

FROM REPRESENTATIONS TO ONTO-SEMIOTIC CONFIGURATIONS IN ANALYSING MATHEMATICS TEACHING AND LEARNING PROCESSES

INTRODUCTION

In this chapter, we reflect on the key role of semiotic representations in constructing and communicating mathematical knowledge, since they are essential components of mathematical practices. However, other mathematical entities, such as concepts, propositions, procedures and arguments intervene in these practices, in addition to problem-situations whose resolution is the ultimate goal of mathematical activity. Therefore, the socio-epistemic and cognitive analyses of the teaching and learning processes should systematically include the conglomerate of ostensive and non-ostensive objects involved in these processes. In this paper, we present the construct, *onto-semiotic configuration*, which generalizes the notion of representation and moves the focus of research towards the system of objects intervening in and emerging from the mathematical activity. In the third section of this chapter, we include a synthesis of the onto-semiotic approach to mathematics education research developed in several previous publications (D'Amore, Font & Godino, 2007; Godino & Batanero, 1998; Godino, Batanero & Font, 2007; Godino, Batanero & Roa, 2005).

SEMIOTIC REPRESENTATIONS, MATHEMATICAL OBJECTS AND PROCESSES

The essential role played in mathematical activity by different representation registers and, more generally, by material and symbolic "artifacts" is widely recognized regardless of whether the activity is carried out by mathematicians when solving new mathematical problems, or if it concerns the diffusion of mathematical knowledge, that is, the teaching and learning processes. The important role that representations play in learning mathematics can explain the wide number of investigations focusing on this topic in mathematics education research (Goldin, 2002).

Different theoretical frameworks assign a representational role (referring to other entities intervening in mathematical practices), as well as an instrumental role (tools for undertaking the mathematical work) to various types of languages (ordinary, symbolic, graphical, gestural, ...). "The part played by signs, or more exactly by semiotic systems of representation, is not only to designate

L. Radford, G. Schubring, and F. Seeger (eds.), Semiotics in Mathematics Education: Epistemology, History, Classroom, and Culture, 157–173.

mathematical objects or to communicate but also to work on mathematical objects and with them" (Duval, 2006, p. 107).

It is also recognized that mathematics should not be reduced to a language and that mathematical objects should not be confused with any of their possible semiotic representations. Therefore, the socio-epistemic and cognitive analysis of mathematical activity should study the role played by material representations, as well as the role and meaning assigned to the system of non-ostensive objects that intervene and emerge from that activity. This is one of the main objectives assumed by the *"onto-semiotic approach"* to mathematical knowledge and instruction (Godino, 2002; Godino & Batanero, 1998; Godino, Batanero & Font, 2007), where the notion of *configuration* of objects and processes is a tool for jointly analyzing the ostensive and non-ostensive objects that intervene in mathematical practice.

Drawing on previous research (Font & Godino, 2006; Font, Godino & D'Amore, 2007; Godino, Font & Wilhelmi, 2006), in this chapter we introduce the notion of onto-semiotic configuration. An example of application shows its potential utility for overcoming some limitations of the idea of representation in the epistemic and cognitive analyses of mathematical learning.

We first describe a teaching episode that will be used as a context for reflection and a synthesis of some basic notions of the onto-semiotic approach. Secondly, three levels of analysis will be applied to the teaching episode: (1) the identification of representations and the practices they make possible; (2) the description of mathematical configurations; and (3) the description of the socio-epistemic and cognitive processes involved in mathematical activity. The chapter finishes with a synthesis and some conclusions.

A TEACHING EPISODE AS A CONTEXT FOR REFLECTION

As a context for reflection, we shall use a teaching episode in which a group of students (17 years of age) were studying the derivative concept. This teaching episode has been selected from an ethnographic observation carried out by the researchers in ordinary mathematics classrooms, that is, without any influence in the design of such lessons.

Task 1, which follows, was given to the students to solve with the help of the "Cabri" software:

Task 1:

If you move point F in figure 1, you will observe that the parabola $f(x) = x^2$ and the tangent line at point P are represented.

- a) Find the relationship between the segments GO and PF.
- b) Find the relationship between the segments PH and FH.

The students solved this task by working in pairs at the computer lab, where they had figure 1 on the screen. They did not have previous knowledge of the Cabri software. Figure 1 was a construction that changed in real time when the students moved point F (the sole point that could be moved with the mouse). By moving point F and activating the *trace* option, the parabola $f(x) = x^2$ and the tangent line at point P were drawn. The teacher asked the students to find some invariant relations between the segments GO and PF, and also between PH and FH. After a period of time, the following properties were recognized and shared in the classroom: (1) in the parabola $f(x) = x^2$ the tangent line at P cuts the axis of ordinates at a point such that the length of the segment that has for extremes this point and the origin of the coordinates is the ordinate of P, (2) the length of segment PH is twice the length of segment PF.

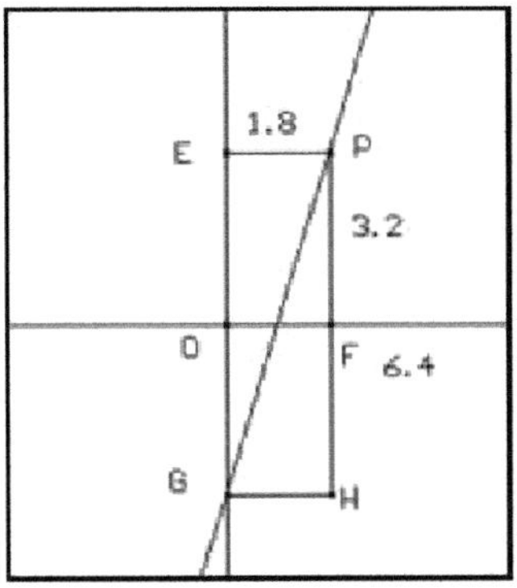

Figure 1. Task 1

They were then asked to use this property to carry out Task 2, as follows:

Task 2:

- a) If $OF = a$, prove that $GH = a$, $PF = a^2$ and $PH = 2a^2$.
- b) If the derivative of the function at a point is the slope of the tangent line, calculate $f'(a)$.
- c) Prove that the derivative of the function $f(x) = x^2$ is $f'(x) = 2x$.

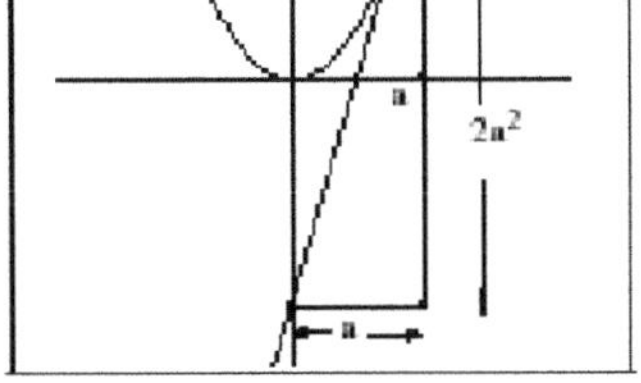

Figure 2. Task 2

The answer given by a student to task 2 is included in Figure 3:

a) $GH = a$ because it is the same distance

$PF = a^2$ *because* the image of a in the function $f(x) = x^2$ is a^2

$PH = 2a^2$ because it is double FP

b) $p = \frac{2a^2}{a} = 2a$

$f'(a) = 2a$

c) $a = x$

$p = \frac{2x^2}{x} = 2x$

$f'(x) = 2x$

Figure 3. Student's answer

We observe that, with the small letter "p" the student indicates the slope of the tangent line.

THE ONTO-SEMIOTIC APPROACH

The onto-semiotic approach to mathematical cognition tackles the problem of meaning and the representation of knowledge by elaborating an explicit mathematical ontology based on anthropological (Bloor, 1983; Chevallard, 1992), semiotic and socio-cultural theoretical frameworks (Ernest, 1998; Presmeg, 1998; Radford, 2006; Sfard, 2000). It assumes a certain socio-epistemic relativity (Cantoral, Farfán, Lezama & Martínez-Sierra, 2006) for mathematical knowledge, since knowledge is considered to be indissolubly linked to the activity in which the subject is involved and is dependent on the institutions and the social context of which it forms a part (Radford, 1997).

In Figure 4, we represent some of the different theoretical notions of the onto-semiotic approach for mathematical knowledge. Here, mathematical activity plays a central role and is modelled in terms of systems of operative and discursive practices. The different types of mathematical objects (problems, languages, concepts, propositions, procedures and arguments) emerge from these practices; these objects are interrelated, forming cognitive or epistemic configurations (hexagon in Figure 1). Lastly, the objects that appear in mathematical practices and those emerging from these practices, depending on the language game in which they participate (Wittgenstein, 1953), might be considered from the five facets of dual dimensions (decagon in Figure 1): personal/institutional, unitary/systemic, expression/content, ostensive/non-ostensive and extensive/intensive. The dualities, as well as the objects can be analysed from a process-product perspective, which leads us to the processes in Figure 4.

The six types of postulated primary objects widen the traditional distinction between conceptual and procedural entities, which we consider insufficient for describing the objects intervening and emerging from mathematical activity. The problem-situations promote and contextualise the activity; languages (symbols, notations, graphics, ...) represent the other entities and serve as tools for action; arguments justify the procedures and properties that relate the concepts. These entities have to be considered as functional and relative to the language game (institutional frameworks and use contexts) in which they participate; they also have a recursive character, in the sense that each object may be composed by other entities, depending on the analysis level—for example, arguments, which may involve concepts, properties, operations, etc. The concept, included as a component of the onto-semiotic configurations, is conceived as "concept – definition"; this view is different from that of Vergnaud (1990), who conceives of a concept as the system formed by the "situations, operative invariants and representations".

In the onto-semiotic approach, the intention is not to give a definition of "process" from the beginning, as there are many different types of processes: one can talk of process as a sequence of practices, as cognitive processes, meta-cognitive processes, processes of instruction, processes of change, social processes, etc. These are very different processes and perhaps the only characteristic many of them may have in common is the consideration of the "time" factor and, to a lesser degree, the sequence in which each member takes part in the determination of the following. For this reason, in the onto-semiotic approach, instead of giving a general definition of the process, the selection of a list of processes considered important in mathematical activity is opted for instead (those of Figure 4), without claiming that it includes all the processes implicit in mathematical activity, because, among other reasons, some of the most important of them (for example, the process of understanding, the solving of problems or modelling) are more than just processes and should be considered hyper or mega-processes.

Meaning is a key notion in mathematics education and also in the onto-semiotic approach, where it is conceived of in a very simple, powerful and operative way by means of the "semiotic function" (Eco, 1978; Hjelmslev, 1943/1963): *Meaning is the content of any semiotic function*, that is to say, the content of the correspondences (relations of dependence or function) between an antecedent (expression, signifier) and a consequent (content, signified or meaning), established by a subject (person or institution) according to distinct criteria or a corresponding code. The content of the semiotic functions, and hence the meaning, could be a personal or institutional, unitary or systemic, ostensive or non-ostensive object; it could be a concept – definition, problem –situation, procedure, argument, or a linguistic element. In agreement with Peirce's semiotics, the onto-semiotic approach also assumes that both the expression (antecedent of a semiotic function) and content (consequent) might be any type of entity.

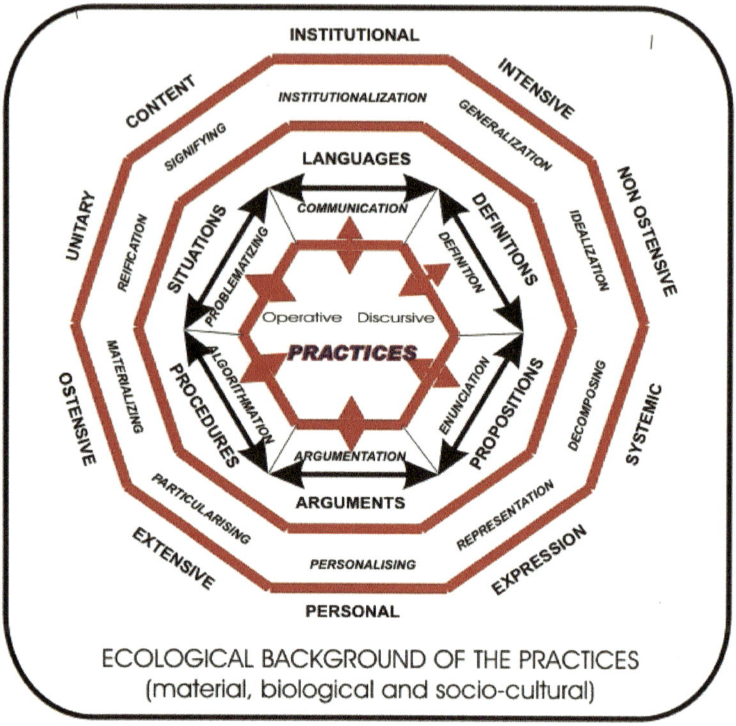

Figure 4. Onto-semiotic configuration for mathematical knowledge

Due to space limitations, we cannot present all the details of the onto-semiotic approach in this chapter. We refer the reader to Godino & Batanero (1998) where the anthropological assumptions are made explicit and to Godino, Batanero & Roa (2005) where the different types of mathematical objects and the dual facets are explained using research on combinatorial reasoning. In Font & Contreras (in press) the materialization-idealization, and particularization–generalization processes are developed and explained.

ANALYSIS OF THE TEACHING EPISODE

In this section, we suggest that the classical dichotomy between external and internal representations is inadequate for analysing the complexity of the epistemological and cognitive phenomena involved in the learning of mathematics, and that the onto-semiotic configuration (Figure 4) allows for a wider and deeper view of such phenomena. Our argument is that the external representations (symbolic, graphical, linguistic, ostensive objects) are inevitable and dialectically accompanied by other non-ostensive mathematical objects and processes. The use of internal representations (conceptions, schemas, competencies) focuses the attention on the individual subject, by dismissing the social context where such non-ostensive objects emerge and are conditioned.

We will apply three levels of analysis to the teaching episode (described in the first section) to show the potential utility of the onto-semiotic approach to mathematical knowledge.

In the first level of analysis, we focus our attention on the linguistic elements (external representations, ostensive objects) and the mathematical practices that these objects make possible. But even in this first step, it is necessary to be aware of the dialectic between ostensive and non-ostensive objects, and the dialectic between the personal (cognitive, internal representations) and the institutional (socio-epistemic) facets. These new categories of analysis are not reducible to the external – internal duality of mathematical knowledge. In a second step, we complete the first level of analysis by making explicit the conglomerate of primary mathematical objects linked to the external representations, and the role played by each of these objects in mathematical teaching and learning processes. The third level of analysis shows new socio-epistemic and cognitive phenomena by focussing attention on the other contextual facets or attributes of mathematical objects and processes. In this chapter, due to space limitations, we restrict the analysis to the materialization– idealization processes (dialectic between ostensive and non-ostensive objects); and particularization–generalization processes (dialectic between extensive and intensive objects) (Figure 4).

Our aim is to show that the application of the onto-semiotic configuration tool offers new analytical insights into the teaching of mathematics that will potentially provide new explanations for the students' difficulties and learning achievements.

First Level of Analysis: Ostensive –Non-Ostensive; Personal - Institutional

It is clear that those representations described in mathematics education bibliography as external (graphics, symbolic expression, etc.) intervene in the teaching episode. Moreover, one student's answer suggests the existence of internal representations related to her answers, usually referred to as conceptions, schemas, competencies, ... So, we can say that the student's conception (Figure 3) of the derivative of a function is the slope of the tangent line to the curve, and that she shows a certain competency in algebraic calculations.

In the onto-semiotic approach, the internal/external classification, in addition to being problematic (Kaput, 1998), is considered to be not very operative, and so, we propose converting it into two dualities or contextual attributes, which, in our opinion are more useful. We are referring to the ostensive non-ostensive and personal-institutional dualities. External representations are ostensive (perceptible) objects and internal representations are non-ostensive objects; nevertheless, not all ostensive objects can be considered as internal representations of individual subjects. Mathematical objects, viewed as cultural entities, cannot be reduced to the material representations used in their generation and communication. Moreover, replacing "external – internal representation" by the expressions "ostensive object", "non ostensive object", is not incidental, because these entities have other roles beyond that of representation.

We feel that the internal/external duality does not explain the institutional dimension of mathematics knowledge, thus confusing, to a certain extent, the said objects with the ostensive resources that are used as support for the creation or emergence of institutional entities. The internal/external duality has serious consequences for understanding learning processes, since the role of human activity and social interaction are not adequately modelled in the production of mathematics knowledge and in learning.

The analysis of the students' responses to Task 2 permits us to suppose relevant differences between the mental processes that "occur" in the mind of each student. At this point, we consider it necessary to take the personal-institutional duality into account before first reflecting on mental processes. It is not enough to reflect on the cognitive processes that have (or have not) permitted these students to answer the questions in the two tasks by carrying out a conversion from a graphic representation to a symbolic representation, when they still did not know what the derivative function of $f(x) = x^2$ was. It is necessary to take into account, above all, the *process of instruction* that these students have followed, if we wish to give an explanation of the achieved learning.

The analysis of the responses, even when it detects important differences between different students, enables us to observe that the students apply the same type of practice to calculating the derivative of the function $f(x) = x^2$. The technique used consists of considering a particular point with the tangent drawn (and so its abscissa and ordinate are not considered to be variables). Then, with the manipulation of dynamic computer programs, like *Cabri Géomètre*, the students find a condition that fulfils all the tangent straight lines (in this case: in the parabola f(x) = x2 the tangent line at P cuts the axis of ordinates at a point such that the length of the segment that has for extremes this point and the origin of the coordinates is the ordinate of *P*), and this permits the calculation of its slope. Finally, students should recognize that the condition they have found and the calculation of the slope from which it is obtained are valid for any point, so the point, which was initially considered as a particular point, is then considered as any point.

In order to answer Task 2, in addition to using the graph of the function, the symbolic expression of the graph, $f(x) = x^2$, should be used. So, this technique relates the following ostensive objects:

$$\text{Graph of } f(x) \text{ and symbolic expression of } f(x) \Rightarrow \text{Symbolic expression } f'(x)$$

With this scheme, we symbolise that the starting point of the students' actions for finding a condition that all tangents fulfil is the graph of the function. The symbolic expression of $f(x)$ is necessary for symbolising the condition that fulfils all the slopes of the tangent straight lines, which enables us to deduce the symbolic expression of $f'(x)$. If the students have practiced the calculation of the slope and the geometric meaning of the derivate at a point, they may obtain the symbolic expression of $f'(x)$ without much difficulty.

One of the relevant aspects of framing the representations within the process of instruction is that it enables us to know that, in this instruction process, the teacher opted to include the institutional intended meaning, and also the implemented and the evaluated meanings, along with practices that form part of the historic-epistemological evolution of the derivative object.

The teacher proposes a sequence of activities to the students that do not correspond to the tangent problem or its inverse. They neither deal with the tangent problem, because the tangent has already been constructed, nor with the inverse tangent problem, because the symbolic expression of the function was known. The use of computers facilitates the students' actions and allows them to find the condition fulfilled by all the tangents (using the triangle formed by the ordinate, the tangent and the sub-tangent). These types of constructions help students compute the derivative of functions without using limits, whenever the students have previously studied the geometrical interpretation of the derivative at a point. This method is suggested by the procedure used to construct the tangent and the normal in the period from Descartes to Barrow.

Second Level of Analysis: Configurations of Primary Mathematical Objects

Identifying the diverse objects that intervene in mathematical practices is essential for understanding the semiotic complexity involved and explaining the learning difficulties. Linguistic elements represent others mathematical entities. But problem-situations make conceptual and procedural entities meaningful; arguments justify the propositions and procedures; definitions and propositions underlay the procedures and arguments. In sum, mathematical activity should be described through the network of objects and relationships involved in solving the problems that motivate such activity. "Mathematical activity is essentially directed towards the exploration, construction and analysis of *conceptual relations* and of *systems of relations*. The main objects of theoretical mathematics are relations" (Steinbring, 2007, p. 104). We describe this network of objects and relations as a configuration, which, in the case of computing the derivative of the function $f(x) = x^2$ in the teaching episode, is summarized in Table 1.

Table 1: Socio-epistemic configuration of primary objects to prove $f'(x) = 2x$

LANGUAGES
Verbal: function, derivative *Graphical:* Graph of the function with the tangent line at a point *Symbolic:* $OF = a$, $GH = a$, $PF = a^2$, $PH = 2a^2$, $f'(a)$, $x = a$, (a, a^2), $f(x) = x^2$, $f'(x) = 2x$, ...

PROBLEM - SITUATION
- It is an internal mathematics problem: to prove that the derivative of $f(x) = x^2$ is $f'(x) = 2x$.

CONCEPTS-DEFINITIONS
Previous
- Graph of a function, coordinate of a point, abscise, ordinate, image, quadratic function, derivative at a point, derivative function, tangent, tangency point, slope, etc.
Emergents

	- Derivative of the function $f(x) = x^2$.
PROCEDURES	PROPOSITIONS
Previous - Calculation of the slope. *Emergents* 1) Find a condition that fulfills all the tangent lines. 2) Symbolize that condition. 3) Find $f'(x)$ from that condition.	*Previous* - The second coordinate of the tangency point is obtained by substituting x by a in the function formula. ... *Emergents* Proposition 1: The tangent line of the function $f(x) = x^2$ at P cuts the axis of ordinates at a point such that the length of the segment that has for extremes this point and the origin of the coordinates is the ordinate of P. Proposition 2: If $OF = a$, then $GH = a$, $PF = a^2$, and $PH = 2a^2$. Proposition 3: The derivative of the function $f(x) = x^2$ is $f'(x) = 2x$.

ARGUMENTS
Argument for proposition 1: It is visually observed. Arguments for proposition 2: - Argument 1: GH = OF because they are parallel sides in the rectangle OGHF. - Argument 2: PF = a^2 because the image of a in the function $f(x) = x^2$ is a^2 - Argument 3: FH = OG because they are parallel sides in the rectangle OGHF; OG = PF (Task 1, a). Arguments for proposition 3: The derivative function of $f(x) = x^2$ is the function $f'(x) = 2x$. Deductive proof in which the following arguments are used: - Argument 1: The tangent line at P cuts the axis of ordinates at a point such that the length of the segment that has for extremes this point and the origin of the coordinates is the ordinate of P. - Argument 2: (a, a^2) is the tangency point. - Argument 3: The slope of the tangent line is $2a$. - Argument 4: The derivative of f'(a) is $2a$ because the derivative is the slope of the tangent line. - Argument 5: The statement about the point (a, a^2) is valid for any other point.

The configuration of objects in Table 1 shows how the problem-situation used in the teaching episode, is the tip of an iceberg that also includes different types of representations, procedures and other mathematical objects. Tasks 1 and 2 place the derivative of $f(x) = x^2$ within a specific context where it is not necessary to use the definition of derivative as the limit of mean variation rates.

Table 1 presents the configuration of primary mathematical objects that the teacher intends to construct in the classroom through a guided teaching process, when solving Tasks 1 and 2 (proving that the derivative of the function $f(x) = x^2$ is $f'(x) = 2x$). We term this conglomerate of objects a "*socio-epistemic configuration*", because it refers to the system of objects and meanings that the teacher wants students to share in the community of practices formed in the classroom. The fact that the mathematical activity was directed at high school students and was based on a specific computer software program determine specific connotations for the linguistic elements, procedures, properties, concepts and arguments, which could be different in other institutional contexts. Put in another words, we recognize some socio-epistemic relativity for mathematical knowledge.

If we apply the onto-semiotic analysis to the student's answer shown in Figure 3, we observe some relevant differences, in particular, in the justification of propositions 2 and 3. That is, the student's *cognitive configuration* shows some agreement with the socio-epistemic configuration, but also some differences, which the teacher should recognize in assessing the achieved learning and making decisions for future implementations of the teaching process.

Third Level of Analysis: Socio-Epistemic and Cognitive Processes

In the final level of analysis, we will focus our attention on the dialectic involved in two pairs of the processes included in the onto-semiotic configuration summarized in Figure 4: materialization–idealization (a complementary look at the ostensive – non-ostensive duality), and the particularization–generalization processes (the extensive–intensive duality).

Processes of materialization – idealization

The ostensive – non-ostensive duality, described in the first level of analysis regarding the dilemma between external and internal representations, permits a complementary insight into the materialization and idealization processes that intervene in building mathematical knowledge. This duality can be applied to each primary entity and, therefore, the materialization – idealization processes take place for the problem – situations, procedures, definitions, etc. We conceive idealization as the generation of non-ostensive objects, while materialization is the linking of ostensive (perceptible) to non-ostensive objects. In both processes, the entities are duplicated, but some students could not recognize this duplication and confused mathematical objects (non-ostensive idealizations) with their related representations (materializations). The distinction between ostensive and non-ostensive is relative to the language game in which they take part. Ostensive objects can also be thought, imagined by a subject or be implicit in the mathematical discourse (for example, the multiplication sign in algebraic notation).

Plato was one of the first thinker to emphasize the relevance of the idealization process, by considering the objects of experience as imperfect copies of mathematical ideas. Since then, the need to take into account the process of idealization in mathematical activity has been noted by many other thinkers. For example, Fischbein (1993) gave a main role to idealization processes in his theory of figural concepts. The idealization process is also emphasized by Kitcher (1984), who assumed empirical and pragmatic origins for mathematics, and adopted a constructivist position by considering mathematics as a science of idealized operations that people are able to carry out on any kind of object. Another example comes from the research program of "embodied cognition" (Lakoff & Núñez, 2000), where a key issue is investigating the way people generate mathematical ideas.

We can recognize processes of idealization and materialization in Tasks 1 and 2 (Figure 5) because the figures (drawn on a sheet of paper or on the computer screen) are materializations of the mathematical objects "graph of the function $f(x) = x^2$" and "tangent line to the function $f(x) = x^2$ at a point".

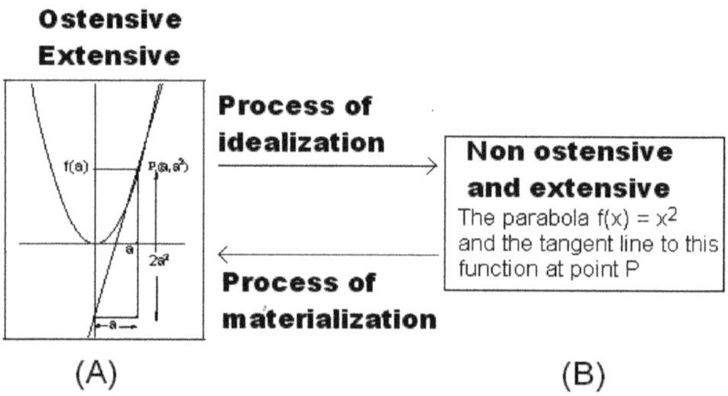

Figure 5. Processes of idealization and materialization

The teacher and students talk about Figure 5(A) as if it were the parabola $f(x) = x^2$ and the tangent line to this function at point *P*. If we look carefully at Figure 5(A) one observes that: (1) the "straight line" is not a straight line, (2) the "straight line" is not the tangent line, (3) the graphic is not a parabola, etc.

It is clear that the teacher hopes the students will go through the same process of idealization of Figure 5(A) drawn on the sheet of paper as he has done. That is to say, Figure 5(A) is an ideal figure, explicitly or implicitly, for the type of discourse the teacher and students makes about it. Figure 5(A), drawn on the sheet of paper, is concrete and ostensive (in the sense that it is drawn with ink and is observable by anyone who is in the classroom) and, as a result of the process of idealization, one has a non-ostensive object (the parabola $f(x) = x^2$ and the tangent line to this function at point *P*) in the sense that one supposes it is a mathematical object that cannot be presented directly. On the other hand, this non-ostensive object is particular, that is, it is the parabola $f(x) = x^2$ and the tangent line to this function at point *P*, and it is not, for example, the parabola $f(x) = 2x^2$ and the tangent line to this function at point *S*. In the onto-semiotic approach, we call this type of "individualized" object an extensive object. Therefore, as a result of the process of idealization, we have moved from an ostensive, which was extensive, to a non-ostensive that continues to be an extensive object.

The other side of the coin is that, to be able to manipulate non-ostensive objects, we need ostensive representations which are the result of a process of materialization (and also of representation). The process of materialization places mathematical knowledge in the "territory of the artefact", (Radford, 2006, p. 107), since its products are cultural artefacts that influence and materialize thought.

Processes of particularization and generalization

The generalization processes and the objects emerging from these processes (generalizations) are essential and characteristic of mathematical work. But, these processes are dialectically linked with the respective particular objects that provide meaning to generalizations. The use of examples and particular cases should be the starting point for a meaningful teaching of mathematics.

In the integrative theoretical framework we intend to build, this characteristic of mathematical work can be described and analysed through the extensive – intensive duality (Figure 4). This duality is applied to the different primary entities: a problem, a definition, a procedure, etc., and can be a particular case of a more general problem, definition, procedure, etc. Moreover, the analysis of mathematical practices from the particularization – generalization processes can be complemented with the application of the other contextual dualities, in particular the expression – content (semiotic function) duality, as we will show in this section.

An extensive object is used as a particular case (a specific example, i.e., the function $y = 2x + 1$), of a more general class (i.e., the family of functions $y = mx + n$), which is an intensive object. The terms extensive and intensive are suggested by the two ways of defining a set, by extension (an extensive is one of the members of the set) and by intension (all the elements are considered at the same time). By extensive we understand a particularized object (individualized) and by intensive, a class or set of objects.

The introduction of the extensive/intensive and the expression/content dualities in the onto-semiotic approach can help to clarify the problem of the use of generic elements (Contreras, Font, Luque & Ordóñez, 2005). Expressed differently, the use of the generic element is associated with a complex network of semiotic functions (and therefore, representations) that relate intensive objects with extensive ones. We will show this with the example of the student's response included in Figure 3.

In writing up the questions for Task 2, much attention was paid to the step from the particular to the general. In item *b*) the teacher asks the students to compute the derivative at a particular value "a", and in item *c*), for any value. We assume that the passage from the extensive to intensive has been carefully considered in the design of the tasks. In this problem, the extensive objects "represent" the intensive objects and, hence, the student has to carry out processes of representation and meaning using semiotic functions.

Without entering into a detailed analysis, as is carried out in Contreras, Font, Luque and Ordóñez (2005), in order to calculate the derivative function of $f(x) = x^2$, the students have to identify the following network of semiotic functions:

> Treat separately the variables related by the formula and the graph of $f(x) = x^2$. To do this, it is necessary to understand this function as a process in which other objects, one being x and the other being $f(x)$, intervene. Here, a semiotic function that relates the object $f(x)$ to the object x, is established.
>
> Associate x to the slope of the tangent line at the point of abscise x. This relation can be considered as a semiotic function that relates the object x with the object "slope of the tangent line at the point of abscise x".
>
> Associate the expression that permits us to calculate the slope of the tangent line at the point of abscise x with $f'(x)$. In this case, we have a semiotic function that relates one notation with another different, but equivalent, one.
>
> Consider x as a variable. In this case, we have a semiotic function that relates an object to the class it belongs to.
>
> Understand the function obtained as a particular case of the "derivative function" class. In this case, we have a semiotic function that relates an object to the class it belongs to.

If we look at the Task 2 handed out to the students, we can observe that the sequence of sections aims at making the establishment of these semiotic functions easier. The use of the letter *a*, in question *b* of Task 2, has the role of introducing a specific element into the students' reasoning and so makes step 1, easier. The reason for including the use of the graph and the symbolic notation together for the point of coordinates (a, a^2) is that the teacher wants the students to carry out steps 2 and 3. Steps 4 and 5 are intended to be achieved from question *c*.

This example permits us to shed light on a phenomenon that we consider to be very relevant: the student, in order to carry out the majority of mathematical practices, has to activate a network of complex semiotic functions and the ostensive objects used are determinant, both for reducing or increasing the complexity of this network, or for carrying out the practice correctly. For example, if we had eliminated question *b* and Figure 2 in Task 2, we would still have wanted the students to apply the technique for calculating the derivative function and we would still use graphs (the ones from the previous activity with the computer) and symbolic expressions (question *c*). However, the complexity of the semiotic functions that the students would have had to carry out would have increased considerably and so too the chances of solving the task.

When we use a representation as a generic element in mathematical practices, we are acting on a specific object, but we are situated in a "language game" where we are interested in the object's general characteristics and we disregard its particular aspects. The analysis of the dialogues between teachers and students as regards the use of generic elements is necessary for knowing the details of this language game and the students' difficulties when taking part in it. The knowing and understanding of the rules of this language game are fundamental to making up the network of semiotic functions associated with the practices in which the generic element intervenes.

If we consider the student's answer to Task 2, we can observe that the student is aware of the rules of using the generic element, since she takes into account the calculation of the derivative function. We see that, in the answer to section *c)*, the equivalence "$a = x$", is expressing that the reasoning of sections *a)* and *b)* is valid for any value of a. This indicates that the student has entered into the language game that governs the use of generic elements.

SYNTESIS AND CONCLUSIONS

In this chapter, we have seen how the analysis of a teaching episode based only on representations is insufficient for including the many aspects involved in mathematical activity. Moving from analysis in terms of representations to analysis in terms of onto-semiotic configurations (mathematical objects and processes linked to mathematical practices) is necessary for obtaining a better understanding of the complexities of the mathematics teaching and learning processes.

Certainly, representation and interpretation processes are crucial, and should be the focus of attention at a first level of analysis. The figures and algebraic notations used in Tasks 1 and 2 include a complex network of semiotic functions; without these semiotic functions, the mathematical work would be impossible or very difficult to carry out. Nevertheless, a socio-epistemic and cognitive analysis that casts light on the conflicts in teaching and learning requires a systematic look toward the diverse type of objects and processes intervening in the activity. At a second level of analysis, we should focus on the configurations of primary objects (languages, problems, definitions, propositions, procedures and argumentations) and the related primary processes. The third level of socio-epistemic and cognitive analysis should be centered on the contextual attributes and secondary associated processes: personalization – institutionalization; particularization – generalization; materialization – idealization; reification – decomposition.

Regarding the problem of the delimitation between the processes of particularization - generalization and the processes of materialization – idealization, our conclusion is that considering the dual facets in the onto-semiotic approach, especially the ostensive/non-ostensive and extensive/intensive facets, allows one to deal separately with both pairs of processes. This is an important distinction, as it permits a more detailed analysis and consequently a better comprehension, of these processes as well as of their combined presence in mathematical activity.

When we use an ostensive object as a generic element in mathematical practices, we are acting on a particular object, but we situate ourselves in a "language game" in which we are interested in its general characteristics and we disregard its particular aspects. Controlling the rules of this game allows the student to activate a complex network of semiotic functions, which is what produces the understanding of the particular – general dialectic. The knowledge of the network is also useful in explaining the students' difficulties. Therefore, we show how the onto-semiotic approach to mathematical knowledge can help us analyse mathematical texts and thus help us to understand students' learning difficulties.

The use of diverse theoretical frameworks is a feature of present research in mathematics education. These frameworks and methodologies come from different disciplines (epistemology, psychology, sociology, pedagogy, semiotics, ...) and various research paradigms. The plurality of approaches may be inevitable and even productive, but also poses a very important theoretical question: How can we take advantage of so many theoretical results produced by so many researchers in mathematics education? This is the crucial issue tackled by researchers who are interested in developing the onto-semiotic approach to mathematical knowledge and instruction.

ACKNOWLEDGEMENT

This research work has been carried out as part of the project, SEJ2007-60110/ EDUC, MEC-FEDER.

REFERENCES

Bloor, D. (1983). *Wittgenstein. A social theory of knowledge*. London, UK: The Macmillan Press.

Cantoral, R., Farfán, R. M., Lezama, J., & Martínez-Sierra, G. (2006). Socioepistemología y representación: algunos ejemplos. *Revista Latinoamericana de Investigación en Matemática Educativa*. Special Issue on Semiotics, Culture and Mathematical Thinking. L. Radford, & B. D'Amore (Guest Editors), pp. 27 – 46.

Chevallard, Y. (1992). Concepts fondamentaux de la didactique: perspectives apportées par une approche anthropologique. *Recherches en Didactique des Mathématiques*, *12*(1), 73–112.

Contreras, A., Font, V., Luque, L., & Ordóñez, L. (2005). Algunas aplicaciones de la teoría de las funciones semióticas a la didáctica del análisis. *Recherches en Didactique des Mathématiques, 25* (2), 151–186.

D'Amore, B., Font, V., & Godino J. D. (2007). La dimensión metadidáctica en los procesos de enseñanza y aprendizaje de la matemática. *Paradigma, 28* (2), 49–77.

Duval, R. (2006). A cognitive analysis of problems of comprehension in a learning of mathematics. *Educational Studies in Mathematics, 61*, 103–131.

Eco, U. (1978). *A theory of semiotics*. Bloomington, IN: Indiana University Press.

Ernest, P. (1998). *Social constructivism as a philosophy of mathematics*. New York, NY: SUNY.

Fischbein, E. (1993). The theory of figural concepts. *Educational Studies in Mathematics, 24*(2), 139–162.

Font, V., & Contreras, A. (2008). The problem of the particular and its relation to the general in mathematics education. *Educational Studies in Mathematics, 69*, 33-52.

Font, V., & Godino, J. D. (2006). La noción de configuración epistémica como herramienta de análisis de textos matemáticos: su uso en la formación de profesores. *Educaçao Matematica Pesquisa, 8* (1), 67–98.

Font, V., Godino, J. D., & D'Amore, B. (2007). An onto-semiotic approach to representations in mathematics education. *For the Learning of Mathematics, 27* (2), 1–7.

Godino, J. D. (2002). Un enfoque ontológico y semiótico de la cognición matemática,. *Recherches en Didactique des Mathématiques, 22* (2/3), 237–284.

Godino, J. D., & Batanero, C. (1998). Clarifying the meaning of mathematical objects as a priority area of research in mathematics education. In A. Sierpinska & J. Kilpatrick (Eds.), *Mathematics Education as a Research Domain: A Search for Identity* (pp. 177–195). Dordrecht: Kluwer, A. P.

Godino, J. D., Batanero, C., & Font, V. (2007). The onto-semiotic approach to research in mathematics education. *ZDM-The International Journal on Mathematics Education, 39* (1–2), 127–135.

Godino, J. D., Batanero, C., & Roa, R. (2005). An onto-semiotic analysis of combinatorial problems and the solving processes by university students. *Educational Studies in Mathematics, 60* (1), 3–36.
Godino, J. D., Font, V., & Wilhelmi, M. R. (2006). Análisis ontosemiótico de una lección sobre la suma y la resta. *Revista Latinoamericana de Investigación en Matemática Educativa.* Special Issue on Semiotics, Culture and Mathematical Thinking. L. Radford, & B. D''Amore (Guess Eds.itors), pp. 131–155.
Goldin, G. (2002). Representation in mathematical learning and problem solving. In Lyn D. English (Ed.), *Handbook of International Research in Mathematics Eeducation* (pp. 197–218). London: Lawrence Erlbaum.
Hjelmslev, L. (1943). Omkring sprogteoriens grundlæggelse. In *Festkrift udg. af Københavns Universitet,* Københavns. (pp. 1–113). English translation by (F. J. Whitfield, English trans.): *Prolegomena to a Theory of Language.* Madison, WI: The University of Wisconsin Press, 1963.
Kaput, J. J. (1998). Representations, inscriptions, descriptions and learning: A kaleidoscope of windows. *Journal of Mathematical Behaviour, 17* (2), 266–281.
Kitcher, P. (1984). *The nature of mathematical knowledge.* Oxford: Oxford University Press.
Lakoff, G. A., & Núñez, R. (2000). *Where mathematics comes from: How the embodied mind brings mathematics into being.* New York: Basic Books.
Presmeg, N. (1998). A semiotic analysis of students' own cultural mathematics. In A. Olivier, & K. Newstead (Eds.), *Proceedings of the twenty-second annual conference of the International Group for the Psychology of Mathematics Education* (vol. 1, pp. 136–151). Stellenbosch, South Africa: University of Stellenbosch .
Radford, L. (1997). On psychology, historical epistemology and the teaching of mathematics: Towards a socio-cultural history of mathematics. *for the learning of mathematics, 17*(1), 26–33.
Radford, L. (2006). The anthropology of meaning. *Educational Studies in Mathematics, 61*, 39–65.
Sfard, A. (2000). Symbolizing mathematical reality into being – or how mathematical discourse and mathematical objects create each other. In P. Cobb, E. Yackel, & K. McClain (Eds.), *Symbolizing and communicating in mathematicsCclassrooms* (pp. 38–75). London, UK: Lawrence Erlbaum.
Steinbring, H. (2007). Epistemology of mathematical knowledge and teacher-learner interaction. *ZDM. The international Journal on Mathematics Education, 39* (1–2), 95–106.
Vergnaud, G. (1990). La théorie des champs conceptuels. *Recherches en Didactiques des Mathématiques, 10* (2–3), 133–170.
Wittgenstein, L. (1953). *Philosophical investigations.* New York: The MacMillan Company.

Vicenç Font
Facultat de Formació del Professorat
Universitat de Barcelona
Barcelona (Spain)

Juan D. Godino
Facultad de Educación
Universidad de Granada
Granada (Spain)

Angel Contreras
Facultad de Humanidades y Ciencias de la Educación
Universidad de Jaén
Jaén (Spain)

STEPHEN J. HEGEDUS AND LUIS MORENO-ARMELLA

ANALYZING THE IMPACT OF DYNAMIC REPRESENTATIONS AND CLASSROOM CONNECTIVITY ON PARTICIPATION, SPEECH AND LEARNING

FROM TOOLS TO SYMBOLIC TECHNOLOGY: AN INTRODUCTION

Crafting a stone tool (an axe, for instance) opened an exploration space: there, it was possible to distinguish between objects with different degrees of softness or hardness. But the softness (or hardness) is not a property we can realize intrinsically from the material under the axe; it is a property we access thanks to the *mediated action* that is possible with the tool. In a certain sense, the tool can be conceived of as a domain of abstraction. The tool *reveals* the idea of softness when we integrate it into a scheme of intentional action. The tool is the medium through which we can realize how a *general* (property) lives in a *particular* (material). Then, in the hands of an expert agent, the axe is more than a device to cut. It becomes a probe into the world of material properties and represents an instance of how human action helps to transform the world into the *human* world.

Jerome Bruner (1990) expounded this argument:

> Cultures characteristically devise "prosthetic devices" that permits us to transcend "raw" biological limits—for example, the limits on memory capacity...The reverse view I am proposing is that it is culture, not biology, that shapes human life and the human mind, that gives meaning to action by situating its underlying intentional states in an interpretive system. (p. 34)

The history of humanity might be described to a considerable measure, as a history of symbolic creativity. Maps, books, novels, mathematical notations are just a few examples of this enterprise that can be named our *symbolic technology*. The deep, transforming effects of symbolic technologies has substantially enriched (and possibly *defined*) human life.

Parietal paintings, incisions in bones, writing, are a few examples of the production of increasingly sophisticated media to externally support our memory and thinking. Of course, the intentionality–the goals of this enterprise–cannot be understood if we forget the social activity (trade, for instance) associated with the solution to real and demanding daily problems. The history of Cuneiform writing is an exceptional, illuminating example (Schmandt-Besserat, 1996). The fact that symbolic technologies and the corresponding semiotic activity transform the functional architecture of conscious mental activity is well documented (Wertsch,

L. Radford, G. Schubring, and F. Seeger (eds.), Semiotics in Mathematics Education: Epistemology, History, Classroom, and Culture, 175–194.

1991). Our thinking is mediated, in every circumstance, by material and symbolic technologies that have modified, for instance, the way we remember and the way we perceive reality.

The semiotic external supports to our biological memory liberate our conscious life of the intrinsic limitations of the biological memory. In fact, much earlier, in the world of orality (before writing), humans had devised strategies to preserve the collective memory. Indeed, *Mythic Culture* (Donald, 1993) was the result of using narrative to preserve the content of collective memory (identity is a case in point). With the advent of writing, since its earlier inceptions, thinkers were given a profound semiotic and epistemological artefact. Thinking gained a more stable and permanent medium that escaped from the burden of presence and simultaneity. The stability of the new medium enabled the possibility to deepen reflections —going beyond the possibilities of orality. Novels, for instance, offered new *models* of social life—let us recall here Wartofsky's discussion on artefacts *as models*. In his celebrated work (Wartofsky, 1979), he introduced *tertiary* artefacts as those that transcend "the more immediate necessities of productive praxis" making feasible to conceive new "possible worlds". A novel, a sonata, a mathematical theory are all instances of tertiary artefacts that, as said before, offer the possibility to transcend the original contexts within which they are created. Tertiary artefacts are the artefacts of creativity.

Speaking of mathematical theories, let us underline that mathematical symbols have never been ancillary to mathematical thinking. Even more, we strongly suggest that there is no mathematics with pre-semiotic objects. For instance, Rotman (1993) has argued in favour of,

> A semiotics which [...] would have to demolish the widely held metaphysical belief that mathematical signs point to, refer to, or invoke some world, some supposedly objective eternal domain, other than that of their own human [...] making. (p. 11)

A vivid example of the awareness of semiotic mediation and its inextricable links with thinking is presented in a recent biography of R. Feynman (Gleick, 1992). Gleick narrates that, during a conversation of Feynman with the historian Charles Weiner, the latter mentioned casually that Feynman's notes were "a record of the day-to-day work" and Feynman reacted sharply (verbatim form Gleick's book, p. 409):

> "I actually did the work on paper", he said.
>
> "Well,'' Weiner said, "the work was done in your head, but the record of it is still here."
>
> "No, it's not a *record*, not really. It's *working*. You have to work on paper, and this is the paper. Okay?"

Feynman experienced no border between his thinking, conceived by himself as a semiotic activity, and the external support of his memory, that is, writing. Thinking is the writing; the writing is not an external image, a posteriori, of a thinking that took place in Feynman's brain. In other words, his cognitive activity is dynamically and inextricably distributed between the poles of the brain and the semiotic space of his writing. As Otte (2006) has clearly stated,

> A mathematical object, such as a number or a function does not exist independently of the totality of its possible representations, but it must not be confused with any particular representation, either. (p. 17)

The problem with the determination of mathematical objects comes from the fact that we never exhaust the set of its possible semiotic representations. The mathematical object is therefore, always, under construction. It is an *unfinished object.* It is not surprising then, that a thinker of the stature of Rene Thom (1972) had expressed in his plenary lecture at ICMI 2:

> The real problem that confronts mathematics teaching is not that of rigor but the problem of the development of *meaning*, of the existence of mathematical objects.

Obviously, we cannot approach this problem without recourse to semiotic representations, as they are the only doors of entrance to (deal with) mathematics objects. Only through semiotic systems of representations can we evolve mathematical ideas in a stable and durable medium. The external memory field can draw mathematical ideas, past and present, from the diverse systems of semiotic representations. Trying to establish connections between two or more semiotic systems used to represent a mathematical object is considered a basic cognitive activity in mathematical learning (Duval, 2006). Up to this point we have been implicitly discussing static semiotic systems.

Today, besides paper and pencil, mathematical learning and teaching have at their disposal digital or dynamic semiotic systems of representations. This statement is not epistemologically neutral. Elsewhere (Moreno & Hegedus, 2008) we have described how static and dynamic technologies differ in terms of the media they are instantiated within and the interactive malleability between the use and the environment. We have described this in 5 stages focused on the inertness of the media, the computational nature of the environment and its executability, and the affordances to co-act with the technology (i.e. the user can guide and be guided by the technology).

We will dedicate the remainder of this chapter to explain and substantiate the assertion that dynamic technologies (more so than static ones) can transform and impact the communicative infrastructure of the mathematics classroom. We focus on technologies that have unique features described above.

SEMIOTIC MEDIATION IN A DYNAMIC ENVIRONMENT: FIRST ENCOUNTER

We first present our classroom studies of the mathematics of variation and variables in a dynamic geometry environment. Dragging, a co-action between the intentional cursor-driven actions of a user, and the guidance of its recipient environment, is a mode well adapted to explore variation and change. It becomes important as well to develop a symbolic level that opens the window to a genuine understanding of these instrumental concepts.

Students will be able to deal with perceptual and cognitive difficulties as dragging allows to identify altogether the elements that change in a Cabri-figure (a dynamic geometry software environment) as well as those that remain invariant while a figure is being dragged. In the context of geometry instruction, this facility has been used as a tool to distinguish between a drawing and the corresponding geometrical object—the deep geometric structure corresponding to what you see on the screen.

The immediacy of perception makes one believe that this is a universal and homogeneous ability, installed ready to help us process information whatever be the environment. Yet, this is not so: one can recognize a familiar face among others, but decoding a functional graphic, for instance, poses a higher cognitive demand on students. Recognizing variation is an even harder problem. We will now review new findings on these research themes. Excerpts from two interviews will exhibit some of the cognitive difficulties students face when dealing with these problems.

The Voice of the Students

Some interesting cognitive behavior was made apparent in the interview with Estela, a 16-year-old student from Mexico. She is a bright student with a good working knowledge in mathematics. The instructor poses the (former) question concerning the largest area of the triangle in these terms:

> An isosceles triangle is inscribed in a circle; one of the triangle's vertices A, coincides with an endpoint of a diameter and the remaining two vertices B, C, are the endpoints of a chord BC that is perpendicular to this diameter.

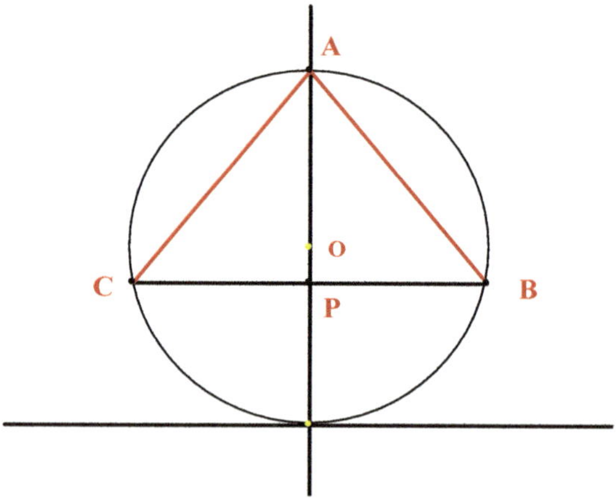

Figure 1. Searching for the largest area

The instructor makes clear that the point P can be moved along the diameter. When the point P moves along this diameter, the chord BC also is moved and remains perpendicular to the diameter.

Instructor: When I move point P what happens to the triangle? Does its area change? Does it remain the same?

Estela: The shape changes…it seems that the triangle remains as isosceles…but its area is the same.

Instructor: Can you explain?

Estela: Because here (pointing to the sides AB and AC) we are making the triangle smaller and here (indicating BC) we are making it larger…

The instructor continues, asking if there is a position of point P wherein the area changes, where the area is smaller or larger, and the student's response is:

Estela: No, the area remains the same…(Then she draws, on the slate, a triangle with BC very short)…for instance in this triangle, BC is smaller but AB and AC are much larger…(Figure 2)

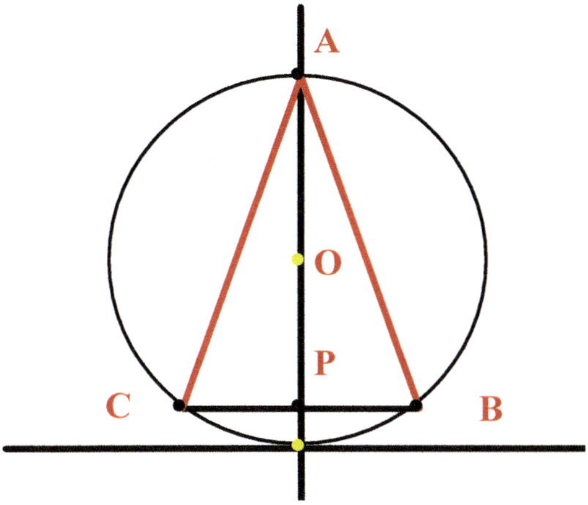

Figure 2. Exploring area changes

The instructor asks Estela to construct the figure on the screen (students were using Cabri Geometry™). Afterwards, she begins displacing point P (upward) along the diameter without reaching the center O of the circle.

Instructor: Does the area change?

Estela: No…it does not.

Instructor: Why?

Before answering, Estela displaces P, very slowly, without reaching O.

Estela: If the side BC grows, the sides AC and AB are decreasing…

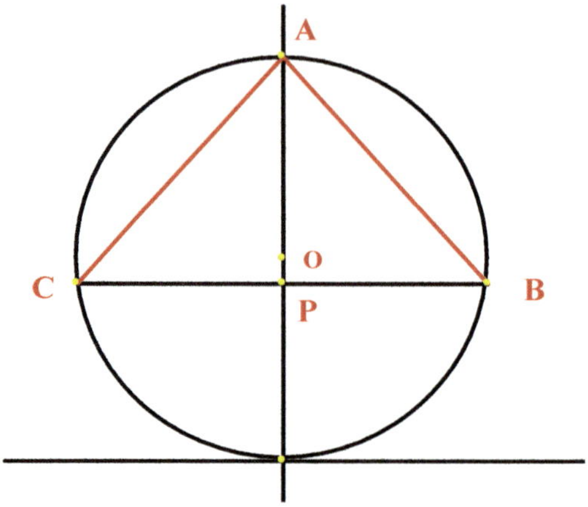

Figure 3. Looking for conditions of area changes

Instructor: The area of the triangle…does it ever change?

Estela: No, because…

Then, Estela moves P towards O and, *for the first time during the interview*, P goes beyond O until almost touching A, and then drags P back beyond O. *At that very moment she discovers something new*, and says:

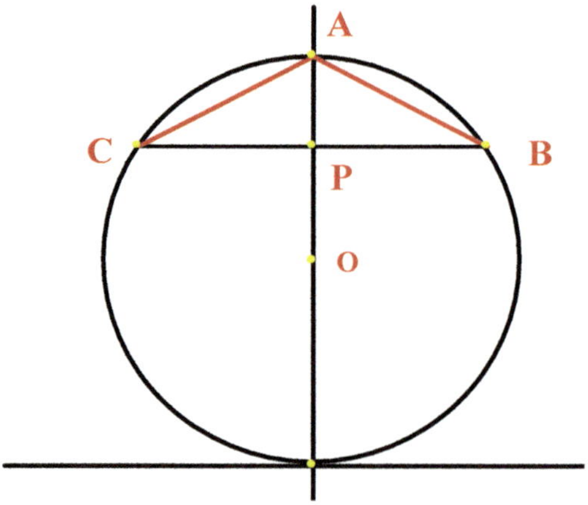

Figure 4. Dealing with a cognitive conflict

Estela: When I drag P beyond O, AB and AC gets shorter and BC also gets shorter…(Estela seems concerned and she drags P again. It is clear that something is disturbing her, attracting her attention. *Her perception is changing...*)

Instructor: What is going on?

Estela: I was looking at the area…it…it decreases, *the area decreases*!

Instructor: When does it decrease?

Looking at the chalkboard she says:

Estela: If I drag P beyond O, the area is smaller.

After the interview, Estela remarked: “I discovered that the area would decrease because, if the base BC and the sides AB and AC are all smaller (than previously), then the area has to be smaller. At the same time, if I drag P from A, without reaching O, the BC gets larger, AB and AC also gets larger, so the area grows.” Estela has seen the simultaneous variation of three magnitudes when she displaces P along the diameter. Yet what she observes on the screen is not the variation of the area itself (which cannot be *seen*) but an *index* (Deacon, 1997, p. 71) of this variation, a sign that triggers an inference: *the area becomes smaller if the sides become shorter*.

What is an index? When a bird sings, an adult, by the side of a child, will indicate the bird and will utter: “bird.” By pointing the bird out, the adult creates a level of reference—an *indexical* level of reference. Next time the child hears the song of the bird, she will utter “bird” even *without seeing the bird*. She has established a connection between the presence of the song and the word “bird.” This level of reference will become key in the process of *symbolizing* the word “bird.” The reference field of a symbol has a potential infinitude of levels of reference. Some are iconic, others are indexical. Sometimes Estela is looking *at* the (static) figure, sometimes, later, she is looking *through* (the movement) the *executable* representation, a figure that finally *reveals* what she has been hiding.

We said that Estela perceived on the screen an *index of the variation of the area* that triggers an inference. This inference is her first level of understanding of the variation of area. Here is where her perception goes beyond a mere act of “encounter the drawing with the eye.” It becomes an *act of visualization*, that is, a sensorial phenomenon controlled by an interpretation.

Within a dynamic-digital environment there are elements that can be used to explore variation and variability with a specific intentionality generating co-action among the different semiotic systems. The mathematical concept under (re)-construction (or appropriation, if you wish) slowly emerges from this semiotic activity. The executability of the representation enriches the pattern of social interaction (there is a significant place for gestures, for instance, when the exploration takes place within a dynamic environment). In this study, the whole dynamic scenario works as a domain of abstraction for the concept of variable. Eventually the students can perceive the general (variation) within the particular (movement). There is a webbing of (partial) meanings that reminds us of C. Geertz concept of culture (Geertz, 1973; Dourish, 2001):

> The concept of culture I spouse...is essentially a semiotic one. Believing with Max Weber, that man is an animal suspended in webs of significance he himself has spun, I take culture to be those webs and the analysis of it to be therefore not an experimental science in search of law but an interpretive one in search of meaning. (p. 57)

The students begin to construe these webbings of meanings from the activity within an environment rich in expressivity and rich in possibilities for all their (embodied) semiotic activities. Students act through the mediation of the dynamic environment. The processing is embedded in the digital instantiations of mathematical objects—opening new ways to conceive of these objects, perhaps opening a different epistemology. It is our conviction that educational problems and conceptions are not independent, in a substantial way, from epistemology. The dynamic nature of the technology environment in the above examples provides an executable representation that enables new forms of knowledge construction and growth, both by the individual learner and the social interaction of learners.

EXPLORATIONS AND INFILTRATION OF PROOF

Imagine a traditional classroom confronted with this problem:

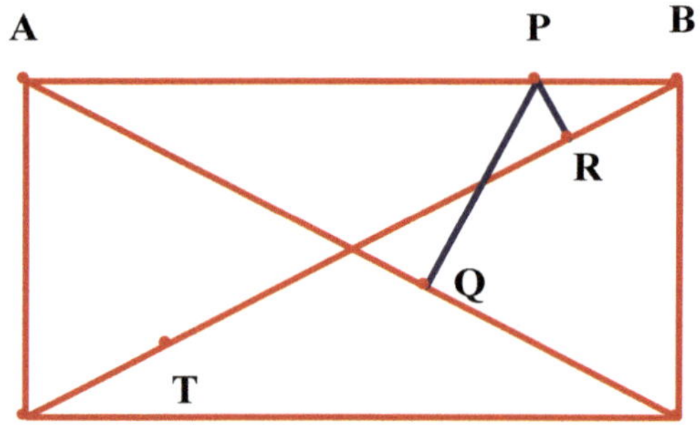

Figure 5. PR+PQ=constant?

Prove that the sum of the length PR and PQ (perpendicular projections to the diagonals of a rectangle) is a constant. This is the kind of problem that leaves students perplexed. A constant? What constant? And there they go. However, when the problem is posed in a dynamic environment (a classroom can become a dynamic environment), the *expressive capacity* with which students are empowered thanks to the executability of the dynamic system of representation, eventually takes them to the realization that dragging the point P along AB can lead to this situation:

There is, consequently, a symmetry hidden in the problem. *They can stay close to B*. This problem was solved, along these lines, by students in an undergraduate Mathematics Inquiry course for seniors at an American University. With a paper and pencil environment, students were not able to formulate a conjecture. The dynamic environment transformed the mathematical identity of the students who were able to share ideas and criticisms openly.

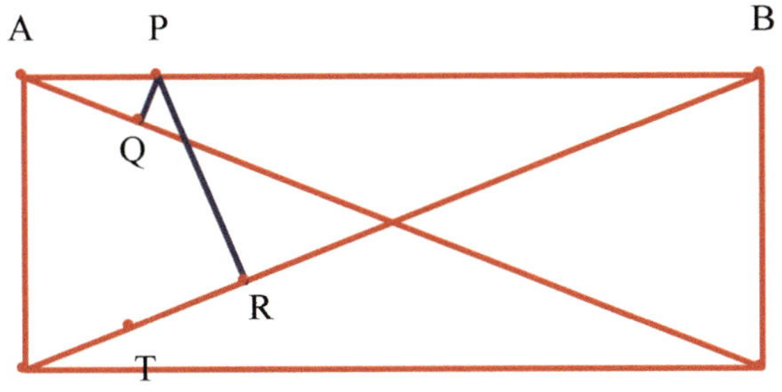

Figure 6. Dragging point P

Now, staying close to B, suggest, eventually, taking P to the position B. This has the effect of revealing a conjecture: the constant is BQ (see figure 7).

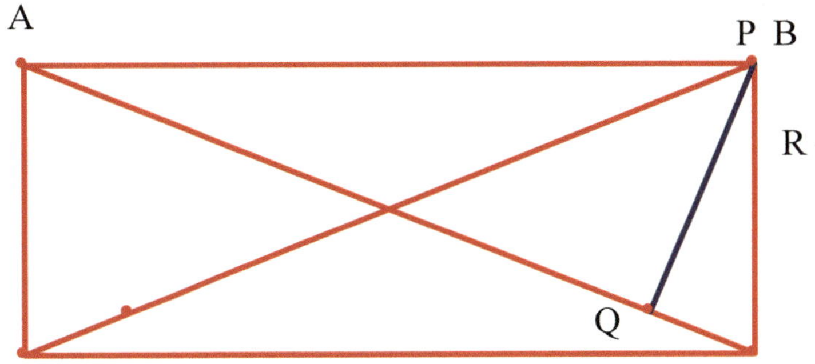

Figure 7. The conjecture is revealed

The expressive capacity and *the plasticity* of a dynamic environment, induce new strategies to explore and conjecture. We could illustrate this assertion with more examples, however, we want to emphasize rather, the mathematical world that is open to dynamic explorations within a dynamic environment. Let us illustrate this with the pedal curves to conic sections. The following figure illustrates the pedal curve (we are simplifying the language here) corresponding to an ellipse, which is the Limaçon (snail) de Pascal:

A whole geometric world is revealed when we drag the focus F2 to the outside, an action whose outcome is *the family of pedal curve*s corresponding to a hyperbola (this way of obtaining the hyperbola is part of the exploration). In a dynamic world, a curve is conceived of as being a member of a family of curves, because this family, as a semiotic representation embodies, crystallizes, an intentional action. It is an instant in an evolutionary process.

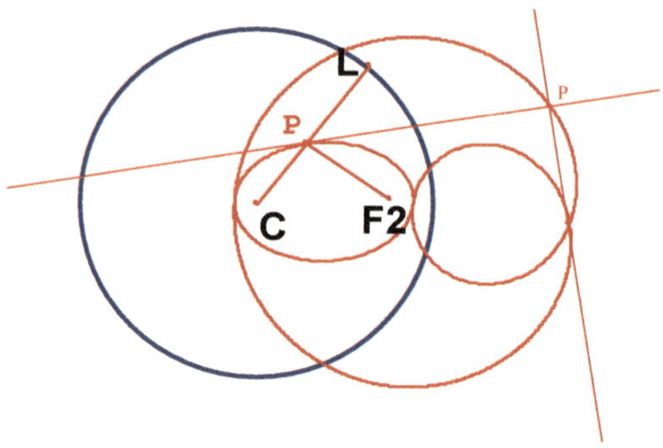

Figure 8. The pedal curve

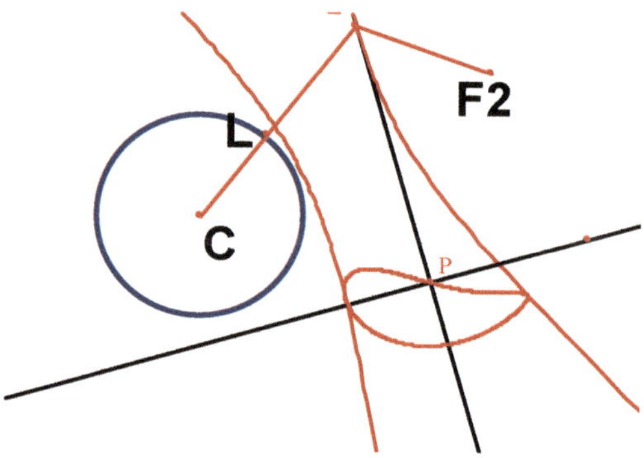

Figure 9. Creating a family of pedal curves

Let us introduce an example to illustrate how we can *infiltrate* the classical methodology of proof. Let us consider the Napoleon Theorem: *Given any triangle, we construct the corresponding equilateral triangles on the sides of the original triangle. The triangle obtained by connecting the centers of the equilateral triangles previously constructed, is always an equilateral triangle.* Such is the content of this classic result. It is very easy to persuade a student of the validity by dragging any vertex (A, B or C) of the original triangle:

The shaded triangle is Napoleon Triangle. Dragging any vertex A, B or C indicates that the napoleon triangle is equilateral; *the activity crystallized in this dragging is indexical.* But there is more knowledge already embodied, infiltrated, in this semiotic representation. To make a long story short, let us describe how to proceed to obtain a valid proof **within** the dynamic environment. Given X and Y, we can construct the third vertex of the equilateral triangle whose side is XY. When we place the pointing device close to point Z, we will receive the question: What point? Because the recently constructed point is occupying the same position

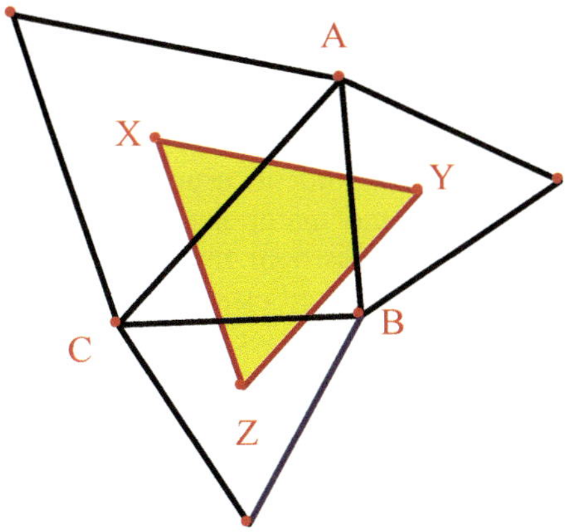

Figure 10. Considering the Napoleon Theorem

as Z. Can we separate these two points *in the environment*? No, we cannot because both are dependent on previous constructions. Being non-separable means they are the same point. This proves that the Napoleon triangle, XYZ is equilateral. This is a new culture of proof and it is up to our community to decide the place this kind of infiltration will have in the future. We might be touching a very delicate epistemological locus. It is timely to remind ourselves of Rotman's (2000) reflections on this issue:

> Is it unnatural or deviant to suggest that immersion in a virtually realized mathematical structure...be the basis for mathematical proof? Would not such proofs, by using virtual experience as the basis for persuasion, add to, and go far beyond, the presently accepted practice of manipulating ideograms and diagrams in relation to an always invisible and impalpable structure?...
>
> Such a transformation of mathematical practice would have a revolutionary impact on how we conceptualize mathematics, on what we imagine a mathematical object to be, on what we consider ourselves to be doing when we carry out mathematical investigations, and persuade ourselves that certain assertions, certain... a "theorem" for example would undergo a sea change (p. 68–69).

So it is important for us to state that the affordances of a dynamic technology environment are not solely confined within the environment in terms of potentials uses, tools and computational functionality, but in the impact of communication, interpretation and reasoning at a social level. We now present a final set of classroom vignettes where the technological infrastructure of dynamic mathematical representations is infused within a specific communication infrastructure yielding an enhanced form of expressivity.

CLASSROOM CONNECTIVITY AND SYMBOLIC DISCOURSE

Earlier we posited that with dynamic media we never exhaust the set of possible semiotic representations, and that a mathematical object is always under construction. But how does it come to exist? And more specifically, what does the existence of a mathematical object mean for a student, especially when they are more actively involved in the construction, and manipulation of mathematical structures? We now offer an exposition of a classroom that uses a different piece of software, SimCalc MathWorlds® in a networked classroom, that is part of the same genre of mathematics software that dynamic geometry environments are situated within. Particularly, SimCalc allows direct, interaction with graphs and function expressions that are connected to animations. Such forms of representations are executable in that a student or teacher can replay, or step through animations to discuss various aspects of the underlying algebraic structure and variation, e.g., what does it mean for a slope to be negative, where do the two motions/graphs intersect? In addition, SimCalc has built its representational infrastructure (Kaput 1994) on top of a new communication system (wireless networks) to allow the passing and sharing of mathematical objects within the classroom, guided by activities that are mathematically meaningful. By this we mean that a student's contribution is part of a larger set of objects, e.g., a family of functions, or a set of varying motions. When these objects are aggregated from the device to a teacher's computer that is projected onto a public display (e.g., a whiteboard), then the actions are transferred from a local, private space, to a public, social work space. In doing this, in a structured way, the communication infrastructure of the classroom is transformed. Not only is the focus of communication or interaction in a public space focused on what a teacher asks a question about, but it is also on the executable representations that a teacher controls in the public space, and that all students have contributed to. In fact, the environment allows for teachers to gradually show or hide contributions (across representations) in order for the structure to emerge in a personalized way. In doing this, we have discovered new forms of expressivity, both verbal and non-verbal, to surface in classroom discussion, that is symbolic in nature. Students create ways of expressing their perception of the general, in ways that shift from informal modes of analysis (metaphors, gestures, deixis) to eventually, formal structures more akin to the traditional study of algebra (e.g., $y = mx + b$).

Let us explain how this can occur with a brief description of the classroom set-up and a mathematically meaningful activity.

The SimCalc software allows students to create mathematical objects on graphing calculators and see dynamic representations of these functions through the animations of actors whose motion is driven by the defined functions. Students are then able to send their work to a teacher computer. Calculators are connected to hubs that wirelessly communicate to the teacher's computer via a local access point. Due to advances in wireless communication and interactivity between desktop PCs and hand-held devices, the flow of data around a classroom can be very fast allowing large iterations of activities to be executed during one class. This is not just an advance in connectivity, but in the development and application of software that maximizes such an innovation.

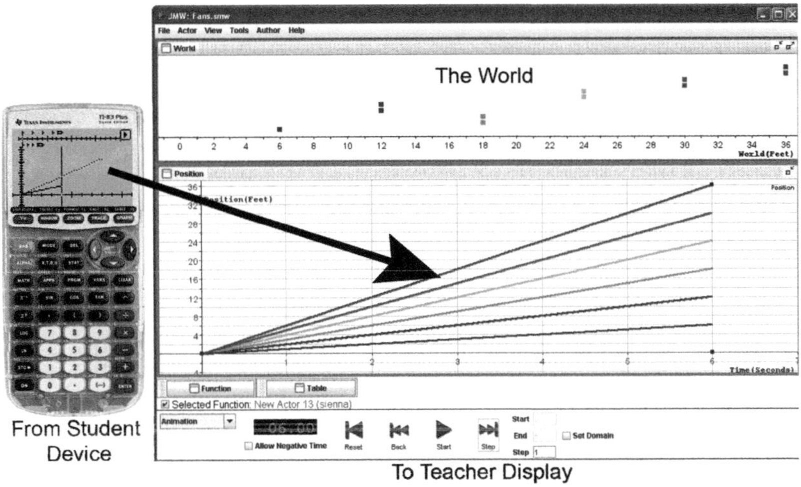

Figure 11. Sample Function in SimCalc MathWorlds®

We have worked to create activities that allow students to create functions in SimCalc on the TI-83 Plus or TI-84 Plus graphing calculator which can then be collected (or "aggregated") by a teacher into the SimCalc software running in parallel on a computer using TI's Navigator Wireless network. The activities are part of a curriculum developed and refined over many years that focus on core high school Algebra ideas such as linear functions, simultaneity, covariation and slope-as-rate vs. slope as m in $y = mx + b$ that utilize Classroom Connectivity (CC) in new ways to supplement or replace existing traditional algebra curriculum (Hegedus & Kaput, 2004). Such curriculum accompanies the software and is structured to take advantage of the natural social setup of the classroom to create variation.

Students, in small groups of three or four, have a group number (e.g., 1 through 5) and are asked to create a position function so that the motion of a character moves for a duration of 6 seconds at a speed equal to their group number. So Groups 1, 2, and 3 create $y = x$, $y = 2x$, and $y = 3x$ respectively for a domain of [0,6]. The important concept is slope as rate—an underlying concept of the mathematics of change and variation—and a family of functions is created by the class via varying the parameter k, in $y = kx$, where k is group number. The group number is an index to naturally vary the contribution set of the students. The variation becomes meaningful for the students; the family of functions is a result of their independent contributions (see Figure 11). There is an *indexical relationship* between the variation of the mathematical objects and the classroom setup (groups of students).

By default, student work is hidden. Student contributions are revealed at the discretion of the teacher; this allows students to conjecture and make generalizations about the class' work across representations, and about how their contributions relate to the class' set of contributions before seeing the variation. The conjectures and generalizations are prompted via curriculum and teacher questions such as, "What do you expect to see in terms of motion (or graphs) for the whole class?" This type of mathematical activity has brought about new forms of participation in the classroom. While analyzing the responses in terms of classroom participation using the SimCalc software, we have noticed a combination of mathematical (sometimes metaphorical) speech accompanied by gesture as students reason, predict, and make sense of the family of functions.

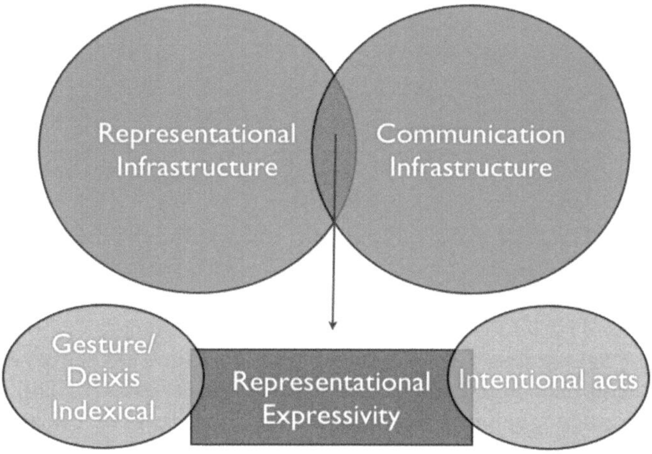

Figure 12: Representational and Communication Infrastructures

In this environment, the intersection of the representational infrastructure and communication infrastructure (each of which can exist independently) is especially of interest because this is where we observe new forms of mathematical expressivity (see Figure 12). These are scaffolded by forms of reference that evolve at rapid rates in the classrooms we have worked in. In such cases, students refer to objects with various registers to point at particular features verbally or attribute ownership of the objects at a public level (a form of deixis). For example, "his dot" (the actor which is animated and connected to a graph that some other student, made), "that actor" (something that the student created but thought of as an external artefact of their work), the "green graph," our graphs, my "actor," or even "me" (as a reference that the student personally thinks that they are "up there" in the public display space. Each of these references can sometimes occur within a few lines of classroom dialogue, with often no apparent confusion for the class, as they continue to discuss the overall mathematical structure that these multiple representations "point at."

We illustrate this phenomenon with a classroom vignette focused on gesture as a means of mathematical symbolization, and flow of discourse to represent the co-action between personal and group-based interpretation.

In this classroom, we have asked students (with the above set-up) to create a motion graphically that starts at twice their group number and ends in a tie with Actor A. After running the animation Actor A runs for 2 feet per second for 6 seconds or is described algebraically as $y = 2x$ on a domain [0, 6].

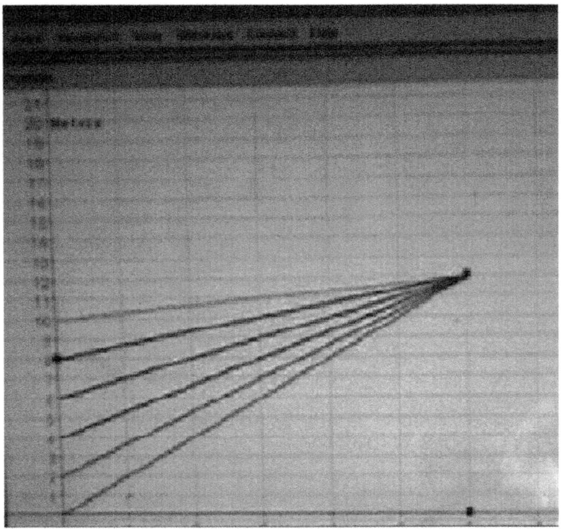

Figure 13. A screenshot of the graphs the students are trying to predict before they are displayed

In the following analysis, JS is the Teacher and we refer to the student as S1... S4. JS, in the beginning of the clip, has the world displayed on the board and asks the students what the graphs, describing the actors' motion, are going to look like. S1 immediately responds using hand gestures. She identifies five lines, each one displayed in the pictures below (there are actually six graphs; one is for Actor A, the target function, and is the only actor that does not represent a student).

It becomes clear in Figure 15 that S1 understands, at this point, the two lines will be stacked upon one another (perhaps an indication that the graphs have different y-intercepts based on the actors starting position in the world) and share a common endpoint. She alternates between her left and right hands in Figures 16-18 (S1's right hand is behind her left in Figure 17) to show the rest of the lines, although she seems to have lost focus of orientation. That is, if you super-impose her gestures on to a coordinate plane her actions do not illustrate what the graph will look like when displayed to the class (refer to picture A at the beginning of this document). Instead, she has accurately displayed and internalized the *shape* of the graph.

Figures 14-18. Hands as Graphs

However, seconds later, S1 alters her gesturing to include both her forearms and hands to describe the orientation of the graphs more accurately:

Figures 19-23. Hands and arms to orient graphs

The basic structure of S1's gesture remains; now she stacks her hands, alternating between left and right, on top of each other where her arms move in a clockwise direction. With each new representation of a graph, S1 is working her way up the y-axis and uses her forearm and elbow in Figure 23 to emphasize it's height. However, she does begin her demonstration in Figure 19 with an elevated arm compared to what is seen in Figures 20-22 when a steeper incline (achieved by lowering the elbow) is expected.

JS proceeds by asking the class what the "start" of the graphs will look like, or their initial positions. S1 responds:

> S1: Different points right there [left hand] and it's gonna go up [draws a line from her left hand four times, all with different heights]. (See Figure 24.)

Figure 24. S1 is now representing the lines by tracing them with her finger in the air

Figure 25. Orienting her hand 5

JS summarizes what the graphs will look like, saying that every two feet there will be another line starting and they all end at the same spot. When prompted by JS, a student says the graphs will look like a sideways triangle. JS asks for more ideas and S1 makes yet another hand gesture to explain the graphs. This time, she uses her fingers to denote each line whose common endpoint is her palm. As she is actualizing this new representation with her fingers, she holds her left hand in front of her face giving JS an accurate prediction of what the graphs will look like. S1 then twists her arm so that the image she is displaying portrays an accurate representation of the graph consistent with her point of view before changing to her right hand, seen in Figure 25. JS prompts S1 for a verbal description of what she's doing with her fingers. S1 mimics the shape of a coordinate axis with her left hand (Figure 26a) and then twists her body to show JS (Figure 26b). JS prompts S1 to identify what each part of her hand represents: index finger and thumb represent the y- and x-axis respectively, and her palm represents "the dot." When JS asks her what dot, S1 replies "12" (the lines meet at the point (6, 12) on the coordinate plane).

Figure 26a. Using hands to "create" a coordinate axis

Figure 26b. Expressive movement of body

JS displays the graphs of the students' lines on the board. It is immediately made public that the graph of the line for Group 4 has the correct initial position but incorrect ending position, and the students were able to instantly identify them. JS proceeds to identify all of the groups by color and/ or pointing (See Figure 27). For example, she starts by asking, "Who is the blue one (Actor A)?" After receiving a response, JS points to the remaining lines in sequential order. Almost simultaneously, upon resting her finger on each line, the students identify which group it belongs to.

JS asks the class if the graph looks like what S1 had predicted and S1 chimed in, "I didn't have enough fingers!" JS continues to review the imagery the class discussed to describe the shape of the graph (sideways triangles) and offers one of her own: a fan. A student argued that there were not enough lines to make it a complete fan. JS, in her overview, makes the point that although people in the class describe the image of the graph differently, all interpretations are correct.

It is interesting to note the progression of S1's gesturing to express the same image. She clearly prefers gesturing to verbal communication. Her first gesture involved her hands to represent the lines without a distinctive sense of direction (Figures 14-18) but she quickly improves upon this in Figures 19-23 where her forearms moving in a clockwise direction signifies the increase along the y-axis for each new line, utilizing her elbow to accentuate the height.

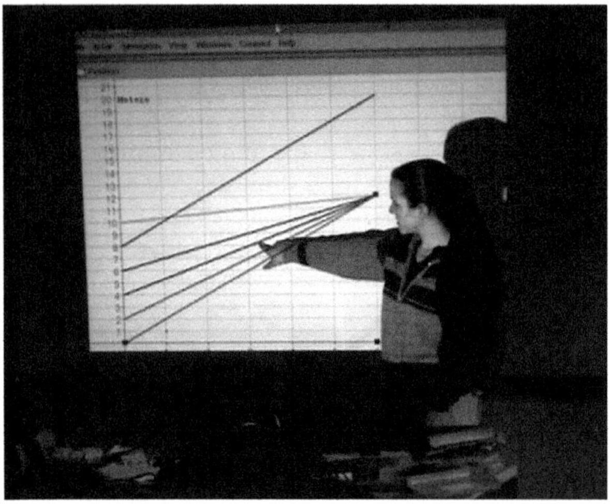

Figure 27. Teacher examining whole class work

It is as if she is not satisfied with this representation, and replaces hand gestures (that represented each line) by tracing the line out with her finger in the air, always starting from the "common endpoint" signified by her fist.

S1's last hand gesture is the most clear in describing the graph of the lines. She mimics the coordinate axis with her left index finger and thumb while using the fingers of her right hand to display the lines where her palm is the common endpoint. Here S1 establishes orientation; it is clear that from her perspective the image is accurate, and she clearly defines each distinct initial position of the line.

In summary, even though students are talking about historical representational forms of functions (i.e. graphs) in this heavily technology-mediated classroom, the implementation of classroom connectivity has enabled the sharing and coordination of personally meaningful objects by students. In transforming the communication infrastructure in the classroom (with dynamic technologies) we have described the impact on speech, action and reference, which in turn impacts some of the most basic expressive acts of humans (e.g. gesture and deixis). Further evidence of the impact on student content knowledge can be found in Hegedus, Kaput, Dalton, Moniz and Roschelle (2007).

SYNOPSIS: NEW APPROACHES TO SEMIOTICS IN A DIGITAL ERA

The evolution of digital media has had an impact on semiotic representations, and hence the mode of representation, inference and interpretation of mathematical ideas. This, quite conceivably, has presented us with a new realm of epistemology with which to make sense of mathematical signs, objects, and concepts as teachers and students, but especially how we share these meanings within a well-appreciated "sign-system". We especially think this is of relevance for new teachers in the 21st Century who aim to help re-conceptualize mathematical ideas in enhanced socio-cognitive ways. The illustrations from our work present some basic tenets for a new perspective on analyzing some fundamental epistemological building blocks in the semiotic culture of mathematics education. This begins with the evolution and conceptualization of the profound potential of media—from static to dynamic. This is not a simple, expected, or linear evolutionary trait, but one that has challenged the very bedrock of how we establish "concrete" ideas such as proof, reasoning, or the establishment of mathematical structure, hitherto an inheritance, or an adoption of beliefs. We believe that the infrastructures of the digital technologies presented here, both in terms of representational land-communicational affordances, offer more than a new transition to helping students understand "concrete" mathematics, but instead allow them entry to a transformed educational environment to discover. Discovery has often been described as a democratic medium, in terms of making "free" a resource to the under-privileged, but we see a digital mathematical media to be one where the medium can allow new knowledge to be discovered in whichever context it can be discovered. Many students have presented solutions to complex "visualization" problems through more malleable media. Just like our ancestors found the ability to express complex ideas through the media around them, we should now appreciate the profound abundance of ideas possible in digital media, not just in terms of re-representing old ideas, but in transforming fundamental ideas in proof and algebraic structure. Our examples have aimed to provide a transitionary story to describe the potential of transforming these basic infrastructures. For us this is an evolutionary perspective. We have not tried to present our ideas as a raison d'être for the future of digital education, but instead consider what, in driving this digital revolution, has impacted the study of signs, and particular symbolic cognition for students in our studies (which range from students in the early grades to teachers in graduate school).

Our model of co-action is a way to express the new medium of expressivity, to allow, through plastic and more elastic modalities, a venue for co-constructing ideas, and building understanding and symbolic meaning. We do not wish to promote a "fluid" media, which relies on computational expertise (e.g., Computer Algebra Systems) but an environment, which allows intentionality to be crystallized in digital actions with and within the interface. We hope that our examples explain that the inter-face is not one between a user/thinker and a screen/programmer but one that is an elaboration of the communicative enterprise that all symbolic meaning is based upon; and that is expression. Our computational ally, the processor, helps us, through the educational intentions of most developers of these environments, to describe opportunities to execute rich representations. These are often built by the users (in our examples) and are not a computational artifact (in a sense that all operations are computational) but semiotic in the sense that give rise to new meaning in the sign structures they allow. These have been exemplified in brief through the basic action of dragging a "rubbery" geometric figure, loose in its connections, but tied to its establisher's connections to understand variation under dragging. And dynamic in the sense that its variance or invariance can be described by a larger corpus of cognition (at the whole class level) as they devise the overall gestalt of an algebraic structure.

REFERENCES

Bruner, J. (1990). *Acts of meaning.* Cambridge, Massachusetts: Harvard U. Press.

Deacon, T. (1997). *The symbolic species.* New York: Norton & Company.

Donald, M. (1993). *Origins of the modern mind: Three stages in the evolution of culture and cognition.* Cambridge, Massachusetts: Harvard University Press.

Dourish, P. (2001). *Where the action is.* Cambridge, Massachusetts: MIT Press.

Duval, R. (2006). A cognitive analysis of problems of comprehension in a learning of mathematics. *Educational Studies of Mathematics, 61*(1–2), 103–131.

Gleick, J. (1992). *Genius.* New York: Pantheon.

Geertz, C. (1973). *The interpretation of culture.* New York: Basic Books.

Hegedus, S., & Kaput, J. (2004). An introduction to the profound potential of connected algebra activities: Issues of representation, engagement and pedagogy. In P. Gates (Ed.), *Proceedings of the 28th Aannual Mmeeting of the International Group for the Psychology of Mathematics Education* (Vol. 3, pp. 129–136). Bergen, Norway: Program Committee.

Hegedus, S., Kaput, J., Dalton, S., Moniz, R., & Roschelle, J. (2007). *Understanding classroom interactions among diverse, connected classroom technologies—Overview of the present findings of a 4-year study.* Technical Report 1:1. Dartmouth, MA: James J. Kaput Center for Research and Innovation in Mathematics Education.

Kaput, J. (1994). Democratizing access to calculus: New routes to old roots. In A. Schoenfeld (Ed.), *Mathematical thinking and problem solving* (pp. 77–156). Hillsdale, NJ: Lawrence Erlbaum Associates.

Moreno-Armella, L., & Hegedus, S., & Kaput, J. (2008). Constitution of symbols and the evolution of the reference field with digital technologies. *Accepted for publication in Special issue of Educational Studies in Mathematics, Volume 68: Democratizing Access to Mathematics through Technology— Issues of Design and Implementation.*

Otte, M. (2006). Mathematical epistemology from a Peircean point of view. *Educational Studies of Mathematics, 61*(1–2).

Rotman, B. (1993). *Ad infinitum... the ghost in Turing's machine. Taking God out of mathematics and putting the body back in.* Stanford, CA: Stanford University Press.

Rotman, B. (2000). *Mathematics as sign.* Stanford, CAalifornia: Stanford University Press.

Schmandt-Besserat, D. (1996). *How writing came about.* Austin, Texas: University of Texas Press.

Thom, R. (1972). Modern mathematics: Does it exist? In A. G. Howson (Ed.), *Developments in mathematical education* (pp. 195–209). Cambridge: CUP.

Wartofsky, M. (1979). *Models, representation and the scientific understanding.* Dordrecht, The Netherlands: D. Reidel.

Wertsch, J. (1991). *Voices of the mind.* Cambridge, Massachusetts: Harvard University Press.

Stephen J. Hegedus
James J. Kaput Center for Research and Innovation in Mathematics Education
University of Massachusetts Dartmouth

Luis Moreno-Armella
Cinvestav, Mexico
James J. Kaput Center for Research and Innovation in Mathematics Education
University of Massachusetts Dartmouth

ADALIRA SÁENZ-LUDLOW AND ANNA ATHANASOPOULOU

THE GSP, AS A TECHNICAL-SYMBOLIC TOOL, MEDIATING BOTH GEOMETRIC CONCEPTUALIZATIONS AND COMMUNICATION

INTRODUCTION

The paper argues the Geometer's Sketchpad (GSP) could be envisaged as a technical-symbolic tool that facilitates the conceptual passage from *drawing* to *figure* (i.e., from the particular to the general) mediating, at the same time, the shared and consensual interaction between teacher and students, the indirect but intentional guidance of the teacher through tasks specially designed for the GSP, and students' conceptualizations of hierarchical inter-relationships between the structural elements of geometric figures to classify them according to certain essential attributes. The argument is extended with the analysis of an interview (from a sequence of interviews) of a pre-service teacher illustrating how and why this tool mediated his geometric thinking.

THE GSP AS A TECHNICAL-SYMBOLIC TOOL

Several researchers (among those Balacheff, 1993; Hoyles, 1996; Laborde, 2003, Mariotti, 2001; Jones, 2001) have argued that dynamic geometric environments are in fact mediational means. For Vygotsky, higher mental processes are generated *in* and *through* social and meaningful mediated activity. For him, the source of mediation is either in *a material tool*, in *a psychological tool*, or in *the behavior of another human being* (Vygotsky, 1986). The concepts of *tool* and *tool mediation* are central to the Vygotskian perspective for the analysis of conceptual development. For Vygotsky, material tools are directed to produce changes in the object(s) on which they are applied and although directed at natural objects they also have a reciprocal influence on the cognitive activity of the individual who uses them. In contrast, psychological tools is a system of symbolic artifacts (e.g., signs, symbols, texts, and graphic-symbolic devices) that helps individuals to master their own natural psychological functions of perception, memory, attention, and so on; that is, psychological tools direct the mind and behavior of the individual (Kozulin, 1998). The GSP, as a graphic device, can be considered as a tool with characteristics of both material and symbolic tools since it mediates the transformation of physical activity (e.g., the construction of drawings on the screen) into conceptualizing activity (e.g., abstracting geometric figures) and subordinating the former to the later.

L. Radford, G. Schubring, and F. Seeger (eds.), Semiotics in Mathematics Education: Epistemology, History, Classroom, and Culture, 195–214.

One of the problems of geometry is that of determining *methods* for drawing instances of geometric figures with certain properties. These methods are called *constructions*. Paper-and-pencil constructions use geometric devices (straightedge and compass) while the GSP and other geometry computer environments use digital compass and some simple implemented constructions to aid in more complex constructions. The product of a construction by either method is called *drawing*. Different kinds of drawings (robust drawings and soft drawings) are widely mentioned in the literature of geometry dynamic environments (Healy, 2000; Laborde, 2001b; Laborde, Kynigos, Hollebrands, & Strässer, 2006): a *robust* drawing is a drawing that passes the dragging test otherwise it is a *soft drawing*. The GSP (through its dragging or animating capabilities) offers the means of constructing dynamic robust drawings and they have the potential of inducing learners to observe their variants and invariants attributes and in the process triggering the conceptualization of a *family* constituted by *all* robust drawings (actual or possible). This family is nothing else than a *geometric figure*. In this conceptualization, each drawing becomes an instance of a geometric figure instead of being confounded with the geometric figure itself. In other words, each robust drawing in the GSP constitutes itself as a perceptual engendering source of a geometric figure in the minds of the learners. It appears, then, that whether in paper-and-pencil or dynamic geometric environments it is important to consider the triad *construction-drawing-figure* differentiating and clustering together geometric elements that are intimately intertwined.

THE GSP AS A MEDIATIONAL TOOL

The GSP as a Tool Mediating the Passage from Construction to Drawing to Figure

Vygostky points out that his principle from "action to thought" should be applied not only to the development of intelligence but also to the functioning of intelligence (Wertsch, 1985). This principle implies a considerable difference between learning *how to operate* with concepts and *becoming aware of* the structure of those concepts and their relations with other concepts or what Vygotsky (1986) calls the degree of generality of a concept. These notions of "how to operate with" and "becoming aware of" seem to be essential in any geometric environment whether static or dynamic. Arsac (1989) and Laborde (1993) brought to our attention the differentiation between *drawing* and *figure*. *Drawing* refers to material objects on sand, paper, or computer screens; in contrast, *figure* refers to a geometric theoretical object. Such a differentiation was first pointed out by Plato in Book VI of The Republic. He clearly says that geometers use material objects or drawings while in their minds they see only conceptual, immaterial objects.

> I suppose you know that students of geometry, arithmetic, and so forth begin by taking for granted odd, and even, and the figures, and the usual figures, and the three kinds of angles, and things akin to these, in every branch of studies; they take them as granted and make them assumptions or postulates, and they think it unnecessary to give any further account of them to

> themselves or to others, as being clear to everybody. Then, starting from these they go on through the rest by logical steps until they end at the object that they set out to consider....And do you know also that although they make use of *the visible figures and give lectures about them, while they are not thinking of these, they can see but the ideas which these are like;* a square *in itself* is what they speak of, and a diameter *in itself*, not the one they are *drawing*. It is always so; the very things that they model or *draw* which have shadows on their own and images in water, they use now as images; but they seek is to see those *ideals* which can be seen only by the mind. (Plato, 1984, p. 310; emphasis added)

In modern terms of set theory, one could say that a *drawing* is a particular element of a class defined by extension or intension while *figure* is the whole class whether it is defined by extension or intension. A numerical example could be useful here. The even numbers could be defined by extension as the set {2, 4, 6, 8, 10, ...} or by intension as $\{x \mid x \text{ is a natural number and } x = 2k \text{ for k a natural number}\}$ or $\{x \mid x \text{ is a natural number and } 2|x\}$. That is, the relation between *drawing* and *figure* is simply the relationship between the *particular/concrete* and the *general/abstract*. In other words, to conceptualize a *figure* it is necessary to observe (in the dragging of robust drawings) the *variation* or change of some attributes (their nonessential attributes) as well as the *invariability* of other attributes (their essential attributes). The dragging and animation facilities of the GSP permit the construction of many robust drawings with the potential of inducing the abstraction of a geometric figure as a family. One special characteristic of this family is that one cannot see a finite number of elements all at once like in the set {2, 4, 6, 8, 10, ...} but one has to "observe the pattern" *in* and *through* motion. Therefore, it could be considered that geometric cognitive processes, in general, depend on elements of perception, visualization, and thought. Kant asserts that in order to abstract a concept we must hold together in one consciousness (as a unit) the manifold of perceptions contained in each representation and calls this *unifying act* of the mind *synthesis* (Wolff, 1973).

Also is important to notice the differentiation between *observation* and *collateral observation*. According to Kant there is in our observation, not only pure sense perception, but also certain amount of inference (Wolff, 1973). That is, observation takes into account only the intrinsic elements of the object being observed and produces an Interpretant (in Peirce's sense) of that object. In contrast, collateral observation links the product of that observation (Interpretant) with other knowledge related to the object being observed. Peirce (1909) clarifies this relationship between the "interpretant" and collateral observation.

> All that part of the understanding of the Sign which the Interpreting Mind has needed *collateral observation* for is outside the Interpretant. I do not mean by "collateral observation" acquaintance with the system of signs. What is so gathered is not COLLATERAL. [Observation] is on the contrary the prerequisite for getting any idea signified by the Sign. But by *collateral*

> *observation*, I mean *previous acquaintance* with what the sign denotes. (Peirce, 1909, p. 494, italics added)

Observation (a blend of perception and reasoning) as well as collateral observation (a blend of perception, reasoning, and prior knowledge) appears to assist the geometer in his abductive, inductive, and deductive reasoning. "The geometer draws a diagram...and by means of *observation* of that diagram he is able to *synthesize* and show relations between the elements which before seemed to have no necessary connection" (Peirce, CP 1.383, 1.888, emphasis added). That is, by collateral observation the learner of geometry could be able to relate the concept/sign represented or instantiated by the drawing with other prior knowledge. Peirce considers ten different signs (not isolated but dynamic and transforming themselves into more sophisticated signs) representing different degrees of iconicity, indexicality, and simbolicity *for* the interpreter according to his levels of generalization and therefore supporting his main hypothesis that knowing is a continuous process of refinement (Sáenz-Ludlow, 2007).

Davidov's dynamic notion of perceptions-conceptions-concept is compatible with Peirce's notion of dynamic sign. "A *concept* is abstracted from the individual features and attributes of particular perceptions and conceptions and is thus the result of a generalization of perceptions and conceptions of a very large number of homogeneous phenomena and objects" (Davidov, 1990, p. 19). The facilities of the GSP, if well used, could get underway the sequence "perceptions-*conceptions*-concept" in the learning of geometry. Through kinesthetic-visual activity the learner constructs and perceives a geometric robust drawing. This drawing represents an instance of a geometric figure with specific angles, shape, size, and position in the plane (e.g., a rectangle) and therefore a particular "conception" of a rectangle. In order to foresee which attributes belong to *all* rectangles, the learner can drag or animate any element of the particular drawing and generate a family of drawings with the same invariant attributes (right angles and opposite parallel sides) and differentiate them from the variant attributes (e.g., size and position in the space and length of sides). By doing this, the learner comes to abstract and conceptualize the indispensable or essential conditions of the whole class of drawings (i.e., the geometric figure called rectangle). As Davidov points out, "the general as something recurring and stable is a definite invariant of the diverse properties of objects of a given sort — that is, it is essential" (1990, p. 15). He also adds that generalization and abstraction are indispensable conditions for the transition from *conceptions* to *concept* and that the group of generalized attributes of an object constitutes the *content* of the concept.

The conceptual move from construction to drawing to figure is in essence the passage from the particular and concrete to the abstract and general. A concept is a symbolic sign, the highest of all signs in Peirce's categorization of signs. In a symbolic sign iconicity and indexicality are still present though superseded by a general type which had been agreed to be significant. What is general has its *being* only in the instances which will determine it. This symbolic sign is consolidated (in the mind of the interpreter) only when the learner recognizes or becomes aware of the relation of this particular sign with other symbolic signs within a particular conceptual system and between different conceptual systems. This is what Vygotsky (1986) called the *degree of generality* of a concept within different structures of generalization.

The conceptualization of a geometric figure is, in essence, a process of de-contextualization afforded by the dragging and animating capabilities of the GSP. This passage from *drawing* to *geometric figure* is nothing else than the passage from *the concrete and particular* to *the general and abstract*. As such, this is an illustration of the Vygotskian *principle of de-contextualization of mediational means* (Werstch, 1985) in which the meaning of signs (in this case *similar drawings* as indexical-iconic signs) becomes less and less dependent on spatiotemporal contexts and become more and more dependent on the meanings conveyed by the structural properties and relations between the constituent elements of the figure (a symbolic sign). In other words, the GSP has the potential of facilitating the endowment of indexical-iconic signs (material objects or drawings) with symbolic meanings to conceptualize the ideal object or geometric figure. That is, the GSP facilitates the emergence of *cognitive continuity* from the empirical, perceptual, and intuitive to the inductive, deductive, and abductive aspects of geometric thinking. This passage from action (*constructing drawings*) to thought *(abstracting-and-generalizing drawings*) is essential to the teaching and learning of geometry.

The GSP as a Tool Mediating Communication

Vygotsky considers that interpersonal relations serve as prototypes for intrapersonal relations; psychological functions, he says, first appear as actual interaction between individuals, and then become internalized as intra-psychological functions (Kozulin, 1998). Thus, the question is how to foster and facilitate the appropriate interpersonal relations based on geometric contextual situations that could later become internalized by the learners of geometry. The GSP is not only a technical-symbolic tool mediating between the student and geometry (as embodied in a digital semiotic medium) but also mediating the communication between teacher and student. Through this mediated communication, the teacher is able to assess the actual knowledge of the student and to hypothesize his potential growth or what Vygotsky (1978) has called a zone of proximal development (ZPD).

When students solve tasks structured for the GSP they become neither the passive recipients of knowledge offered by the teacher nor independent thinkers that arrive on their own at the conceptualization of geometric objects; rather, they participate in a shared learning activity with other students and the teacher. It can be said that the GSP is a technical-symbolic tool which enhances the establishment of a three-way communication between the students, geometry, and the teacher. This three-way communication allows not only for the sharing of interpretations and discussion of ideas but it also gives the teacher the opportunity to challenge students' interpretations and misinterpretations or misconceptions (usually called errors) as steps or missteps of their conceptualizing activity. Leont'ev and Brousseau consider errors and successes essential to the continuity of sense-making in the process of conceptualization. In fact, errors and successes are stopping-to-reflect opportunities in the refining process of knowing. "Man learns from errors — and still more from the successes — of other people" (Leont'ev, 1970, p. 123). "Errors are not erratic or unexpected; the error is a component of the meaning of the acquired piece of knowledge" (Brousseau, 1997, p. 82). Peirce argues that all cognitions draw upon former ones for their significance and therefore knowledge is a self-corrective process (Davis, 1972). Thus, implicitly he admits the value of errors in the process of knowing.

The GSP, as a mediating technical-symbolic tool, re-defines the channels for dialogical interaction between teacher and students and encourages students to modify their subjective contents and meanings of their conceptions and to re-organize their geometric knowledge in a continuous manner. The re-definition of the channels of communication also implies a shift from the teacher's monologue to the student-teacher dialogues and the concomitant shift of authority from "domination and obedience to negotiation and consent" (Amit & Fried, 2005, p. 164). This new way of looking at the teacher-student communication mediated by the GSP shifts the conceptual agency from the teacher to the conceptual agency of the student, and from the *passive* to the *active* participation of the students in their own learning. That is, the GSP makes possible a teaching-learning environment that is student-centered though directed by the teacher. Students' active participation in their own learning activity contributes to transform their geometric reasoning and, like any transformation, it must be developed through a consistent reasoning activity in a variety of geometric situations in order to consolidate and self-organize one's geometric concepts.

In summary, the GSP can be used as an effective technical-symbolic tool to promote student-centered teaching environments that focus on the development of independent learners who come to re-organize their concepts and to trust the authority of their own reasoning. The teaching of geometry, after all, should develop students' reasoning capability and their ability to build up logical arguments, to make and to investigate geometric conjectures, to develop and to evaluate geometric arguments and proofs as well as to recognize reasoning and proof as fundamental aspects of mathematics (NCTM, 2000).

The GSP as a Tool Mediating the Classification of Geometric Figures

As it was discussed above, a class of drawings is generated when dragging or animating vertices or sides of a robust geometric drawing in the GSP. Do all the drawings generated in this process have the exactly same structural properties as the initial robust drawing? It depends on the construction of the drawing being considered. When the robust drawing being dragged or animated is the drawing of a square, the class generated is constituted only by drawings of squares. When the robust drawing being dragged or animated is the drawing of a parallelogram constructed with two pair of parallel sides, the class generated is constituted by drawings of parallelograms, rectangles, rhombuses, and squares. When the robust drawing being dragged or animated is the drawing of any general quadrilateral, the class generated is constituted by drawings of convex and non-convex quadrilaterals. Among the convex quadrilaterals, parallelograms, rectangles, rhombuses, squares, general trapezoids, isosceles trapezoids, kites, and other quadrilaterals can be observed. How can students make sense of this variety of shapes and the relationships among them? Classification of figures requires putting them into groups or subgroups according to certain similarities and differences. With respect to classification, in general, Davydov says, "Correctness and completeness of classifications depend, naturally, on the precision and completeness with which the concept's essential attributes are *single out* and *coordinated.* One of the basic methods of classification is the establishment of *generic-specific* relationships, the delineation of the genus and the specific differences in concepts" (1990, p. 21, emphasis added). The following chart presents a classification of quadrilaterals with respect to a particular essential attribute, the number of pairs of parallel sides.

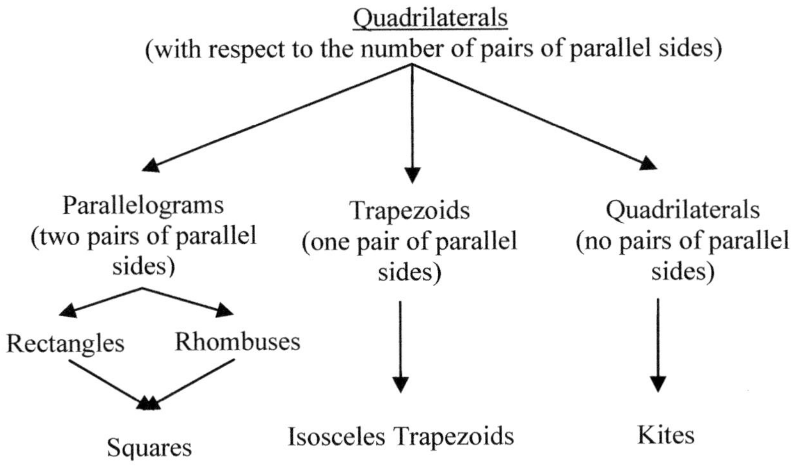

Figure 1. A classification of quadrilaterals

Therefore, in the context of geometry, the Aristotelian notions of *genus*, *species*, and *individual* are important for classification; this concepts are relative concepts. *Genus* is a class of figures with general properties while the *species* of a particular genus is a subclass of the genus with the same distinctive properties as the genus plus other specific and more restrictive properties. An *individual* is one element of a genus or a species, or one element of a class of things. In the classification of quadrilaterals shown in Figure 1, the figure square is a species of the figure rectangle and also of the figure rhombus. Going up in the hierarchy, the figure rectangle and the figure rhombus is a species of the figure parallelogram. Going up again in the hierarchy, the figure parallelogram, the figure trapezoid, and the class of other quadrilaterals with no parallel sides are different species of the genus class of quadrilaterals. Since genus, species and individuals are relative concepts, the position of the class one wants to consider could be seen as a genus of a lower class. For example in Figure 1, parallelograms could be considered as genus of lower order whose species are the class of rectangles and the class of rhombuses, thus the class of squares will be an individual of the class of rectangles as well as an individual of the class of rhombuses. And so on we can continue conceptualizing the chart.

A classification of geometric figures as a result of a classifying activity is a very important mental process that allows for the conceptualization of similarities, differences, and co-relationships among them to impose a hierarchy on a set of figures. The GSP facilitates the conceptualization of classifications through the dragging and animating capabilities. The differentiation between drawing and figure as well as between genus and species allows for the transformation of the commonly used expression "the square *is* a rectangle" into "the square *is a species* of the rectangle". In the first expression the student does not know if the definition and properties of the square are the same as those of the rectangle regardless of their different appearance. With the second expression, they come to understand that squares inherit the right angles from the rectangle but they have an additional condition on the lengths of the sides.

TASKS FOR THE GSP AS MEDIATIONAL TOOLS

The GSP, as argued in the prior sections, is a technical-symbolic tool that (a) mediates the passage from construction to drawing to figure, (b) mediates the communication about geometry between teacher and students, and (c) induces and facilitates classification processes. Nonetheless, the geometric tasks, proposed to the students also constitute a mediatory instructional tool to make significant their conceptual activity. The dragging and animation features of the GSP facilitate the construction of a plurality of robust drawings each of which comes to stand, metonymically, for a singular class (i.e., figure, or geometric concept). For example, the concept of triangle is the singular class or family of all triangles (a genus figure). By dragging or animating a robust drawing of a triangle and by observing the sequential drawings, the student is left with the traces of kinesthetic-visual experiences that constitute themselves into the perceptual genesis of the conceptualization of triangle as a geometric figure or geometric concept. Thus, one of the most important means of cognitive mediation of the GSP is its dynamic and versatile construction of singular drawings, that each time, present the student with a replica or representation of the class or family (e.g., a representative of the class called *triangle* or a representative of the geometric object *triangle*).

Robust drawings on the screen are sought out because of their usefulness to explore the properties of geometric figures, to make conjectures, and to gain insights into the solution of geometric problems. Hoyles (1996) and Jones (1999) have observed that manipulating *drawings* on the screen does not necessarily mean that the properties of geometric objects have been understood and appreciated. This observation is very similar to the one from Vygotsky about the differentiation between "how to operate with and become ware of" (Wertsch, 1985). The question is whether or not a robust drawing could be re-formulated as a semi-robust drawing to help the learner to become aware of properties and definitions of a particular geometric figure.

Tasks for the GSP could be unstructured, semi-structured, or structured. By *unstructured* tasks for the GSP we mean any task that gives no directions to the student of how to interact with the GSP. For example, "In a given triangle ABC, construct the median AM. In the extension of the median AM from the point M we take the point D such that AM is congruent to MD. Prove that triangles ABC and BCD are congruent". These tasks are in essence traditional textbook tasks designed to be solved by means of paper-and-pencil.

How can the teacher utilize the mediatory capabilities of the GSP to pose geometric tasks to guide students' geometric conceptualizations? To make the most of the GSP as a cognitive mediator, the tasks for the GSP cannot be identical to tasks designed to be solved by means of paper-and-pencil. A geometric task for the GSP should be a task that allows for exploration and open-ended questions. These tasks could be semi-structured and structured.

A *semi-structured task* is based on a *drawing of a lesser degree of robustness.* By a drawing of a lesser degree of robustness we mean a drawing in which some of its elements maintain a particular relationship while the rest of the elements are free to change. The drawing in Figure 2 has sides AB and CD parallel and O is the intersecting point of segments AD and BC. When any of the free elements of this drawing (vertices of sides) are dragged or animated, a family of quadrilaterals is generated. The task asks to *drag points* and *segments* of the figure to see *which* kinds of quadrilaterals could be constructed and *how* they could be transformed into isosceles trapezoids. Another purpose of the task is to investigate and prove properties of isosceles trapezoids and to inquire *how* the knowledge of these properties could be used in such a transformation. It is also worthwhile to notice that Figure 2[1], could also be transformed into a parallelogram or in any of the other species-figure with two pairs of parallel sides such as rectangle, rhombus, or square. In other words, this task could be multipurpose to make students aware of the importance of definitions and properties in the construction of robust drawings in the GSP environment.

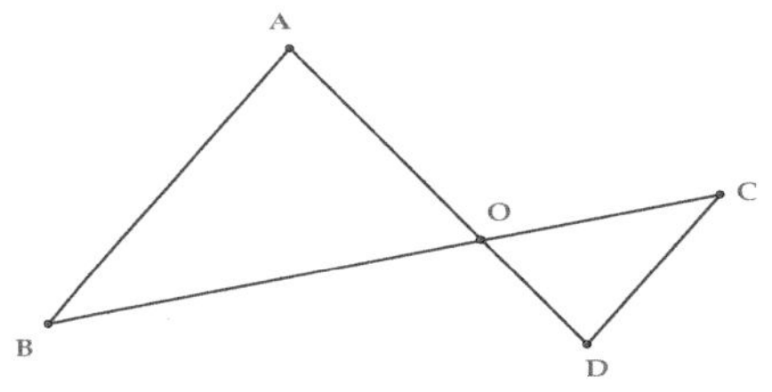

Figure 2. Semi-structured task

In contrast, a *structured task* is based on a *robust drawing.* As said before, dragging any particular robust drawing generates only one singular class or a plurality of drawings that belong to the same class. An example of structured task is the following.

(a) Construct a parallelogram ABCD with <A > <B.
(b) Construct the diagonals AD and BC.

(c) Measure <A, <B, BD, and AC.
(d) Drag any vertex of the parallelogram and tabulate all your measurements for different parallelograms with <A > <B.
(e) According to your measurements and observations, write your conjecture about the length of the diagonals.
(f) Prove your conjecture.

We concur with Laborde (1993, 2001a) that geometric tasks for computer geometric environments should be different from geometric tasks for paper-and-pencil environments. She argues that "changing tools leads to changing the way tasks are performed and the facilities offered by computers may offer a strong interaction between visualization and knowledge in geometry" (p. 49). Mariotti (2001) and Choi-Koh (1999) also emphasize the role of tasks as fundamental to enhance mathematical discussion to guide the evolution of students' construction of geometric meanings in a dynamic environment and their ability to analyze geometric objects in order to generate conjectures and to produce geometric proofs.

THE GSP MEDIATING GEOMETRIC CONCEPTUALIZATION AND COMMUNICATION— AN ILLUSTRATION

Several research studies have concentrated on elementary and secondary school pre-service teachers to give them the opportunity to experiment by themselves ways of learning geometry using dynamic environments. Among these studies are those conducted by Mariotti, 2001; Jiang, 2002; Christou, Mousoulides, Pittalis, and Pitta-Pantazi, 2004; and De Villiers, 2004. Other studies have concentrated on the geometric progress of individual students. The conclusions of these studies indicate that both appropriate guiding questions on the part of the teacher and the use of tasks for the GSP provide valuable contexts for discoveries, explanations, justifications, and proofs that could be incorporated in the high-school geometry curriculum.

Teaching Experiment

A pilot teaching-experiment on the teaching and learning of geometry lasted one academic semester and was conducted with one pre-service teacher, one lateral entry teacher, and two in-service teachers. The purpose of the pilot teaching-experiment was to understand *how* learners, who already have some knowledge of geometric objects, are able to interact with the GSP to reorganize and broaden their knowledge of geometry. The guiding principle of the teaching-experiment was to use unstructured, semi-structured, structured tasks especially designed for the GSP in order to foster the three-way-interaction among students, the tasks in the GSP, and the teacher. The objective of each task was to allow each participating teacher the freedom to explore, to make conjectures, to investigate them, and to prove or reject them.

Twenty one (21) one-to-one interviews were conducted; teachers were interviewed on a weekly basis and each interview lasted 90 minutes. Two researchers conducted the experiment who acted in consensual agreement; one was the interviewer and the other played the role of active participant observer. All the tasks for the GSP were prepared before hand and some of them re-designed according the conceptual needs of the students. Here we concentrate on the 14th interview of the middle school pre-service teacher (with the pseudonym Michael) to analyze *how* he solved a task whose purpose was to investigate properties of isosceles trapezoids. The first thirteen interviews of the teaching-experiment focused on congruence and similarity of triangles.

Michael's Unsuccessful Attempt to Transform a Drawing into a Robust One

The task analyzed here was the first task in which student-teachers had to link their knowledge about triangles to their knowledge about quadrilaterals. In this particular task, the objective was to make the connection between triangles and trapezoids. The interviewer presented Michael with the drawing in the GSP (see Figure 2) and he was asked what kind of plane geometric figures he could foresee.

27 MICHAEL: If you connect AC and BD then you have a quadrilateral.
28 INTERVIEWER: OK, make the connections.
29 MICHAEL: OK
30 INTERVIEWER: And this is what?
31 MICHAEL: A quadrilateral.
32 INTERVIEWER: Could you animate one of the points A, B, C, or D?
33 MICHAEL: Yes.
34 INTERVIEWER: Please do it.
35 MICHAEL: Umhum, I see all kinds of quadrilaterals.
36 INTERVIEWER: Please stop the animation and then transform the quadrilateral on the screen into a trapezoid with bases AB and CD.
37 MICHAEL: [He connects A and C as well as B and D and makes the following drawing]

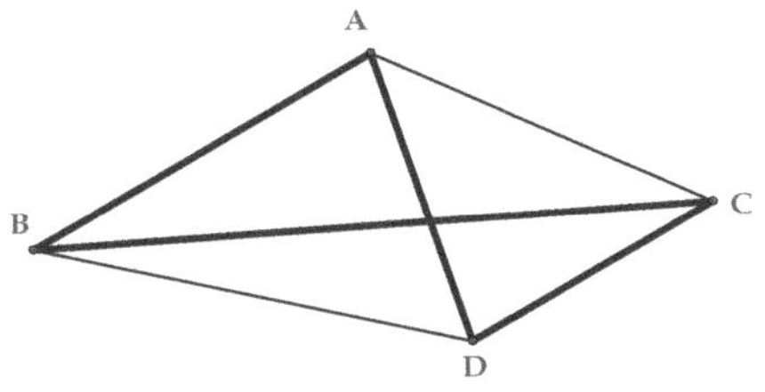

38 INTERVIEWER: What kind of trapezoids do you know?
39 MICHAEL: Isosceles trapezoids.
40 INTERVIEWER: Can you transform the trapezoid ABDC into an isosceles trapezoid?

Michael tried to transform the given figure into an isosceles trapezoid dragging sides and vertices of the trapezoid in Figure 2, but he was not sure that it was really an isosceles trapezoid or a trapezoid that looks like an isosceles one. In order to construct an isosceles trapezoid, Michael would need to know the properties of isosceles trapezoids and use one of them to construct a robust drawing. He did not use any particular construction strategy to make the given trapezoid an isosceles trapezoid. As Michael was dragging he was not talking; thus, the interviewer asked him to explain how he was trying to transform the trapezoid into an isosceles one. He said, "I am trying to make the base angles congruent". He continued dragging the vertices without a particular strategy in mind.

The above dialogue and Michael's silent activity clearly indicates the affordances of the GSP and Michael's ability to generalize the drawings on the screen by anticipating the whole class of quadrilaterals (a genus-figure or the geometric concept of quadrilaterals). The following question, in essence, asked Michael to put geometric conditions on a representative of the class of quadrilaterals to generate a representative of the class of trapezoids (species-figure), and from there, to generate a representative of another subclass, the class of isosceles trapezoids (a species-figure generated from the class of trapezoids which now should be seen as a genus-figure of lower order). To do this, Michael had to know the properties of isosceles trapezoids, before hand, and use them as geometric conditions for the transformation to take place.

The interviewer advised him to construct an isosceles trapezoid on his own. He constructed an angle and then he said, "I try to mirror this angle to have the other base angle …If I measure the angle…". Clearly his goal was to mirror the constructed angle and get an equal angle at the other endpoint of the horizontal line segment; however, he did not know which line should be the mirror and where the mirror should be in order to mirror the angle. So he constructed a segment and angles at each of its endpoints using line segments. Then, he measured both angles, decided to take an angle as a referent, and move the non-common side of the other angle until the two measures were roughly the same. Of course, such a construction was a drawing which did not pass the dragging test. This construction indicated that Michael did not know how to make elementary constructions in Euclidean geometry and was not a matter of not knowing how to do it in the GSP since he had good knowledge on how to manage the program. Thus, the interviewer guided him to make the construction of congruent angles and, after that, he was able to finish the construction of an isosceles trapezoid.

Michael's first attempt to construct an isosceles trapezoid indicates that he was not well aware of the congruence of the non-parallel sides or any other of its properties. Thus, the interviewer asked Michael which properties of the isosceles trapezoid he knew.

66 MICHAEL:	First for any trapezoid you have two parallel lines. And then the angles on one parallel line…they are equal, they are congruent, these are congruent and those two are congruent [*Michael shows the two pairs of base angles in his trapezoid*]

The interviewer helped Michael to establish a connection between an isosceles triangle and an isosceles trapezoid by means of the congruent base angles. He was asked to construct an isosceles triangle. He constructed a triangle, measured two of the angles, and by dragging a vertex he made them congruent forming an isosceles triangle. Again he made a drawing which did not pass the dragging test. Nonetheless, this construction directed him to remember the definition of isosceles triangles. This definition prompted Michael to give, for isosceles trapezoids, the analogous property of isosceles triangles. He said, "The non-parallel sides of isosceles trapezoids are congruent". Establishing this property for isosceles trapezoids gave Michael no insight into how he could use it to transform the given drawing (Figure 2) into an isosceles trapezoid. Thus, the interviewer continued asking Michael if he knew other properties of isosceles trapezoids.

Proving a property of isosceles trapezoids

Michael, with the help of the interviewer, constructed a drawing of an isosceles trapezoid (Figure 3). The interviewer asked again if he new about other properties of isosceles trapezoids. Michael gave no answer so the interviewer recommended him to join the midpoints of the parallel sides of the trapezoid.

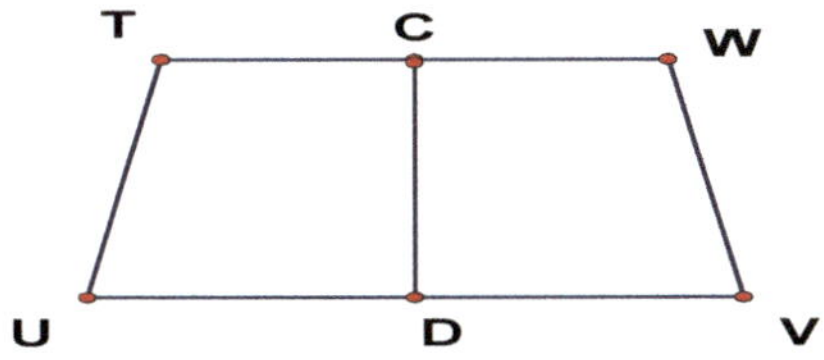

Figure 3. The mid-segment joining the bases of isosceles trapezoids

Michael observed the drawing, measured the angles formed between CD and the bases UV and TW. Then he said, "line CD, joining the midpoints of the bases, is perpendicular to the bases". In trying to explain why CD was perpendicular to the bases, he gave a description of the perpendicular bisector as a line of reflection. He also said, "If CD were not perpendicular to the bases, then the quadrilateral would not be an isosceles trapezoid". The interviewer asked if he could prove this property. Michael did not know how to proceed. The interviewer asked him to give the definition of perpendicular bisector. He saw the relationships between the points C and U as well as the points C and V, constructed the segments CU and CV and, even further, saw the relationship between the definition of perpendicular bisector and anticipated the isosceles triangle UCV (Figure 4).

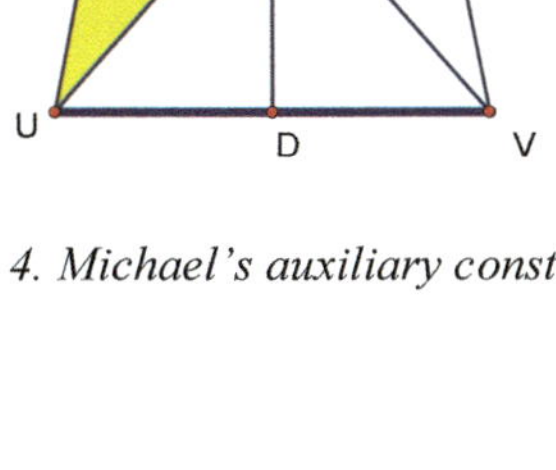

Figure 4. Michael's auxiliary construction

182 MICHAEL: I need to prove that CU and CV are congruent.
200 MICHAEL: C is the midpoint of TW, TC is congruent to CW. Sides TU and WV are congruent. Angles UTC and CWV are congruent because they are base angles. ΔUTC and ΔVWC are congruent by SAS. Therefore CU is congruent to CV.

The interviewer and Michael sustained a dialogue so that Michael could see how the congruence of the segments CU and CV could be used to prove that CD was perpendicular to UV.

249 MICHAEL: I know CU is congruent to CV. D is midpoint of UV, UD is congruent to DV. CD is congruent to CD. Thus ΔCUD is congruent to Δ CVD by SSS.
250 INTERVIEWER: What are the implications of this congruence?
252 MICHAEL: Angle CUD is congruent to CVD, angle UCD is congruent to VCD, and angle UDC is congruent to CDV.
254 MICHAEL: Since angle UDC is congruent to angle CDV and they are also supplementary, then they are right angles.
255 INTERVIEWER: What does this mean?
256 MICHAEL: That the line joining the midpoints of the bases of an isosceles trapezoid is perpendicular to both basis because a similar argument could be made with respect to the other base.

In the above episode, the interviewer hypothesized Michael's conceptual geometric level and guided him to prove his initial empirical observation that in an isosceles trapezoid, the line joining the midpoints of the bases is also perpendicular to the bases. That is, the interviewer was working within Michael's zone of potential development.

Proving another Property of Isosceles Trapezoids

Once Michael finished with the above proof, the interviewer guided Michael to continue the inquiry of other properties of isosceles trapezoids.

260 INTERVIEWER: Now, what about the diagonals of isosceles trapezoids? Please construct them.
261 MICHAEL: [*He uses the GSP to copy and paste the prior figure and constructs the diagonals TV and UW*]. The point of intersection of the diagonals is not the midpoint of any of them, they do not bisect each other.
262 INTERVIEWER: If they bisected each other, then what kind of quadrilateral this would be?
263 MICHAEL: It could be parallelogram, or a rectangle, or a rhombus, or a square.
264 INTERVIEWER: Do you observe any other kind of relationship between TV and UW?
265 MICHAEL: [He measures them] They are …equal.
266 INTERVIEWER: Why are they equal? Can you prove this?

267 MICHAEL:	I think I need to compare triangles TUV and WVU.
268 INTERVIEWER:	What do you know about them?

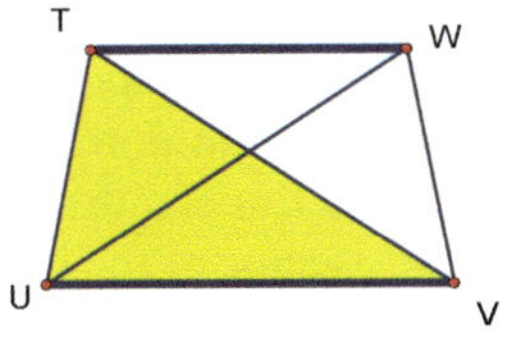

Figure 5. Property of the diagonals of isosceles *trapezoids*

269 MICHAEL:	TU and WV are congruent because they are the non-parallel sides of the isosceles trapezoid. UV is congruent to UV. The base angle TUV is congruent to the base angle WVU. Thus triangles TUV and WVU are congruent by SAS.
270 INTERVIEWER:	What are the implications of this congruence?
271 MICHAEL:	TV and UW are congruent. Angle UTV is congruent to angle UWV. And angle UVT is congruent to angle WUV. What I need is the congruence between TV and UW and now I have it. Thus the diagonals are congruent.

The above episode shows that Michael needed less and less guidance from the interviewer. It also manifests that Michael had knowledge of the properties of the diagonals of certain quadrilaterals. His knowledge about congruence of triangles allowed him to produce, on his own, a logical argument about the congruence of the diagonals of isosceles trapezoids; property he observed through the measuring capability of the GSP.

Back to the Initial Question and Discovering a New Property of Isosceles Trapezoids

The interviewer suggested Michael to go back to Figure 2 and try to transform the trapezoid into an isosceles trapezoid using its properties. He tried to make the diagonals congruent by means of measuring. He measured each of the diagonals and then dragged a vertex in an effort to make them congruent. This construction was analogous to the one he did at the beginning of the interview using the congruence of the base angles of isosceles trapezoids; this construction produces only a drawing that fades away as soon as the drawing is dragged (i.e., soft drawing).

The interviewer asked Michael to figure out a method to make the trapezoid isosceles without measuring. The interviewer suggested him to construct the perpendicular bisectors of bases AB and CD. Michael constructed the midpoints of the bases AB and CD and constructed the perpendicular bisectors of each one. The two perpendicular bisectors did not coincide. Michael dragged a vertex until the two perpendicular bisectors coincided. He immediately observed that at the moment the two perpendicular bisectors became one and the same, the intersection of the diagonals fell over it and the trapezoid became an isosceles one (Figures 6 and 7). Therefore, Michael answered the initial question about the transformation of a trapezoid into an isosceles trapezoid. In addition, he discovered a new property of isosceles trapezoids (i.e., in isosceles trapezoids, the point of intersection of the diagonals lies on the perpendicular bisector of the bases).

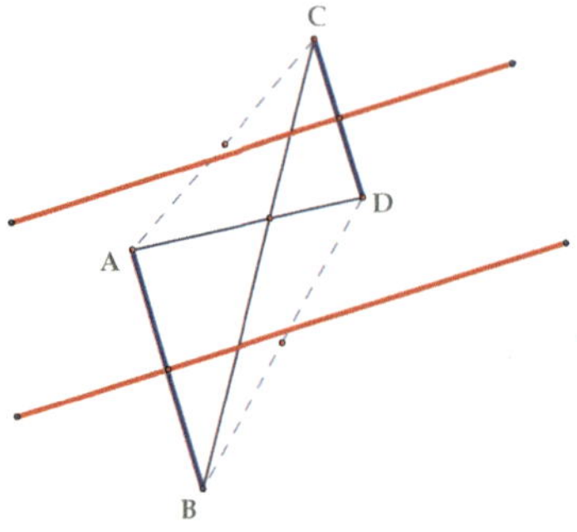

Figure 6.

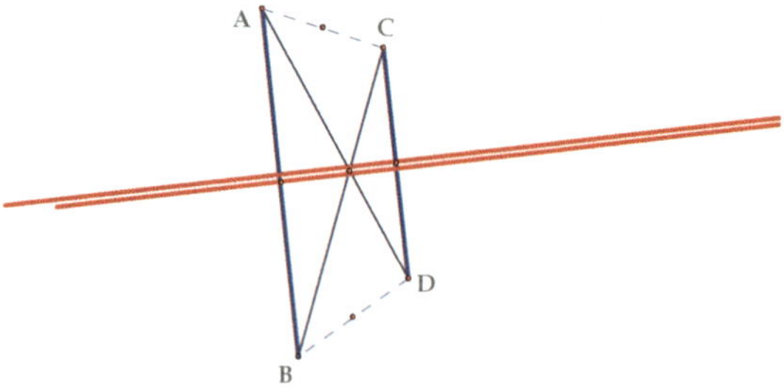

Figure 7.

Perpendicular bisectors to the basis of non-isosceles and isosceles trapezoids

The investigation of the properties of isosceles trapezoids allowed Michael to transform Figure 2 into an isosceles trapezoid. In the process of this transformation, Michael's conjecturing level and his knowledge on isosceles trapezoids increased throughout his interaction with the interviewer. The communication between the interviewer and Michael was possible due to both the cognitive invariants of the concept managed by the interviewer to guide the interaction, and Michael's kinesthetic-visual and conceptual actions on the drawing of a trapezoid to transform it into an isosceles one.

Proving the New Property

The interviewer continued the dialogue with Michael to follow up his new discovery and asked him if he could prove the following proposition: "If in a trapezoid the perpendicular bisector of one of the bases passes through the intersection of the diagonals, then the trapezoid is an isosceles trapezoid". He thought for some minutes, made the perpendicular bisector of one base to pass through the intersection of the diagonals and colored a right triangle formed by the perpendicular bisector and one of the diagonals (Figure 8). Then, he said the following.

MICHAEL: Triangles COP and OPD are right triangles because OP is the perpendicular bisector of the base CD. OP=OP, P is midpoint of CD so CP=PD. Triangles COP is congruent to triangle OPD by leg-leg. Thus CO=OD. Triangle COD is an isosceles triangle. Then,

> angle OCP is congruent to angle ODP. Points C and D are equidistant from O. Thus O, the point of intersection of the diagonals, is on the perpendicular bisector of CD. Now, since CD is parallel to AB [ABDC is a trapezoid] when extended segment OP, the line OP is perpendicular to the base AB. Also, angle OCP is congruent to angle OBM and angle PDO is congruent to angle MAO [alternate interior angles] and because angle OCP is congruent to angle ODP then the angle MAO is congruent to the angle MBO. Triangle AOB is an isosceles triangle with MO perpendicular to AB [line OP perpendicular to base AB] thus MO is bisector of AB. Now, the perpendicular bisector of CD is also perpendicular bisector of AB. Therefore, the trapezoid is an isosceles trapezoid.

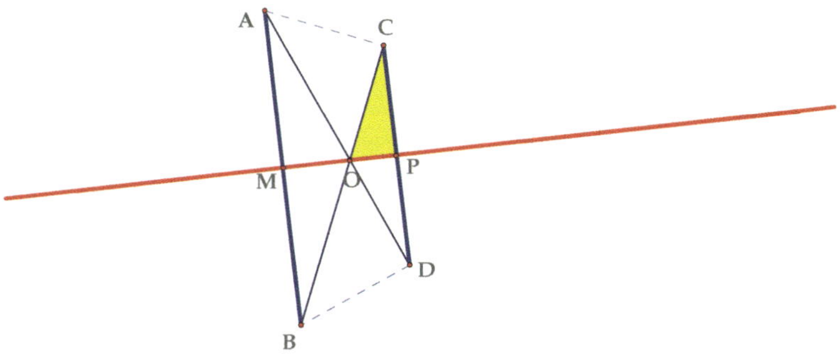

Figure 8. The intersection of the diagonals and the perpendicular bisector of the bases of isosceles trapezoids

The above dialogue indicates how Michael's knowledge about the congruence of triangles, isosceles triangles, and angles between parallel lines allowed him to validate his empirical conjecture. In contrast to the other proofs, he showed independence about his choice of triangles in order to produce a logical argument to confirm his assertion. This last proof indicates Michael's improvement on his geometric thinking due to the nature of the task, the communicative actions between interviewer and interviewee mediated by the GSP, and his interface with the GSP which mediated his conceptualization of isosceles trapezoids.

CONCLUDING REMARKS

The analysis of Michael's interview illustrates how a learner, who uses the GSP as a technical tool, comes around, with the help of the interviewer, to use it as a symbolic tool to mediate conceptualization. This indicates that although the GSP is a tool of technical nature, it has the potential of being a symbolic tool that directs the mind and behavior of the individual to mediate the understanding of geometric figures as a class of drawings with particular characteristics. It also indicates that the GSP mediates the communication between learner and interviewer (who uses the cognitive invariants of the concept) to guide Michael from his actual knowledge of isosceles trapezoids to a deeper understanding of this concept.

A significant source of Michael's conceptualizing activity was also the semi-structured task for the GSP that allowed the main features of the definition of trapezoid (exactly one pair of opposite sides parallel) to remain fixed. When the free elements of the drawing were dragged, quadrilaterals with two pairs of opposite parallel sides were generated (i.e., parallelograms). The same drawing could have been used to investigate the properties of parallelograms proper or any of the species-figures of parallelograms (i.e., rectangles, rhombuses, or squares). In other words, the semi-structured task, the GSP, and the interviewer's intervention were elements that mediated Michael's passage from kinesthetic-visual actions on soft drawings or drawings of a lesser degree of robustness to kinesthetic-visual and conceptual actions on robust drawings as well as the transformation of a generic trapezoid (a representative of the genus-figure trapezoid) into an isosceles trapezoid (a representative of a species-figure of the genus trapezoid).

NOTES

1 In this paper the word *figure* as in "Figure 1, Figure 2, etc." is used in reference to the writing convention to document pictorial representations in a text. In the rest of the paper, the word *figure* plays a very important role in reference to geometric conceptual objects. Therefore, each "Figure n" in this paper may depict *a drawing* of a particular *geometric object or figure*.

REFERENCES

Amit, M., & Fried, M. (2005). Authority and authority relations in mathematics education: A view from and 8th grade classroom. *Educational Studies in Mathematics, 58*, 145–168.

Arsac, G. (1989). La construction du concept de figure chez les élèves de 12 ans. In *Proceedings of the 13th Conference of the International Group for the Psychology of Mathematics*, (pp. 85–92). Paris: Editions GR Didactique et acquisition des connaissances scientifiques.

Balacheff, N. (1993). Artificial intelligence and real teaching. In C. Keitel and & K. Ruthven (Eds.), *Learning from computers—Mathematics education and technology*, (pp. 131–158). New York: Springer-Verlag.

Brousseau, G. (1997). *Theory of didactical situations in mathematics*. Dordrecht, The Netherlands: Kluwer.

Choi-Koh, S. S. (1999). A Student's learning of geometry using the computer. *The Journal of Educational Research. 92* (5), 301–311.

Christou, C., Mousoulides, N., Pittalis, M., & Pitta-Pantazi, D. (2004). Proofs through exploration in dynamic geometry environments. In *Proceeding of the 28th Cconference of the International Group for the Psychology of Mathematics Education* (Vol. 2, pp. 215–222)., Bergen-Norway, *2*, 215–222.

Davidov, V. V. (1990). *Types of generalization in instruction: Logical and psychological problems in the structuring of school curricula*. Translated by (Joan Teller, Trans.). Reston, Virginia: National Council of Teachers of Mathematics.

Davis, W. H. (1972). *Peirce's epistemology*. La Hague, Netherlands: Martinus Nijhoff.

De Villiers, M. (2004). Using dynamic geometry to expand mathematics teachers' understanding of proof. *International Journal of Mathematical Education in Science and Technology, 35* (5), 703–724.

Healy, L. (2000). Identifying and explaining geometrical relationship: Interactions with robust and soft Cabri constructions. In *Proceeding of the 24th Cconference of the International Group for the Psychology of Mathematics Education* (Vol. 1, pp. 103–117), Hiroshima-Japan, Vol. 1, pp. 103–117.

Hoyles, C. (1996). Modeling geometrical knowledge: The case of the student. In J.-M. Laborde (Ed.), *Intelligent Learning Environments: The case of geometry*, (pp. 94–112). Berlin: Springer-Verlag.

Jiang, Z. (2002). Developing preservice teachers' mathematical reasoning and proof abilities in the geometer's sketchpad environment. *Proceedings of the Annual Meeting [of the] North American Chapter of the International Group for the Psychology of Mathematics Education.*, Vol. *2*, 717–729.

Jones, K. (2001). Providing a foundation for deductive reasoning: Students' interpretations when using dynamic geometry software and their evolving mathematical explanations. *Educational Studies in Mathematics.*, *44* (1), 55–85.

Jones, K. (1999). *Student interpretations of a dynamic geometry environment.* Retrieved September 01, 2007, from European Society of Research in Mathematics Education I.I + I.II Web site: http://www.fmd.uni-osnabrueck.de/ebooks/erme/cerme1-proceedings/cerme1-proceedings.html

Kozulin, A. (1998). *Psychological tools: A sociocultural approach to education.* Cambridge, MAassachusetts: Harvard University Press.

Laborde, C. (1993). The computer as part of the learning environment: The case of geometry. In C. Keitel and & K. Ruthven (Eds.), *Learning from computers: Mathematics education and technology.* Berlin: Spring Verlag.

Laborde, C. (2001a). Integration of technology in the design of geometry tasks with Cabri-geometry. *International Journal of Computers for Mathematical Learning, 6*, 283–317.

Laborde, C. (2001b). Dynamic Geometry Environments as a Source of Rich Learning Contexts for the Complex Activity of Proving. *Educational Studies in Mathematics, 44*, 151–161.

Laborde, C. (2003). Technology used as a tool for mediating knowledge in the teaching of mathematics: The case of Cabri-geometry. In W. C. Yang, S. C. Chu, T. de Alwis, & M. G. Lee (Eds.), *Proceedings of the 8th Asian Technology Conference in Mathematics* (Vol. 1, pp. 23–38). Hsinchu, Taiwan ROC: Chung Hua University.

Laborde, C., Kynigos, C., Hollebrands, K., & Strässer, R. (2006). Teaching and learning geometry with technology. In A. Gutiérrez, P. and & P. Boero (Eds.), *Handbook of research on the psychology of mathematics education: Past, present and future.* (pp. 275–304).

Leont'ev, A. A. (1970). Social and natural in semiotics. In J. Morton (Ed.), *Biological and social factors in psycholinguistics.* Urbana, IL: University of Illinois Press.

Mariotti, M. A. (2001). Introduction to proof: The mediation of a dynamic software environment. *Educational Studies in Mathematics.*, *44* (1), 25–53.

Nacional Council of Teachers of Mathematics. (2000). *Principals and Standards for School Mathematics: An overview.* Reston, VAirginia: NCTM.

Peirce, C. S. (1909). *The essential Peirce: Selected philosophical writings (EP)*, Vol. 2. (1893–1913), edited by the (Peirce Edition Project, Ed.). Bloomington, IN: Indiana University Press.

Peirce, C. S. (1931–1935). *Collected Papers (CP).* (Volumes 1–6 edited by Charles Hartshorne and & Paul Weiss; (Eds.), volumes 7–8 edited by Arthur Burks (Ed.)). Cambridge, Massachusetts: Harvard University Press.

Plato. (1984). *The Republic.* New York: Mentor, Penguin Work.

Sáenz-Ludlow, A. (2007). Sign and the process of interpretation: Sign as an object and as a process. *Studies in Philosophy and Education, 26*, 205–223.

Vygotsky, L. (1978). *Mind and society.* Cambridge, Massachusetts: Harvard University Press.

Vygotsky, L. (1986). *Thought and language.* (Translated by Alex Kozulin, Trans.). Cambridge, Massachusetts: The MIT Press.

Wertsch, J. V. (1985). *Vygotsky and the social formation of the mind.* Cambridge, Massachusetts: Harvard University Press.

Wolff, R. P. (1973). *Kant's theory of mental activity: A commentary on the transcendental analytic of the Critique of Pure Reason.* Gloucester, Massachusetts: Peter Smith.

Adalira Sáenz-Ludlow
Department of Mathematics and Statistics
University of North Carolina ast Charlotte
USA

Anna Athanasopoulou
Department of Mathematics and Statistics
University of North Carolina at Charlotte
USA

LUIS RADFORD

THE ETHICS OF BEING AND KNOWING: TOWARDS A CULTURAL THEORY OF LEARNING[1]

This chapter sketches a theory of teaching and learning that takes its inspiration from some anthropological and historico-cultural schools of knowledge—the theory of knowledge objectification. Within this theory, the problem of learning is formulated in such a way that rationalist or individualists views of cognition and social interaction are avoided. The theory of knowledge objectification posits, indeed, the problem of learning as a social process through which students become progressively conversant with cultural forms of reflection. Arising in the course of sensuous mediated cultural praxes embedded in historically formed epistemes and ontologies, learning, it is argued, is not just about knowing something but also about becoming someone. The formulation of learning as a process where knowing and being are mutually constitutive leads to a non-utilitarian conception of the classroom: entrenched in unerasable ethical concerns, the classroom appears as a space for the growth of intersubjectivity and the nurturing of what is called here the *communal self*.

The chapter is divided into six sections. In the first section, I discuss some problematic assumptions often adopted by many contemporary theories of teaching and learning, in particular assumptions related to the learner, the content to be learned and process of learning. In Section 2, I introduce a non-mentalist, culturally embedded, concept of thinking that neither reduces the thinking subject to the mere product of discursive structures, nor posits it as a culturally-detached *res cogitans*. Section 3 is devoted to a discussion of the epistemological and ontological bases of the cultural theory here advocated. The concepts of learning and the mathematical classroom portrayed in this theory are presented in Sections 4 and 5, respectively. The main ideas of the previous sections are brought together in Section 6, where the educational questions surrounding the ethics of being and knowing are discussed.

1. THEORIES OF TEACHING AND LEARNING

Theories of teaching and learning differ from each other mainly in their conceptions about: (a) the content to be learned; (b) the learner; and (c) how learning actually occurs.

Concerning the third point, most contemporary theories have adopted the view according to which the student constructs his or her own knowledge (Lesh, Doerr, Carmona, & Hjalmarson, 2003). Although, in their account of learning, these theories do not necessarily exclude the role of the social, the social dimension of

L. Radford, G. Schubring, and F. Seeger (eds.), Semiotics in Mathematics Education: Epistemology, History, Classroom, and Culture, 215–234.

knowing is often reduced to a kind of external environment to which the cognitive activity of the student has to adapt. In these theories, assumed universal mechanisms of knowledge formation—e.g., the logical-mathematical structures of thinking in Piaget's genetic epistemology—account for the supposedly universal patterns of conceptual development. However, recent research in psychology, anthropology and other disciplines has pointed out the contextual nature of knowing and being (de Haan, 1999; Lave, 1988; Radford, 1997, 2003a, 2008a; Shweder and LeVine, 1984).[2] What this research makes clear is that cognition is much more complex than standard adaptive epistemologies intimate: cultural environments do indeed play a significant role in the ways we come to know and to be.

As to the second point, more often than not, theories that explain learning in adaptive terms share the same idea of the cognizing subject—an intrinsic rational auto-sustained individual maturing as it interprets and refines allegedly ethically neutral environmental feedback. This conceptualization of the cognizing subject leads to a narrow idea of the learner that Canadian psychologist Jack Martin describes in the following terms: a self-regulated individual whose "most vital resources are apparently available within its detached internality . . . a self that already knows its business, one that requires only a facilitative grooming to become more fully socialized and intellectually engaged" (Martin, 2004, p. 197). In short, what these theories convey is the problematic idea of a learner who "naturally" acts in a scientific, rational mindful manner.

Last but not least, adaptive explanations of learning mechanisms lead to important difficulties concerning the type of knowledge produced by adaptation. Indeed, the cognitive regulatory mechanisms of adaptation are usually conceived of in biological terms, with peripheral room for cultural considerations, leading to ahistorical and acultural accounts of knowledge. Often, biological premises are supplemented with a subjectivist interpretation of the individual's realm of experience. The result is a subjectivist interpretation of knowledge production. The best example is perhaps radical constructivism. In this theory, knowledge is merely made up of *personal* viable constructs. For many, however, this move is unconvincing: on the one hand, the idea of personal viability of knowledge leads to the unavoidable problem of solipsism (Lerman, 1996); on the other hand, radical constructivism gives up ontology (of any kind) and posits the *subjective* experiential realm as the limits of reason and knowledge.

At the educational level, radical constructivism has been criticized, among other things, for failing to account for the dissymmetric distribution of knowledge in the classroom. In a recent plenary lecture, Brousseau (2004) argued that "In didactics, radical constructivism is an absurdity." What Brousseau finds absurd in the radical constructivist position is not the claim that legitimate knowledge can only be the result of the individual's own achievement and deeds. What he finds erroneous is the idea that students' constructions necessarily lead to the institutional form of mathematical knowledge (*le savoir savant*). As Brousseau was able to observe over and over again in the classrooms of the Michelet School in Bordeaux, the students' subjective conceptual constructs require that an external perspective, among other things, *institutionalize* the knowledge arising from classroom mathematical activity.

The students cannot become aware of the cultural epistemic status of, say, a method arising as the result of their enquiring activity or, as Brousseau puts the matter, the students may not know that they know. The teacher hence has to encourage and highlight the kind of reasoning and the methods valued by the mathematicians' community.

These few comments provide an idea of some of the theoretical differences in current perspectives in mathematics education. Of course, the differences between theories are subtler than hinted at here. My interest is not to delve into these differences.[3] Rather my aim is to mention some focal points from where theoretical differences arise. In the rest of this paper, I present some elements of a theory of teaching and learning that takes its inspiration from some anthropological and historico-cultural schools of knowledge. This theory—the *theory of knowledge objectification*—relies on a non-rationalist epistemology and ontology, which gives rise, on the one hand, to an anthropological conception of thinking, and on the other, to an essentially social conception of learning.

2. A NON-MENTALIST CONCEPTION OF THINKING

2.1 Thinking as a Mediated Praxis Cogitans

Typically, thinking is understood as a kind of interior life, a series of mental processes on ideas carried out by the individual. This conception of thinking, as "mental activity" (de Vega, 1986, p. 439), comes from Augustine's interpretation of Greek philosophy at the end of the fourth century. For Augustine, ideas refer to something situated *inside of the individual*, contrary to the Greek tradition, where the term idea (*eidos*) referred to something external. Influenced by the Augustinian transformation of the Greek term, seventeenth-century rationalists such as Descartes and Leibniz believed that mathematics could be practiced even with one's eyes closed. As Leibniz put the matter, the principles that we need to understand objects or see their properties, the internal rules of reason, are "interior principles" that is, they are within our interior (Leibniz, 1704/1966, pp. 34–37). Anthropologists such as Clifford Geertz have demonstrated the limitations of the conceptualization of ideas as "things in the mind" or of thinking as an exclusively intracerebral process. Geertz (1973) claims that "The accepted view that mental functioning is essentially an intracerebral process, which can only be secondarily assisted or amplified by the various artificial devices which that process has enabled man to invent, appears to be quite wrong" (p. 76). He argues that "the human brain is thoroughly dependent upon cultural resources for its very operation; and those resources are, consequently, not adjuncts to, but constituents of, mental activity" (p. 76).

The conception of thinking as a kind of interior life has been very influential in the investigation of cognition in mathematics education. Written questionnaires, interviews, and drawing exercises have often been used to get a glimpse of what is going on *in* the head. To avoid the pitfalls of this mentalistic approach, some theories have simply discarded any psychological considerations. They simply avoid any talk about psychological constructs.

The theory of knowledge objectification adopts a non-mentalist position on thinking and intellectual activity. This theory suggests that thinking is a type of a social practice (Wartofsky, 1979), a *praxis cogitans*. To be more precise, thinking is considered to be *a mediated reflection in accordance with the form or mode of the activity of individuals*.

The Mediated Nature of Thinking The mediating nature of thinking refers to the role, in the Vygotskian sense, played by artifacts (objects, instruments, sign systems, etc.) in carrying out social practice (Bartolini Bussi & Mariotti, 2008). Artifacts are neither merely aids to thinking nor simple amplifiers, but rather constitutive and consubstantial parts of thinking. We think with and through cultural artifacts. The following example will help clarify this idea.

In a first-grade class in elementary school, the pupils had to solve a problem about a numeric sequence. The teacher introduced the problem through a story in which a squirrel, at the end of the summertime, brings two nuts to his new nest every day in preparation for the coming winter. In one part of the problem, the pupils had to determine how many nuts the squirrel had collected in his nest by the end of the tenth day, given the fact that there were already 8 nuts in the nest when the squirrel found it and that the squirrel never ate nuts from his winter provision. Christina, one of the pupils, began counting two by two: "ten, twelve, fourteen, sixteen." When she noticed that she was not keeping track of the number of days that had passed, she started the count again. However, doing things simultaneously ended up being quite a difficult task. Addressing herself to Michael, her group mate, Christina said, "let's do it together!" While the rest of the class continued working on the problem in small groups, Christina and Michael went to the blackboard and, using a large wooden ruler, Christina began counting two by two while Michael counted the days out loud.

In Figure 1 (left), when Michael says "nine," Christina points with a wooden ruler to the number 26 on a number line placed above the blackboard, being the number of nuts the squirrel had collected by day 9. In Figure 1 (right), Michael, who continued counting the days, says "ten," while Christina moves the ruler to the right and points to the number 28, finding the answer to the question in this way.

Figure 1. (Left picture) Michael says 9, and Christina points to number 26. (Right Picture) Michael says 10, and Christina points to number 28

The wooden ruler, the number line, the mathematical signs on the piece of paper that Michael holds up as he reads behind Christina, all are artifacts that *mediate* thinking. They are not merely aids: their mediating role is such that they *orient* and *materialize* thinking and, in so doing, become an *integral* part of it. Indeed, according to the theory advocated here, Christina and Michael's thinking is not something occurring merely in the students' mental plane. Thinking also occurs along the social plane, in a region that, paraphrasing Vološinov (1973), I want to call the *territory of artifactual thought*. It is within this territory that subjectivity and cultural objectivity mutually overlap and where the mind extends itself beyond the skin (Wertsch, 1991).

Thinking as Re-flection The reflexive nature of thinking means that the individual's thinking is neither the simple assimilation of an external reality (as the Empiricists and Behaviorists suggested) nor an *ex nihilo* construction (as certain constructivist schools claim). Thinking is a *re-flection*, that is, a dialectical movement between a historically and culturally constituted reality and an individual who refracts it (as well as modifies it) according to his/her own subjective interpretations, actions and feelings.

One of the roles of culture is to suggest ways of perceiving reality and its phenomena to students: literally, ways of intending (*manières de viser*), as Merleau-Ponty (1945) would say, or ways of intuiting, as Husserl (1931) might have it. In more general terms, the *re-flexivity* of thinking, from the phylogenetic point of view, consists in individuals giving rise to thinking and to the objects that thinking creates. However, at the same time, from the ontogenetic point of view, the thinking of individuals is, from the outset, subsumed by their cultural reality and by the historically formed concepts that they encounter in their environment. This is why we originate thinking, but at the same time become subsumed by it (Eagleton, 1997).

2.2 The Anthropological Dimension of Thinking

Thinking is not merely generated in the course of human activity. The *form* of the activity imprints its mark on thinking and in its product–i.e., knowledge. Now, the form that all activity takes depends on *symbolic superstructures*. These symbolic superstructures, which elsewhere I have called *Semiotic Systems of Cultural Signification* (Radford 2003a), include cultural conceptions surrounding mathematical objects (their nature, their way of existing, their relation to the concrete world, etc.) and social patterns of meaning production (see Figure 2). In their interaction with activities (their objects, actions, division of labour, etc.) and with the territory of artifactual thought, the *Semiotic Systems of Cultural Signification* give rise, on the one hand, to forms or *modes of activities*, and, on the other hand, to specific *modes of knowing* or *epistemes* (Foucault, 1966). While the first interaction gives rise to the particular ways in which activities are carried out at a certain historical moment, the second interaction gives rise to specific modes of knowing which allow for the identification of "interesting" situations or problems and the methods, reasoning, evidence, etc. that will be considered culturally valid.

Here is an example. The difference between the thinking of a Babylonian scribe and that of a Greek geometer cannot be reduced only to the kinds of problems with which they were respectively occupied or to the artifacts they used to think mathematically. The difference between their modes of thinking cannot be reduced to the fact that the Babylonian scribe was reflecting in a context tied to political and economic administration, whereas the Greek geometer was thinking within an aristocratic and philosophical context. The difference between the thinking of the Babylonian mathematician and that of the Greek one has to do with the fact that each one of these forms of thinking was underpinned by a particular *symbolic superstructure*. The thinking of the Babylonian scribe was framed by a realist pragmatism where mathematical objects such as "rectangle," "square," and so forth—objects which the Greek geometer of Euclid's time conceptualized in terms of Platonic forms or Aristotelian abstractions—acquired their meaning. The manner

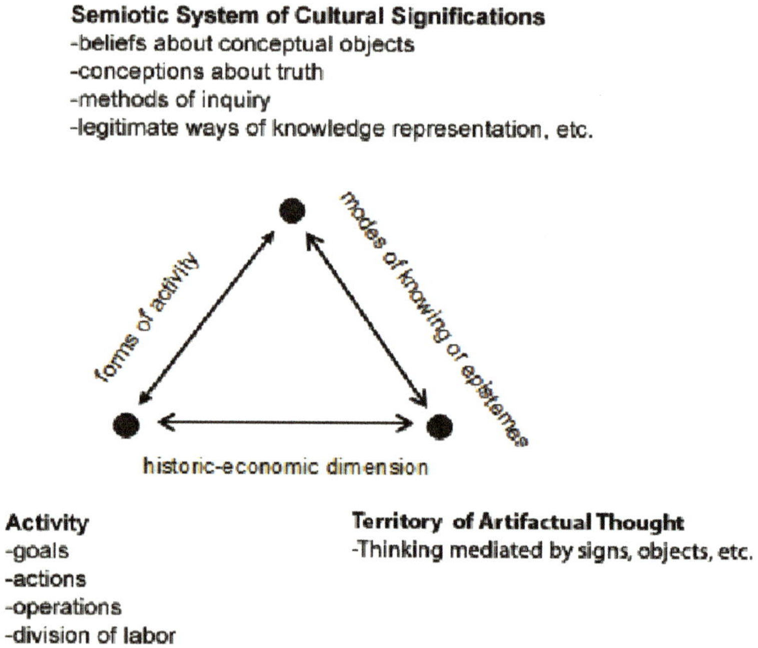

Figure 2. The arrows show the interaction between a Semiotic System of Cultural Significations, Activity and the Territory of the Artifactual Thought. The interaction generates the forms of activity and the modes of knowing on the base of the specific historic-economic dimension. In a dialectic process, forms of activity, modes of knowing, and the historic-economic dimension alter the triangle's vertices

in which the Babylonian scribe, the Greek geometer and the Renaissance abacist ended up thinking about and knowing objects of knowledge, the way in which they tackled their problems and considered them to be solved, all were framed by the very form of their activity and the corresponding cultural episteme (Radford, 1997, 2003a, 2003b).

Rather than seeing these (and other) historical and contemporary forms of mathematical thinking as "primitive" or "imperfect" versions of current mathematical thought (ethnocentrism), the anthropological dimension of the theory of knowledge objectification considers these forms as belonging to particular, genuine types of mathematics in their own right.

3. THE EPISTEMOLOGICAL AND ONTOLOGICAL BASES OF THE THEORY OF KNOWLEDGE OBJECTIFICATION

Any didactic theory, at one moment or another (unless it voluntarily wants to confine itself to a kind of naïve position), must clarify its ontological and epistemological position. The *ontological* position consists in specifying the sense in which the theory tackles the question of the nature of conceptual objects (in our case, the nature of mathematical objects, their forms of existence, etc.). The *epistemological* position consists in specifying the way in which, according to the theory, these objects can (or cannot) end up being known.

Often, contemporary didactic theories that start from an application of mathematics to the experiential world, adopt—even if it is not mentioned explicitly—a realist ontology and deal with the epistemological problem in terms of abstractions. Naturally, the situation is not that simple, as Kant himself recognized. As for Realism—which, in an important way, is the Platonist version of the instrumental rationalism (Weber, 1992) which emerged during the Renaissance—the existence of mathematical objects precedes and is independent from the activity of individuals. Like the Platonist, the Realist believes that mathematical objects exist independently of time and culture. The difference is that, whereas Platonic objects do not mix with the world of mortals, the conceptual objects of the Realist govern our world through "natural laws." According to realist ontology, this explains the miracle that is the applicability of mathematics to our phenomenal world (Colyvan, 2001). Naturally, in order to achieve this, Realism makes a leap of faith that consists in believing that the abstractive ascent from the concrete objects of sensuous experience to general, pre-existing objects is certainly possible. The faith that Plato placed in reasoned social discourse (*logos*) and which Descartes placed in cogitating with oneself are subjected to scientific experimentation by Realism.

The ontological and epistemological position of the theory of knowledge objectification moves away from Platonist and realist ontologies and their corresponding conception of mathematical objects as eternal objects preceding the activity of individuals. It also moves away from Rationalist ontologies and their conception of mathematical objects as products of a mind that works folded in onto itself working in accordance to the laws of logic. The theory of knowledge objectification suggests that mathematical objects are historically generated during the course of the mathematical activity of individuals. More precisely, mathematical objects *are fixed patterns of reflexive human activity incrusted in the ever-changing world of social practice mediated by artifacts*.

The conceptual object "circle," for example, is a fixed pattern of activity whose origins cannot be found in the intellectual contemplation of (more or less) round objects that the first individuals would have encountered in their surroundings. The conceptual object "circle" must rather be found in the *sensual* and *practical* activity that led individuals to notice the emergent object:

> People could see the sun as round only because they rounded clay with their hands. With their hands they shaped stone, sharpened its borders, gave it facets. (Mikhailov, 1980, p. 199)

This sensual experience of labour has remained fixed in language which encapsulates original meanings, so that

> the meaning of the words "border," "facet," "line" does not come from abstracting the general external features of things in the process of contemplation. (p. 199)

but rather comes from the activity of labour that has been taking place since the origins of humanity. Far from surrendering itself completely to our senses, our relationship with nature and the world is filtered through conceptual categories and cultural significations so that

> man could contemplate nature only through the prism of all the social work-skills that had been accumulated by his predecessors. (p. 199)

4. LEARNING

In the previous sections we have seen that, from a phylogenetic point of view, conceptual objects are generated in the course of human activity. From an ontogenetic point of view, the central problem is to explain how acquisition of the knowledge deposited in a culture can be achieved: this is a fundamental problem of mathematics education in particular and of learning in general.

As mentioned in the Introduction, classical theories of mathematical education posit the problem of learning in terms of a construction or re-construction of knowledge on the part of the student. I already mentioned some of the difficulties that arise from such a perspective at the ontological and epistemological levels (for a detailed discussion, see Radford, 2008d). Here, I just want to add that the idea that knowledge has to be "constructed" by each individual has its own history. Actually, such an idea only became thinkable after the Renaissance and did not receive a full and explicit articulation before the 18th century. It was in Kant's work that the idea was expressed with unprecedented detail. For Kant, the individual is not only an introspective thinker whose mental activity brings him mathematical truths, as upheld by the rationalists (Descartes, Leibniz, etc.); nor is he only a passive individual who receives sensory information in order to formulate ideas, as proposed by the Empiricists (Hume, Locke, etc.). For Kant, the individual is a craftsman of his/her own knowledge (Radford, 2006a). In formulating the relationship between subject and object in this way, Kant expressed, in a coherent and explicit manner, the epistemological change that had been gradually taking place since the appearance of manufacturing and the emergence of capitalism during the Renaissance. The reasons of this epistemological change can be summarized in the following way. The modern era is marked by a displacement in the conception of knowledge clearly manifested in a shift that went from a focus on "the what" (the object of knowledge) to "the how" (the process). Unlike the medieval individual, the modern individual can only know that which he/she himself/herself has made (Arendt, 1958). This idea arose out of the political and economic climate surrounding the Enlightenment, one of its chief characteristics being the opposition it presented to tradition and religion. From the 18th century onward, knowledge was no longer conceived of as something to be received or passed on, but something made by the autonomous, rational, auto-sufficient, culturally-detached Enlightened self: within this individualist tradition, knowledge was hence considered as a purely personal "construction".

Although historically interesting, the idea that every piece of knowledge is necessarily a personal construction has been subjected to a series of critiques, in particular because of the tremendous subjectivism that it entails (Adorno, 2001). If we were really meant to construct everything we know, we would still be trying to light some fire in front of a dark cavern, as the French biologist and philosopher Henri Laborit (1985) once suggested. And, of course, the idea of knowledge as a personal construction is even more problematic in education (Lerman, 1996; Radford, 2008d). Constructing knowledge by oneself is certainly *one* form of knowing among others. But stating that this is the only possible one fails to capture the diversity of cognitive forms of learning, knowing and being that can be found in the mathematics classroom.

According to the theory of knowledge objectification, learning does not consist in constructing or reconstructing a piece of knowledge. It is a matter of actively and imaginatively endowing the conceptual objects that the student finds in his/her culture with meaning. It is what we will later call a process of objectification. For the moment, we need to discuss two important sources of meanings that underlie all forms of learning.

4.1 The Knowledge Deposited in Artifacts

One of the sources of learning results from our contact with the material world, the world of cultural artifacts which surrounds us (objects, instruments, etc.). The Grade 1 example discussed previously illustrated the artifact mediated nature of thinking and led us to argue that thinking occurs in a zone that was called the Territory of Artifactual Thought. But the role of artifacts is more than materializing thinking and making it thinking-with-and-through-artifacts. Artifacts, indeed, are bearers of historically deposited knowledge from the cognitive activity of previous generations. Although it is true that some animals are able to use artifacts (Boesch & Boesch, 1990; Torigoe, 1985), nevertheless, for animals, artifacts do not end up acquiring a durable meaning. The wooden stick that a chimpanzee uses to reach a piece of fruit looses its meaning after the action has been executed (Köhler, 1951). It is for this reason that animals do not preserve artifacts.[4] Furthermore—and this is a fundamental element of human cognition—unlike animals, the human being is profoundly altered by the artifact: by making contact with it, the human being restructures his/her movements (Baudrillard, 1968) and new motor and intellectual skills are formed such as anticipation, memory, and perception (Vygotsky & Luria, 1994).

The world of artifacts appears, then, to be an important source for the process of learning, but it is not the only one. Objects cannot make clear the historical intelligence that is imbedded in them. This requires that they be used in activities as well as in contact with other people who know how to "read" this intelligence and help us to acquire it. Symbolic-algebraic language would otherwise be reduced to a group of hieroglyphics. The intelligence that symbolic-algebraic language carries would not be noticed without the social activity that takes place in the school. It is this social dimension which constitutes, for the theory of knowledge objectification, the second essential source for learning.

4.2 Social Interaction

Even though the importance of the social dimension has been underlined by a great number of recent studies on classroom interaction, there are subtle differences with regards to its cognitive contribution (see, e.g., Kidron, Lenfant, Bikner-Ahsbahs, Artigue, and Dreyfus, 2008; Yackel and Cobb, 1996; Sierpinska, 1996; Steinbring, Bartolini Bussi and Sierpinska, 1998). Often, interaction is considered as a negotiation of meanings occurring in an environment that simply offers the stimuli of adaptation that are required for students' cognitive development. The problem is that the classroom is not merely a material space where the students negotiate and find an environment to adapt themselves; it is not only a matter of "external" conditions to which the subject must accommodate his/her activity. The crucial point is that the classroom is a *symbolic space*. It is a space that conveys scientific, aesthetic, ethical and other historically constituted cultural values impressed in "social languages" (Bakhtin, 1986) such as the scientific, the artistic, etc. that end up affecting individuals' actions and reflections. As was mentioned in the first part of this article, the actions that individuals carry out are submerged in cultural modes of activity. It is for this reason that the classroom cannot be viewed as an enclosed space, folded over against itself, where knowledge and rules of interaction are negotiated anew. In fact, knowledge and rules of social interaction have a whole cultural history behind them and therefore pre-exist the interaction that takes place in the classroom.

According to the sociocultural perspective advocated here, interaction plays a different role. Rather than performing a merely adaptive function—a catalyzing or facilitating one—interaction is consubstantial to learning. Therefore, we see that the material world and the social dimension play a basic role in learning. The allocation of meaning that rests on these dimensions has profound psychological importance inasmuch as it is both a progressive immersion into cultural forms of thinking as well as the process of development of the specific capacities of the individual—cognitive, ethical, subjective, etc. It is for this reason that learning is not merely appropriating something or assimilating something; rather, it is the very process by which our human capacities are formed (I shall come back to this point below).

4.3 Objectification and Subjectification

In the previous sections it was suggested that learning consists of endowing conceptual cultural objects with meaning. In fact, learning is much more than that. Learning rests on an attitude of open-mindedness: it is an opening movement towards others and the objects of culture. It is worth noticing that this is, in fact, the etymological sense of the term *acquisition.* Acquisition comes from the Latin *adquaerere,* which means *to seek.* In this context, to learn is not merely to acquire something in the corrupted sense of possessing it or mastering it, but to go to culture to find "something" in it. This is why the outcome of the act of learning is not the construction, re-construction, re-production, re-invention or mastering of concepts: its true outcome is to be found in the fact that, in this encounter with the other and cultural objects, the seeking individual *finds herself.* This creative process of finding or noticing something (a dynamic target) is what I have termed elsewhere a process of *objectification* (Radford, 2002).

As understood here, objectification thus is more than the connection of the two classical epistemological poles, subject and object: it is in fact a transformative and creative process between these two poles, where, in the course of learning, the subject objectifies cultural knowledge and, in so doing, finds itself objectified in a reflective move that can be termed *subjectification.* The making of the subject, the creation of a particular (and unique) subjectivity is thus a process of subjectification that is made possible by the activity in which objectification takes place, and by the *re-flective* nature of thinking and the possibilities that e.g. language and other cultural instruments of thought offer to distinguish between an "I" and its surroundings (I/non-I; I/you; I/it; we/them, the impersonal discourse of science, etc.). In the culturally mediated experience that a subject '*s*' makes of an object '*o,*', '*s*' comes to know '*o*' within the possibilities and constraints offered by the dynamic and ever-changing cultural-normative sphere of knowledge. In knowing '*o*', '*s*' enters into a historically mediated relationship with, '*o*' and other subjects 's_i'. This historically mediated relationship not only makes the object '*o*' *noticeable* to '*s*' but also '*s*' to itself through the available forms of subjectivity and agency of the culture. This is why learning is both a process of knowing and a process of becoming.

5. THE MATHEMATICS CLASSROOM

5.1 Learning Activity

A central element of the concept of activity is its objective (Leont'ev, 1978). Even though the objective of classroom activity is clear for the teacher, this is not necessarily the case for the students. If the objective were to be clear to them, then there would be nothing left for them to learn. Within the didactic project in the class, the teacher suggests a series of mathematical tasks to the students so that a given objective can be achieved. Solving these problems becomes an end that directs the actions of the students. However, from the perspective of the theory of knowledge objectification, doing mathematics is more than doing tasks and solving problems. Without devaluing the role of problems in knowledge formation (see, for example, Bachelard, 1986), for us, problem solving is not the end but rather one of the means for achieving the type of *praxis cogitans* or cultural reflection that we call mathematical thinking. Behind the objective of the lesson, there lies a greater and more important objective—the generally held objective for the teaching and learning of mathematics—namely, the elaboration on the part of the student of a reflection defined as a *communal* and active relationship with his/her cultural-historical reality. Unfortunately, the learning of mathematics has often been reduced to merely obtaining a certain conceptual content. Knowledge has been reduced to a sort of *commodity*. This fetishist conception of learning operates a separation between knowing and being and ends up favouring an alienating form of being. The cultural theory of teaching-and-learning advocated here resists such a separation and argues for a reconnection between knowing and being. In other words, learning mathematics is not simply learning *to do* mathematics (problem solving), but rather it is learning *to be* in mathematics. This theoretical stance has important consequences, not only for the designing of activities, but also for the organization of the class itself and the roles that students and teachers play within it.

5.2 Layers of Generality

Teaching consists of generating and keeping in movement contextual activities which are heading toward inter-subjectively engaging the students with conceptual objects—fixed patterns of reflexive activity incrusted in the culture. This movement has three essential characteristics. First, the conceptual object is not a monolithic or homogenous object. It is an object made up of layers of generality. Second, from the epistemological point of view, these layers will be more or less general depending on the characteristics of the cultural meanings of the fixed pattern of activity in question (for example, the kinaesthetic movement that forms a circle; the symbolic formula that expresses it as a group of points at an equal distance from its centre, etc.). Third, from the cognitive point of view, the layers of generality are noticed in a progressive way by the student.

For the student, the learning process consists in becoming receptive to others and fluidly conversant with the various layers of generality of the object and their enabling forms of action—e.g. techniques and reflections on these techniques

(Bosch & Chevallard, 1999), modelling (Lesh et al., 2003), etc. Now, in order to get to know objects and products of cultural development, it is "necessary to carry out a determined activity around them, that is to say, a kind of activity that produces its essential characteristics, embodied, 'accumulated' in said objects" (Leontiev, 1968, p. 21). Thus, for the teacher, the teaching process consists in offering the students rich activities featuring, in a suitable manner, the encounter with other voices and the various layers of generality of the cultural object, making sure that this encounter is supported by the two sources of meaning discussed in Section 3—namely, meaning in artifacts, and meaning arising out of social interaction between students and between students and the teacher.

Because of the artifactual and embodied nature of thinking (Arzarello, Edwards, and Radford, 2008; Nemirovsky, 2003; Roth, 2001; Seitz, 2000), in the course of their objectification of knowledge, students and teachers use signs and artifacts of different sorts (mathematical symbols, graphs, words, gestures, calculators and so on). We call these artifacts and signs used to objectify knowledge *semiotic means of objectification* (Radford, 2003c). In previous works, we have discussed the prominent role of gestures and language in students' processes of knowledge objectification. We have provided evidence of the key role of deictic activity, both at the level of gestures, such as pointing, and at the level of language, such as when students use indexical terms such as *this* and *that* (Radford, 2002) and showed how, through various types of semiotic means of objectification, the students reach different layers of generality (Radford, 2000, 2003c, 2006b, 2008e; Radford, Bardini, & Sabena, 2007; Sabena, Radford, & Bardini, 2005). The investigation of the students' and teachers' interaction and use of semiotic means of objectification is indeed a methodological strategy to account for the processes of learning in the classroom. It provides a broad, but sufficiently specific, frame with which to track students' progressive acquisition of cultural forms of mathematical being and thinking.

6. THE ETHICS OF BEING AND KNOWING

The classroom is the symbolic space in which the student elaborates a communal and active relation with his/her historical-cultural reality. It is here that the aforementioned encounter between the subject and the object of knowledge occurs. The objectification that allows for this encounter is not an individual process but a social one. The sociability of the process, nevertheless, cannot be understood as a simple business "negotiation" during which each stakeholder invests some capital (e.g. some meaning) in the hopes of ending up with more of it. Here, sociability means the process of the formation of consciousness which Leont'ev (1978), paraphrasing Vygotsky, characterized as *co-sapientia*, that is to say, as knowing in common or knowing-with others.

Naturally, these ideas imply a re-conceptualization of the student and his/her role in the act of learning. Insofar as current theories in mathematics education draw on the concept of the individual as formulated by Kant and other Enlightenment philosophers, education justifies itself by guaranteeing the formation of an autonomous subject (understood in the sense of being able to do something for oneself without the help of others). Autonomy is, in effect, a central theme of modern education (Piaget, 1973) that has served as a basis for the theorizing of socio-constructivism (see, for example, Yackel and Cobb, 1996) and the theory of didactic situations (Brousseau, 1986; Brousseau and Gibel, 2005). The rationalism that weighs heavily on this concept of autonomy comes from its alliance with another key Kantian concept: that of freedom. There can be no autonomy without freedom and, for Kant, freedom means the convenient use of Reason according to its own universal principles (Kant, 1797/1974).

Since the Enlightenment did not put forward the possibility of there being a multiplicity of reasons, but rather postulated that western reason was The Reason, community coexistence implies respect for a duty which, in the end, is nothing but a manifestation of that alleged universal reason, whose epitome is mathematics. It was this supposed universality of reason that led Kant to fuse together the ethical, political and epistemological dimensions of life and to affirm that "to do something for the sake of duty means obeying reason" (Kant, 1803, p. 37).

For the theory of knowledge objectification, classroom functioning and the role of the teacher are not meant to promote the Enlightened individualistic idea of autonomy. The theory of knowledge objectification pleads for an idea of subjectivity and the self that goes beyond the ahistorical individualism inherited from the Enlightenment. It seeks rather to promote a concept of the autonomous person that is sensitive to the importance of history, the context and others, and where autonomy is both self-fulfilment and social commitment.

In her studies of Ancient Greek and Roman cultures, Arendt has shown that, in opposition to the modern idea of autonomy as something that comes from within—a personal and individual attribute coterminous with free will—autonomy for the Greek and Roman citizen had a social-civic connotation: it was related to action in the public sphere; it was a characteristic of human existence in the world. Unfortunately, Arendt comments, "Our philosophical tradition is almost unanimous in holding that freedom begins where men have left the realm of political life inhabited by the many, and that it is not experienced in association with others but in intercourse with one's self" (Arendt, 1993, p. 157).

Despite its legendary endurance in Western thought, non-individualistic conceptions of autonomy and freedom are frequently reported in contemporary anthropological research. Anthropologist Richard Shweder (1991) notes that

> not all cultures socialize autonomy or redundantly confirm the right of the individual to projects of personal expression, to a body, mind, and room's of one's own. . . . Linked to each other in an interdependent system, members . . . take an active interest in one another's affairs and feel at ease in regulating and being regulated. Indeed, others are the means to one's functioning, and vice versa. (p. 154)

It is indeed along the lines of a communal engagement, displayed in the public sphere, that autonomy might be better conceptualized, for, as the French philosopher Emmanuel Lévinas (1976) reminds us, "It is not through a relationship with the self, but through a relationship with another self, that man can become complete" (p. 31) and also, certainly, free and autonomous.

Instead of the idea of the self-regulated Enlightened individual common to many contemporary theories in education, the theory of knowledge objectification suggests the idea of a communitarian self, one busy with learning how to live in the community that is the classroom, learning how to interact with others, to opening oneself up to understanding other voices and other consciousnesses, in brief, *being-with-others*.[5]

The intrinsic social nature of knowledge and mathematical thinking has brought us then to conceiving of the classroom as an ethical and political space—the ethico-political space of the continuous renewing of being and knowing.[6] In our research with teachers, we encourage the students to work towards the creation of opportunities for the personal achievement of each individual, fostering respect, critique and mutual understanding (which includes understanding disagreements). We encourage them to show commitment to others and their community. Their community has to be flexible in its ideas and its forms of expression and be open to resistance and subversion in order to insure: modification, change and its transformation.

Being a member of the community, however, is not something that comes as a matter of course. Thus, in one of the classes we have worked with (a Grade 5 class), we encouraged the students to discuss mathematical problems in small groups (usually groups of 2 to 4 students). It was not unusual to invite the small groups to contrast their mathematical arguments in order to end up with more sophisticated ideas. During these inter-group encounters, the students were supposed to listen to the other groups' arguments, make sense of them and explain whether or not the argument could be improved. A student from one group whispered to a student from another group: "We'll let them fight for it!" It takes time (a long time indeed!) to make students aware of other non self-interested and alienating modes of classroom participation. In order to be a community member, students are encouraged to: share in the objectives of the community; involve themselves in the classroom activities; communicate with others (Radford and Demers, 2004). The abovementioned guidelines are not simply codes of conduct. On the contrary, they are indexes of forms of being in mathematics (and, as a consequence, of knowing mathematics) in the strictest sense of the term.

CONCLUDING REMARKS

The theory of knowledge objectification suggests a view of teaching and learning anchored in the idea that learning is a social activity (*praxis cogitans*) deeply rooted in historically constituted cultural forms of thinking and being. Its fundamental principles are articulated according to five interrelated concepts. The first of these is a concept of a psychological order: the concept of *thinking*, elaborated in non-mentalist terms. The second concept of the theory is of a socio-cultural order. This is the concept of learning. The third concept of the theory is of an epistemological nature and deals with those super-epistemic aspects that frame learning in the form of *semiotic systems of cultural signification*—cultural systems that "naturalize" the ways that one questions and investigates the world. The aforementioned concepts come to be completed by a fourth concept of an ontological nature—that of mathematical objects, which we have defined as *fixed patterns of reflexive activity incrusted in the ever-changing world of social practice mediated by artifacts*. To render the theory operational in its ontogentic aspect, it was necessary to introduce a fifth concept of a semiotic-cognitive nature—that of objectification, or a subjective awareness of the cultural object. In this context, and in light of the previous fundamental concepts, learning is defined as the social process of *objectification* of those external patterns of action fixed in the culture. Objectification entails another process, the process of *subjectification*—i.e., the becoming of the self. Subjectification has received very little attention in the literature on mathematics education (in addition to M. Fried's chapter in this volume, some works are: Brown, in press; Lerman, 1996; Popkewitz, 2004). However, its importance is easy to grasp as soon as we resist the temptation to reduce mathematics to its technical dimension and become aware that learning is much more than constructing logico-mathematical mental structures or picking up ready-made knowledge, that is to say, if and when we become aware of the fact that learning is about knowing *and* being. Objectification and subjectification should in fact be seen as two mutually constitutive processes leading to students' engagement with cultural forms of thinking and a sensibility to issues of interpersonal respect, plurality, inclusiveness and other main characteristics of the *communitarian self* (Radford, 2006a).

Such a view requires us to move away from traditional epistemologies where learning and knowing are the outcomes of a detached self, moved by its own business and interests—as, e.g., in radical constructivism, where the Self reduces the Other to a practical concern: one grounded in the individual subject's need for other people to corroborate its own constructions (see von Glasersfeld, 1995, p. 126). Moving away from traditional epistemologies means acknowledging that there is much more to others than the echoes they refract of our own cogitations. In order to teach and learn mathematics, we need to move beyond the standard noetic-noematic correlations between subject and object and to acknowledge that alterity, that is, the relationship to the Other, is not, as Levinas (1989) noted, of a conceptual order; for, the other cannot be dealt with through the same methods and forms of representation as conceptual objects. The relation to the other is one of solidarity, commitment and engagement, a relationship that challenges the reduction of the other to the same.

ACKNOWLEDGMENTS

A part of this paper was presented at the Working Group of Theories in Mathematics Education at the *Fifth Congress of the European Society for Research in Mathematics Education* (CERME 5), Cyprus, 2006. A partial version of it appeared in *Revista Latinoamericana de Matemática Educativa* (2006). I am grateful to the CERME Working Group participants and the reviewers for their insightful comments and critiques.

NOTES

1 This article is a result of a research program funded by The Social Sciences and Humanities Research Council of Canada / Le Conseil de recherches en sciences humaines du Canada.

2 See also Crombie (1995), Høyrup (2007) and Lizcano (1993).

3 For some recent discussions on these differences see e.g. Bikner-Ahsbahs and Prediger (2006), Cobb (2007), Lerman (2006), Lester (2005), Niss (1999), Radford (2008b, 2008c), Silver and Herbst (2007), and Sriraman and English (2005).

4 New Caledonian crows are an exception (see Bluff, Weir, Rutz, Wimpenny, & Kacelnik, 2007). I am indebted to Michael Roth for bringing this fact to my attention.

5 The essence of Being, argues Jean-Luc Nancy, is *to-be-with*. "The one/the other is neither 'by', nor 'for', nor 'in', nor 'despite', but rather "with'." (Nancy, 2000, p. 34).

6 The moral and ethical dimensions of the classroom and everyday human praxis have been recently emphasized by Roth (2007a, 2007b). Valero and Zevenbergen (2004) have stressed the importance of taking into account the social and political dimensions in mathematics education.

REFERENCES/BIBLIOGRAPHY

Adorno, T. W. (2001). *Kant's critique of pure reason.* Stanford, CA: Stanford University Press.

Arendt, H. (1958). *The human condition.* Chicago: The University of Chicago Press.

Arendt, H. (1993). *Between past and future.* (Reprint of the 1968 ed. published by The Viking Press). New York: Penguin Books.

Arzarello, F., Edwards, L., & Radford, L. (2008). Gesture and multimodality in the construction of mathematical meaning. Special Issue, *Educational Studies in Mathematics* (forthcoming).

Bachelard, G. (1986). *La formation de l'esprit scientifique.* Paris: Vrin.

Bakhtin, M. (1986). *Speech genres and other late essays.* Austin, TX: University of Texas Press.

Bartolini, B.ussi M. G., & Mariotti, M. A. (2008). Semiotic mediation in the mathematics classroom: Artefacts and signs after a Vygotskian perspective. In L. English (Ed.), *Handbook of international research in mathematics education* (2nd ed.ition). Mahwah, N.J.: Lawrence Erlbaum Associates. (in press).

Baudrillard, J. (1968). *Le système des objets [The system of objects].* Gallimard, Paris: Gallimard.

Bikner-Ahsbahs, A., & Prediger, S. (2006). Diversity of theories in mathematics education - How can we deal with it? *ZDM - The International Journal on Mathematics Education, 38*(1), 52–57.

Bluff, L. A., Weir, A. A. S., Rutz, C., Wimpenny, J. H., & Kacelnik, A. (2007). Tool-related cognition in new Caledonian crows. *Comparative Cognition & Behavior Reviews*, 2: 1–25.

Boesch, C., & Boesch, H. (1990). Tool use and tool making in wild chimpanzees. *Folia Primatologica, 54,* 86–99.

Bosch, M., & Chevallard, Y. (1999). La sensibilité de l'activité mathématique aux ostensifs. *Recherches en Didactique des Mathématiques, 19*(1), 77–124.

Brousseau, G. (2004). *Une modélisation de l'enseignement des mathématiques.* Plenary Lecture delivered at Convegno di didattica della matematica, September 24–25, 2004, Locarno, Switzerland.

Brousseau, G. (1986). Fondements et méthodes de la didactique des mathématiques,. *Recherches en Didactique des Mathématiques, 7*(2), 33–115.

Brousseau, G., and & Gibel, P. (2005). Didactical handling of students' reasoning processes in problem solving situations. *Educational Studies in Mathematics, 59,* 13–58.

Brown, T. (in press). Lacan, subjectivity, and the task of mathematics education research. *Educational Studies in Mathematics.*

Cobb, P. (2007). Putting philosophy to work. Coping with multiple theoretical perspectives. In F. K. Lester (Ed.), *Second Handbook of Research on Mathematics Teaching and Learning* (pp. 3–38). Greenwich, CT: Information Age Publishing.

Colyvan, M. (2001). The miracle of applied mathematics. *Synthese, 127,* 265–277.

Crombie, A. C. (1995). Commitments and styles of European scientific thinking. *History of Sciences, 33,* 225–238.

de Haan, M. (1999). *Learning as cultural practice*. Amsterdam: Thela/Thesis.

de Vega, M. (1986). *Introduccion a la psicologia cognitiva [Introduction to cognitive psychology]*. Mexico: Alianza.

Eagleton, T. (1997). *Marx*. London: Phoenix.

Foucault, M. (1966). *Les mots et les choses*. Paris: Éditions Gallimard.

Geertz, C. (1973). *The interpretation of cultures*. New York: Basic Books.

Høyrup, J. (2007). The roles of Mesopotamian bronze age mathematics tool for state formation and administration. *Educational Studies in Mathematics, 66*(2), 257–271.

Husserl, E. (1931). *Ideas*. London: George Allen & Unwin.

Kant, I. (1797/1974). *Anthropology from a pragmatic point of view*: (Translated by Mary J. Gregor, Trans.). The Hague: Martinus Nijhoff.

Kant, E. (1803). *On Eeducation*. Boston: D. C. Heath. (Retrieved, on June 11, 2006, from: http://oll.libertyfund.org/EBooks/Kant_0235.pdf)

Kidron, I., Lenfant, A., Bikner-Ahsbahs, A., Artigue, M., & Dreyfus, T. (2008). Social interaction in learning processes as seen by means of three theoretical frameworks. *ZDM - The International Journal on Mathematics Education*. (in press).

Köhler, W. (1951). *The mentality of apes*. New York: Routledge and Kegan Paul.

Laborit, H. (1985). *Éloge de la fuite*. Paris: Gallimard.

Lave, J. (1988). *Cognition in practice*. Cambridge: Cambridge University Press.

Leibniz, G. W. (1704/1966). *Nouveaux essais sur l'entendement humain*. (published 1765). Paris: Flammarion.

Leont'ev, A. N. (1978). *Activity, consciousness, and personality*. New Jersey: Prentice-Hall.

Leontiev [or Leont'ev], A. N. (1968). El hombre y la cultura. In *El hombre y la cultura: problemas téoricos sobre educación* (pp. 9–48). México: Editorial Grijalbo.

Lerman, S. (1996). Intersubjectivity in mathematics learning: A challenge to the radical constructivist paradigm? *Journal for Research in Mathematics Education, 27*(2), 133–150.

Lerman, S. (2006). Socio-cultural research in PME. In A. Gutiérrez & P. Boero (Eds.), *Handbook of research on the psychology of mathematics education: Past, present and future*. Rotterdam: Sense Publishers.

Lesh, R., Doerr, H., Carmona, G., & Hjalmarson, M. (2003). Beyond constructivism. *Mathematical Thinking and Learning, 5*(2–3), 211–233.

Lester, F. (2005). On the theoretical, conceptual, and philosophical foundations for research in mathematics education. *ZDM - The International Journal on Mathematics Education, 37*(6), 457–467.

Lévinas, E. (1976). *Noms Propres*. Paris: Fata Morgana.

Levinas, E. (1989). *The Levinas reader*. (Edited by (S. Hand, Ed.)). Oxford: Blackwell.

Lizcano, E. (1993). *Imaginario colectivo y creación matemática [Collective imaginary and mathematical creation]*. Barcelona: Editorial Gedisa.

Martin, J. (2004). The educational inadequacy of conceptions of self in educational psychology. *Interchange: A Quarterly Review of Education, 35*, 185–208.

Merleau-Ponty, M. (1945). *Phénomenologie de la perception*. Paris: Gallimard, Paris.

Mikhailov, F. (1980). *The riddle of the self*. Moscow: Progress Publishers.

Nancy, J.-L. (2000). *Being singular plural* (R. D. Richardson & A. E. O'Byrne, Trans.). Stanford, CAalifornia: Stanford University Press.

Nemirovsky, R. (2003). Perceptuo-motor activity and imagination in mathematics learning. In N. Pateman, B. Dougherty, & J. Zilliox (Eds.), *Proceedings of the 27th Cconference of the International Group for the Psychology of Mathematics Education* (Vol. 1, pp. 103–135). University of Hawaii.

Niss, M. (1999). Aspects of the nature and state of research in mathematics education. *Educational Studies in Mathematics, 40*, 1–24.

Piaget, J. (1973). *To understand is to invent. The future of education*. New York: Grossman.

Popkewitz, T. (2004). School subjects, the politics of knowledge, and the projects of intellectuals in change. In P. Valero & R. Zevenbergen (Eds.), *Researching the socio-political dimensions of mathematics education* (pp. 251–267). Boston: Kluwer.

Radford, L. (1997). On psychology, historical epistemology and the teaching of mathematics. *For the learning of mathematics, 17*(1), 26–33.

Radford, L. (2000). Signs and meanings in students' emergent algebraic thinking: A semiotic analysis. *Educational Studies in Mathematics, 42*(3), 237–268.

Radford, L. (2002). The seen, the spoken and the written. A semiotic approach to the problem of objectification of mathematical knowledge. *for the learning of mathematics, 22*(2), 14–23.

Radford, L. (2003a). On culture and mind. A post-Vygotskian semiotic perspective, with an example from Greek mathematical thought. In M. Anderson, A. Sáenz-Ludlow, S. Zellweger, & V. Cifarelli (Eds.), *Educational perspectives on mathematics as semiosis: From thinking to interpreting to knowing* (pp. 49–79). Ottawa: Legas Publishing.

Radford, L. (2003b). On the epistemological limits of language: Mathematical knowledge and social practice in the Renaissance. *Educational Studies in Mathematics, 52*(2), 123–150.

Radford, L. (2003c). Gestures, speech and the sprouting of signs. *Mathematical Thinking and Learning, 5*(1), 37–70.

Radford, L. (2006a). Communication, apprentissage et formation du *je communautaire*. In B. D'Amore & S. Sbaragli (Eds.), *20th Italian national conference Incontri con la Matematica*, November 3–5, 2006. (Plenary Lecture available at: http://laurentian.ca/educ/lradford/).

Radford, L. (2006b). Algebraic thinking and the generalization of patterns: A semiotic perspective. In S. Alatorre et al. (Eds.), *Proceedings of the 28th Conference of the International Group for the Psychology of Mathematics Education, North American Chapter* (Vol. 1, pp. 2–21), Mérida, Vol. 1, pp. 2–21.

Radford, L. (2008a). Culture and cognition: Towards an anthropology of mathematical thinking. In L. English (Ed.), *Handbook of International Research in Mathematics Education (2nd Edition)* (pp. 439–464). New York: Routledge, Taylor and Francis.

Radford, L. (2008b). Connecting theories in mathematics education: Challenges and possibilities. *ZDM - The International Journal on Mathematics Education, 40*(2), 317–327.

Radford, L. (2008c). Theories in Mathematics Education: A Brief Inquiry into their Conceptual Differences. Working Paper. Prepared for the ICMI Survey Team 7. The notion and role of theory in mathematics education research. Retrieved from http://www.laurentian.ca/educ/lradford/.

Radford, L. (2008d). Di Sé e degli Altri: Riflessioni su un problema fondamentale dell'educazione [The self and the other: Reflections on a fundamental problem in education]. *La Matematica e la sua didattica, 22*(2), 185–205.

Radford, L. (2008e). Iconicity and contraction: A semiotic investigation of forms of algebraic generalizations of patterns in different contexts. *ZDM - The International Journal on Mathematics Education, 40*(1), 83–96.

Radford, L., Bardini, C., & Sabena, C. (2007). Perceiving the general: The multisemiotic dimension of students' algebraic activity. *Journal for Research in Mathematics Education, 38*, 507–530.

Radford, L., & Demers, S. (2004). *Communication et apprentissage. Repères conceptuels et pratiques pour la salle de classe de mathématiques.* Ottawa: Centre franco-ontarien des ressources pédagogiques.

Roth, M.-W. (2001). Gestures: Their role in teaching and learning. *Review of Educational Research, 71*(3), 365–392.

Roth, W.-M. (2007a). Identity in scientific literacy: Emotional-volitional and ethico-moral dimensions. In W.-M. Roth & K. Tobin (Eds.), *Science, learning, and identity: Sociocultural and cultural historical perspectives* (pp. 153–184).

Roth, W.-M. (2007b). The ethico-moral nature of identity: Prolegomena to the development of third-generation cultural-historical activity theory. *International Journal of Educational Research, 46*, 83–93.

Sabena, C., Radford, L., and & Bardini, C. (2005). Synchronizing gestures, words and actions in pattern generalizations,. *Proceedings of the 29th Conference of the International Group for the Psychology of Mathematics Education* (Vol. 4, pp. 129–136), Australia, Vol. 4, pp. 129–36.

Seitz, J. A. (2000). The bodily basis of thought. *New Ideas in Psychology, 18*, 23–40.

Shweder, R. A. (1991). *Thinking through cultures*. Cambridge: Harvard University Press.

Shweder, R., & LeVine, R. (1984). *Culture theory. Essays on mind, self, and emotion*. Cambridge: Cambridge University Press.

Sierpinska, A. (1996). Interactionnisme et théorie des situations: Format d'interaction et contrat didactique. In D. Grenier (Ed.), *Didactique et technologies cognitives en mathématiques, Séminaires 1996* (pp. 5–37). Grenoble: Université Joseph Fourier.

Silver, E., & Herbst, P. (2007). Theory in mathematics education scholarship. In F. K. Lester (Ed.), *Second handbook of research on mathematics teaching and learning* (pp. 39–67). Greenwich, CT.: Information Age Publishing.

Sriraman, B., & English, L. (2005). Theories of mathematics education: A global survey of theoretical frameworks/trends in mathematics education research. *ZDM - The International Journal on Mathematics Education, 37*(6), 450–456.

Steinbring, H., Bartolini Bussi, M., and & Sierpinska, A. (1998). *Language and communication in the mathematics classroom*. Reston, Virginia: NCTM.

Torigoe, T. (1985). Comparison of object manipulation among 74 species of non-human primates. *Primates, 26*(2), 182–194.

Valero, P., & Zevenbergen, R. (Eds.). (2004). *Researching the socio-political dimensions of mathematics education: Issues of power in theory and methodology*. Boston: Kluwer.

Vološinov, V. N. (1973). *Marxism and the philosophy of language*. Harvard: Harvard University Press.

von Glasersfeld, E. (1995). *Radical constructivism: A way of knowing and learning*. London: The Falmer Press.

Vygotsky, L. S., and & Luria, A. (1994). Tool and symbol in child development. In R. van der Veer and & J. Valsiner (Eds.), *The Vygotsky reader* (pp. 99–174). Oxford: Blackwell.

Wartofsky, M. (1979). *Models, representation and the scientific understanding*. Dordrecht: D. Reidel.

Weber, M. (1992). *The protestant ethic and the spirit of capitalism*. London: Routledge.

Wertsch, J. V. (1991). *Voices of the mind*. Cambridge, MA: Harvard University Press.

Yackel, E., and & Cobb, P. (1996). Sociomathematical norms, argumentation, and autonomy in mathematics. *Journal for Research in Mathematics Education, 27*(4), 458–477.

Luis Radford
École des sciences de l'éducation
Université Laurentienne
Sudbury, Ontario
Canada

TANIA M. M. CAMPOS, VERA HELENA GUISTI DE SOUZA AND ROSANA NOGUEIRA DE LIMA

AN ATTEMPT TO ACHIEVE REIFICATION IN FUNCTIONS – A STUDY BASED ON SEVERAL SEMIOTIC REGISTERS

INTRODUCTION

As part of a research project entitled, "Mathematics teaching-learning processes at Basic School" (6- 14-year-old), we have worked, between 2000 and 2006, with a group of seven Mathematics teachers from public schools in São Paulo, Brazil. This group was assisted by researchers with a Mathematics or Mathematics Education PhD or MsC degrees. We have held weekly three and a half hour meetings to discuss the pedagogical and mathematical aspects of a large variety of school mathematics topics, such as Pythagoras' Theorem, Thales' theorem, fractions, equations, similitude, functions and statistics.

The group of teachers in question has chosen to discuss the topic of functions, since, by the National Curriculum Standards (PCN) elaborated by Brazilian Education Ministry, it must appear at the end of secondary school and at the high school level with a focus on:

- plotting, analyzing and reading graphs;
- analyzing different phenomena in daily life, physics, geography, economy;
- modeling with functions and graphs;
- connecting subjects inside and outside mathematics.

Primary and Secondary School levels are compulsory for all children aged respectively 6-10 and 11-14 years old, but High School is not compulsory for students aged 16-18 years old. So, these teachers are interacting with many students that need to learn the concept of function with the focus described above.

When we started the discussion about functions under the guidance of a lecturer, we engaged in a brainstorming session about the word "function". On this occasion, the teachers started saying things such as,

> "I am scared of graphs …"; (1)

> "I usually play with the location of points in order to introduce the subject to my students …"; (2)

> "I understand when the horizontal variable is *t*; if not … I can't see …"; (3)

> "I immediately do a table of values"; (4)

L. Radford, G. Schubring, and F. Seeger (eds.), Semiotics in Mathematics Education: Epistemology, History, Classroom, and Culture, 235–248.

"The manuals for teaching at the basic level do not explore graphs ..."; (5)

"I only know a function if it is linear or quadratic ...". (6)

We believe that, for the teachers in question, the concept of function continues to be just a process and not an object (Linchevski & Sfard, 1991; Sfard & Linchevski, 1994; Kieran, 1993) and they are at a level of *condensing*, somewhere between seeing the function as a process and seeing it as an object. Therefore, we felt a very strong need to try and change this situation, at least initially, with these particular mathematics teachers.

FOCUS

It became a challenge to make these particular teachers develop a more elaborated understanding of function. We wanted to help them advance from an *interiorizing* to a *condensing phase* and, if possible, to help them then evolve from *interiorizing* and *condensing phases* to a *reifying phase* (Sfard, 1991).

We believed that, in order to do so, it was necessary to work with several semiotic registers of representation, because they would make it possible for the teacher to have a broader view of the concept, which, in turn, would most likely broaden his or her understanding of function. Hence, we have elaborated three activities based on several semiotic registers of representation, to be solved by using Cabri-géomètre II software dynamics, to strongly explore graphs and their relation to algebraic and natural language registers, as recommended by Duval's (1995) Semiotic Register of Representation Theory.

It is important to emphasise that we were not trying to teach how to convert from one kind of register to another. Our goal was to make the teachers work with different registers, so that they could analyse the different properties of each register.

From this, we have posed the following question:

"Can a didactical sequence involving the use of graphic, algebraic and natural language registers help to achieve a reification of the concept of function?"

THEORETICAL FRAMEWORK

To Sfard (1991) and Sfard & Linchevski (1994), there is a duality process-object in most mathematical concepts. They call it Reifying Theory, meaning that there is an *operational conception* (as an oriented process), which comes first, and it evolves to a *structural conception* (as a mathematical object), by means of *reifying* the process. They also say that this evolution is not easy; it can be challenging and it happens in three continuous phases: *interiorization, condensation and reification.*

In the *interiorization phase*, the process is made over familiar mathematical objects; for example, in the case of functions, using input values in the "function machine" to calculate the output values. If we need to look at specific values, as in a table of values, for example, we are in an *interiorization phase*. It is exactly what we do when we have a new problem and we look for some kind of pattern.

In the *condensation phase*, the process has general characteristics and it is compacted to turn into operational units. For example, one is in this phase when one sees that when there is something going into and something going out of the "function machine", this is because there is a relationship between input and output values, independent of specific values. The better we can deal with a function as a whole, the more advanced we are in the *condensation phase*. At this level, we are able to investigate, to draw graphs of, compound the function and even determine its inverse.

In the *reification phase*, the process is solidified and it is transformed into an object. In the case of functions, *reifying* means that we can deal with a function as an unknown, either in equations with parameters or in differential equations, for example.

We want mathematics teachers to be able to see the mapping as a whole (Sfard, 1991), meaning that they must be in an advanced *condensation phase*, somewhere near the *reification phase*.

Based on this, we have designed our research instruments in such a way as to help subjects achieve a global view of graphs of more general functions, and be able to relate them to their algebraic law, in order to guide them to a more evolved level of condensation, or even to reification.

RESEARCH METHOD

We have developed a qualitative research based on the following steps from didactical engineering (Artigue, 1995): an overview of previous conceptions from the researched group; creating, applying and observing a didactical sequence; analysing teachers' protocols. These steps and the observers' notes were thoroughly analysed, in an attempt to find the validation argument each subject has used to justify his answers. Such argumentation may provide evidence for which phase each subject is in, and whether or not he or she is evolving to the next phase.

In order to get an overview of teachers' previous conceptions, we carried out a brainstorming session with the word *function*. It was during this session that the teachers expressed their feelings of fear and concern about graphs, variables, table of values and the safe use of algebraic registers.

Faced with the results from the brainstorming session, we realised that our subjects seemed to be in the *interiorization* phase regarding the concept of function. In our view, when one evolves through *interiorization* and *condensation phases* and is in the *reification phase*, one has learned to deal with a mathematical object as it is, not just as an oriented process. We believe that this evolution occurs if one does not mistake an object with its representation and, to achieve this, it is necessary to discriminate and use at least two representation systems (Duval, 1995; 2000). In this way, the activities for this study were developed using three representation systems: natural language, algebraic systems and graphic systems.

We chose an *algebraic system*, as it is the most effective one for dealing with functions, at least when one can master it; also, because it is the most common one in Brazilian Mathematics textbooks and the teachers are more familiar with it. We also chose a *graphic system*, not only because it has visual characteristics, but also because we believe it can lead to ascertainable learning, since it entails different and complementary properties from the algebraic one. Finally, we also made use of *natural language*, as it is in natural language that we think and speak when we try to understand either register, that is, algebraic or graphic systems.

We designed three activities, with which the teachers worked in six meetings. In all the activities, the teachers were to deal with algebraic, natural language and graphic registers. The teachers worked in pairs, in an environment with computers and using dynamical software Cabri-géomètre II, with a neutral observer watching each pair. The protocols and written observations were carried out at the end of each session, and, at the beginning of the next one, we did an institutionalization, mainly based on doubts and questions posed about the protocols and/or the observers' notes.

We must make note of the fact that the variable "use of an environment with computers" was not taken into account because the teachers were used to it. We used the Cabri-géomètre II, in particular, because the teachers were already familiar with it, as this group typically works with such software. In addition, it is possible to personalise the software's menu with macro-constructions. In this way, the teachers would not need to construct any of the graphs, as they were already obtained by a change in the numerical edition of the software, but would only need to make the proposed changes in either the independent or dependent variable, which in turn would make changes in the graph of the function itself.

In order to analyse teachers' protocols and observers' notes, we have classified teachers' ways of validating their argumentation into four different categories: *visual*, *numerical*, *graphic*, and *algebraic*. These categories were inspired by the phases of Reification Theory: *visual*, if the teacher concluded something by just looking at the graph; *graphic*, if the teacher had used any tool from the software to visualize what was going on; *numerical*, if the teacher produced some kind of calculation such as a table of values or if he or she had substituted a particular value in the function or in the graph; *algebraic*, if the teacher did a formal verification by using function's algebraic expression.

As a way of exemplifying these categories, we provide possible answers for a question like, "What is the relation between the graphs of $h(u)=u^2$ and $g(u)=u^2+1$?" When considering both graphs, justification is *visual* if a subject just looks at the graphs and does not take any action to give the answer; *numerical* if one takes numerical values such as (2, 4) in the graph of h and (2, 5) in the graph of g to decide what happens; *graphic* if an individual analyses mathematical characteristics of the graph looking at the changes in coordinates of a moving point over both graphs; *algebraic* if one makes use of both algebraic expressions to conclude anything.

It is our view that the first two categories, *visual* and *numerical*, may indicate an interiorization phase, while *graphic* and *algebraic* categories may indicate condensation and/or reification, depending on the level of sophistication given to them. In this way, we have tried to verify if the teachers' vision evolved from "function as a process" to "function as an object". We present, in this text, some observations we have made, but, before this, we think it is necessary to bring in some ideas about the three activities.

ABOUT THE ACTIVITIES

If indeed we reflect carefully about the comments (1), (2), (3), (4), (5) and (6) at the beginning of this paper, we can ask ourselves: (1) what is so scary about graphs? Is it because they need a *formal definition*? Aren't the students ready for just a little bit of formal Mathematics? (2) Is points' location a good approach to *visualize* a graph as a whole? (3) Is time *t* variable an *easier* one? Is it because the time gives the idea of movement? Is it possible to learn the significance of independent variables if the teacher just uses the label *x*? (4) Does the table of values give us relevant information about the graph, if we have the *algebraic expression*? Is it possible for the pupils to *generalize* and to perceive that three points on the Cartesian plane are not always on a parabola? (5) Why don't mathematics textbooks explore other graphs besides straightedges and parabolas? Why is the gap between all kinds of functions so big that first-degree and second-degree polynomial functions have one chapter each and the others types do not appear for discussion? Nowadays, we have a lot of graphic software, which allows us to bring a large variety of graphs to the classroom. We can also use graphic calculators or computers. (6) If we only bring first and second-degree polynomial functions to the classroom, will the pupils really learn the concept of function? Or will they say "this parabola ..." about the graph of $g(x) = x^3$?

In search of answers to these questions, we designed a group of activities that aimed to overcome some of the obstacles classified by Sierpinska (1992). This was to understand the notion of function and was directly related to the teachers' comments.

– (1) *Graphs scaring* may be explained by a strictly algebraic focus on the learning of functions.

  In her list of epistemological obstacles related to the concept of function, Sierpinska suggests the following one: "(A conception of function) Only relationships describable by analytic formulae are worthy of being given the name of functions" (Sierpinska, 1992, p. 46).

Taking this obstacle into consideration, we decided to use a great variety of graphs in the activities and this was made easier by Cabri-géomètre II software.

– (2) Importance given to *points' location* seems to be connected to two causes. The first one is well described by Sierpinska (1992):

> Another epistemological obstacle (EO) in Sierpinska's list (the seventh one) is the following: "EO(f)-7: (An attitude towards the notion of number) A Pythagorean philosophy of number: everything is number" (Sierpinska, 1992, p. 41).

This obstacle, we believe, is motivated by an underlying preoccupation, coming from our textbook writers and their emphasis on the idea that Mathematics is concerned with practical problems.
Considering these causes, in our activities we stressed situations in which the teachers were invited to *visualize* graphs (Duval, 1999).

- (3) *t variable* seems to be easier to understand if we accept Sierpinska's arguments about her 16th obstacle:

> "EO(f)-16: (A conception of variable) The changes of a variable are changes in time" (Sierpinska, 1992, p. 55).

In order to prevent subjects from being confined to using just the *t* variable, we used several names for both dependent and independent variables.

- (4) Mathematics textbooks, in general, have given too much importance to *tables of values* as soon as the algebraic expression is completed, without the necessary discussion about function as a relationship between physical or other quantities.

To trigger a vision of function as a relationship between variables, we used several kinds of functions' graphs in an effort to guide subjects to a global visualization, rather than the point-to-point one given by the table of values.

- (5) and (6) Emphasis on *first-degree polynomial function* seems connected to Sierpinska's 9th obstacle, which reads as follows:

> "EO(f)-9: (An unconscious scheme of thought) Proportion is a privileged kind of relationship" (Sierpinska, 1992, p. 43).

During the activities, subjects explored unusual functions (at least for them), such as rational and trigonometric ones, as a way to break this unconscious scheme of thought. This approach was possible due to the use of graphic software like Cabri-géomètre II.

In our case, since we are dealing with Mathematics teachers with at least seven years in classroom practice, we are focusing on the *development of a more elaborate level of understanding functions* and the *use of various representations*. This choice is due to their previous knowledge of the subject of function and their experience in teaching it. Each of the three activities is twofold:

- They use several semiotic registers, such as graphic, algebraic and natural language.
- They stimulate the global visualization of functions' graphs, not as a set of points but as a trajectory in which one can see relations between independent and dependent variables.

For each of the activities, we consider two *reference functions* from among the following:

$$f(x)=x\,;\;\; g(x)=x^2\,;\;\; h(x)=x^3\,;\;\; m(x)=\sin(x)\,;\;\; n(x)=\frac{1}{x} \qquad (3)$$

as well as *associate functions* which were obtained from the reference ones by "summing up a constant in" (Activity 1), "multiplying by a constant" (Activity 3) and "applying modulus in" (Activity 2) the independent variable as well as the dependent variable. One of these operations was proposed in each activity.

New tools were added to Cabri's menu by means of macro-constructions, and were ready for use. With them, the teachers could, firstly, obtain the reference function graph, then, several of the associated ones, by using a tool for the intended operation, which was proposed in natural language, first in the dependent variable, and then in the independent one. This order is due to our experience working with functions with undergraduate students (Da Silva et al., 2002).

At the beginning of the activity, the teacher opens the menu with the new configuration and uses the new tools to get one of the reference functions' graphs.

To exemplify this, we present the first task of activity 1: *"Open file paref.fig to get the graph of* $g(x)=x^2$ *and the file OficProf.men. What is the relation between the graph of g and the graph of the associated function which is obtained when one sums up 2 to g?"*

Afterwards, he or she chooses the tool to perform the asked-for operation, which is described in natural language and makes reference to a constant choice that can be made by the numerical edition that appears on the screen (see Figure 1).

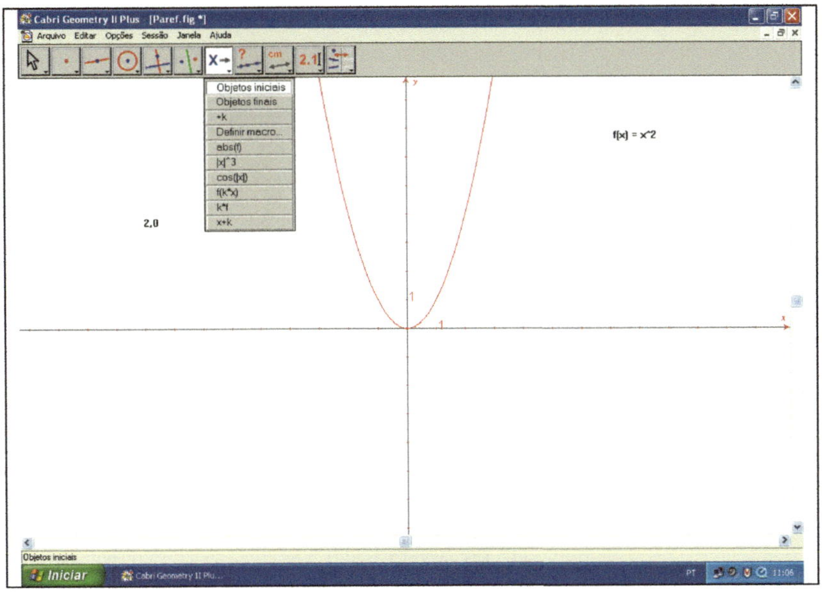

Figure 1. New tools for the development of activities

The resulting graph from this operation, which appears on the screen, is that of the associate function. The teacher can change the value of this constant by "clicking" on the numerical edition, and this act changes the graph as well. In this way, the teacher can see what happens to the graph of a function when either of its dependent or independent variables is summed up as a constant (see Figure 2). The same can be done when multiplying one of the variables by a constant or applying modulus.

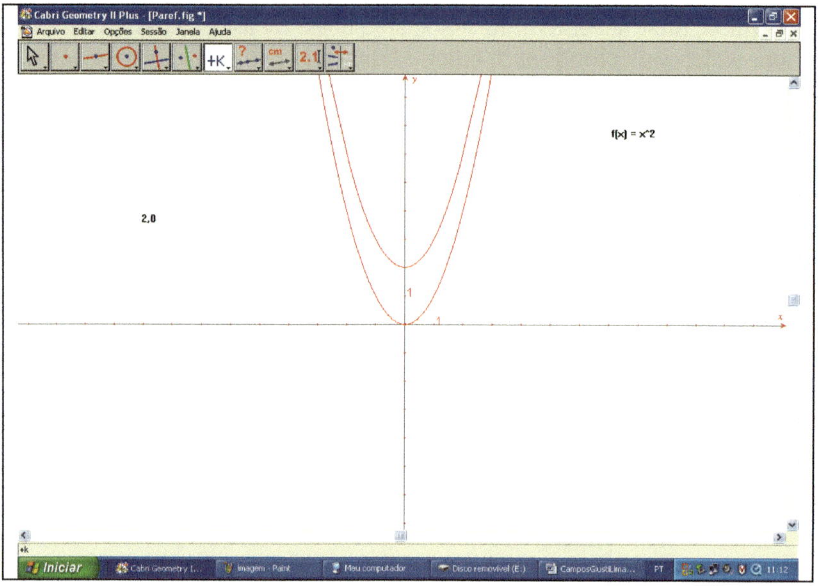

Figure 2. Graphic of the associated function obtained by summing up a constant to g

While analysing changes in the graphs from the reference function to the associated ones, teachers can relate them with the algebraic laws, and see the new function's domain and range as well as the graph's symmetries. From this analysis and the activity instructions in natural language, they are asked to write down their reasoning and answers for all the characteristics of the functions.

One of the tasks involved in each of the activities is as follows: *"What is the algebraic expression, the domain and the image of the associated function?"*

Our analysis is mostly based on the justifications they use for questions such as this one.

DATA AND RESULTS

We have analyzed teachers' protocols as well as observers' notes, trying to find out which validation arguments, *visual, numeric, graphic* and *algebraic*, each teacher uses to answer the proposed questions, because we think this is a good parameter for measuring the level of condensation and/or reification the individual is in.

Analysing written observations, we have found that, while working on the activities, all seven teachers used *visual* validation, meaning that they just looked at the graphs to reach their conclusions, as we can conclude from two subjects' justifications.

> *"When summing up 2 to the original function, the associated one translates in the y-axis 2 units without changes in the graph's shape."*

> *"Yes. Because the function vertex is on the y-axis, therefore values of x to the left and to the right of the y-axis are equidistant."* (When asked whether the graph has symmetry.)

None of them used *graphical* validation; although two frequently made comments related to *numerical* validation as follows.

> *"Yes, related to y-axis parabola it [the parabola] is divided into two equal pieces. Ex.: (2; 4.01) and (-2; 4.01)."* (When asked whether the graph has symmetry.)

> *"I think it is going to have symmetry. Wait! Let me think! If I take -1 to the third, it is going to be -1, and the absolute value is 1. Then, it will be a parabola."*

And six of the teachers never referred to *algebraic* validation (see Figure 3). Only one made an attempt to do so, but he did not succeed

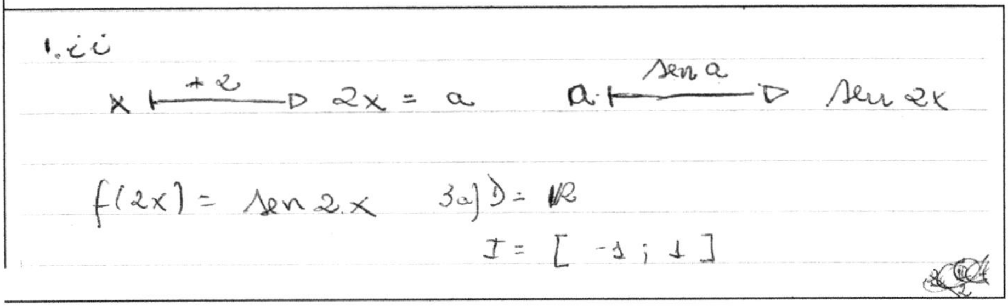

Figure 3. Subject's attempt to justify based on algebraic validation

The fact that the software produced a graph and the teachers could see it seemed to be enough for them to conclude something about the characteristics of the function. Thus, they used a visual validation, which, in our view, stresses the fact that they, apparently, are in the interiorization phase when dealing with the graph register.

From teachers' protocols and observers' notes, we can perceive that visual validation was and still is very, very strong. All of them have accepted, by just looking at the graph, that translation occurs when we "sum up a constant in" either variable. Three of them showed some kind of *oral* preoccupation with graphic validation, asking us if it was possible to verify symmetry by using Cabri II tools, but none did so. One of them made an attempt to justify it saying

"How to explain to a pupil that there is symmetry [in the graph]*? They are going to find the coordinates of a point* [with Cabri II tools]*"*.

He marks point (…; 8.89) in the graph and says: *"I have to find -8.89."*

From this, we can see the worry involved in a search for a different kind of validation, not only the kinds of validation related to the interiorization phase. Hence, we believe that the subjects are not just in the interiorization phase, but they have started to reach a level of understanding of function in the condensation phase. This shows the beginning of a sophistication they did not demonstrate before their work with different registers.

One of the teachers refused the computer help, saying, *"I always do a table of values* [numerical validation] *in my draft"*. This shows that, although he is using a validation still related to the interiorization phase, he needs a more sophisticated way to validate than just looking at a graph. He has also done good work with the conversion from natural language to algebraic registers.

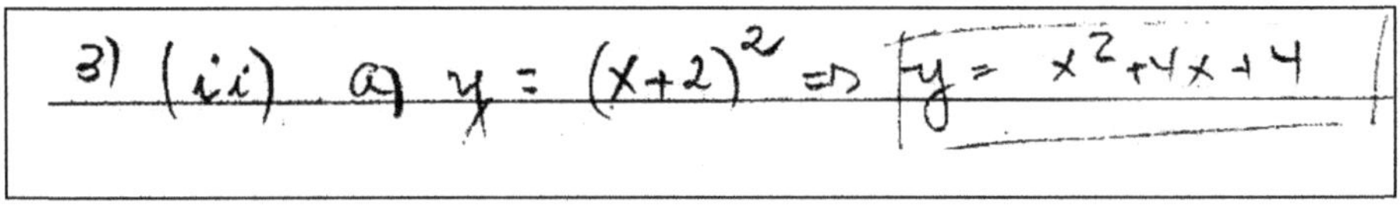

Figure 4. The work of one of the teachers in finding the algebraic register from natural language, when asked to "sum up 2" to independent variable

As we see in Figure 4, in the case of the algebraic law, he could easily find the associate function law, but he always uses $y = \ldots$ and never $f(.) = \ldots$. In other words, we can say that he seems to mistake the mathematical object with its representation, and he cannot see the graph as a set of coordinate pairs, in which the second coordinate, y, has a specific value, and not a random one.

Another teacher always uses numerical validation to evaluate coordinates for some point over the graph, like the parabola's vertex. Sometimes, he has compared the coordinates given by the software with the ones he has calculated in the draft. This teacher has shown that he has experienced a lot of difficulty in converting the natural language register into the algebraic one. At least two more teachers have had the same difficulty, especially when the association was made over the dependent variable.

No one has tried an algebraic validation in any case. Just one time one of them wrote something comparable to Figure 5. Before trying the algebraic expression for the associate function: he gave one or two values for x; used software to confirm values; and gave up saying that he could not get the generic case. This is an attempt to use an algebraic validation, again evidence that these teachers are growing in sophistication while dealing with the proposed activities, tending to the condensation phase.

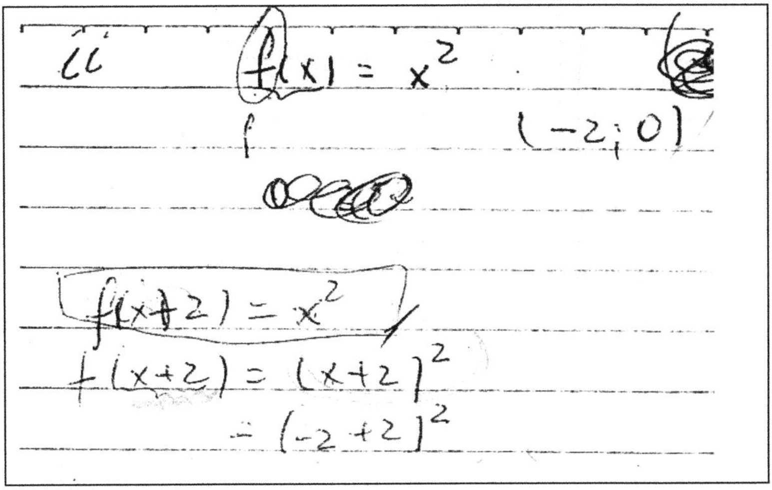

Figure 5: The work of one of the teachers in finding the algebraic register from natural language

All of the teachers have tried to write algebraic expressions for the associate functions – another attempt to use algebraic validation. For the associate function, they always used the same name as the reference one, although no one wrote "$f(x)=f(x+a)$".

Symmetry of all kinds was troublesome because it is not usual for them and needs a global visualization. At first, they were looking for a symmetry related to the y-axis. During the group discussion, two of them had *seen* that a translation of the graph did not change symmetry but rather the symmetry axis; no one had tried some kind of graphical or algebraic validation in this case.

It was difficult to achieve some response in all operations with modulus. When asked about the meaning of it, two of them said *"modulus is distance"*; one answered *"it is positive"*; another, *"it is x if x is positive and it is –x if x is negative"*, but had doubts about the meaning of $|x^2|$; another one, gave us an example: |-3|=3, but said that $\sqrt{(-3)^2}$ is –3 and no one recognized $\sqrt{(-3)^2}$ as |-3|. All of them had difficulty generically operating modulus.

It is important to emphasise that working with such functions enlarged the teachers' universe of functions. One of them said,

> *"When I saw a function as $f(x)=\dfrac{2x+3}{3x-4}$, I used to look at it as a linear function, since the denominator is linear. Now I know it is not".*

Another said,

> *"I have been teaching linear and quadratic functions only. Now I will try some other ones"*;

and still another said that he can bring other functions to the classroom, such as sine, cosine, exponential.

All of them accepted the independent variable as the horizontal one in the graph, regardless of its name, *x*, *a*, *b* or *t*, even the teacher who said,

> *"I understand when the horizontal variable is t; if not ... I can't see ..."*

at the beginning of the project.

CONCLUSION

In our data analysis, we can see that visual validation is much too strong and it was not possible to completely break it. We believe that this happened because the graph register was not usual for them and more time and discussion on the subject is necessary to make teachers understand that the graph brings different properties not easily seen only with the algebraic expression, but we do not mean that they should replace one with the other. It is important to discuss and work with both of them, as well as with natural language. In this way, this confirms Duval saying (1995, 2000) that we learn to use many semiotic representation systems by using them, and the teacher is responsible for creating opportunity and means for this to happen in the classroom. It seems that these teachers themselves never learned by means of different representations, so they lack the ability to do so. In this sense, the work with these activities was the first time they had seen such an approach, which may be a reason why they could not intertwine the different properties of algebraic and graphic registers. And this was one of the reasons why we believe they could not reach a more sophisticated level of condensation phase.

In reference with the goal of this study, namely, to "help this group advance in the *condensation phase* and, if possible, evolve from the *interiorization* and *condensation phases* to the *reification phase* in the subject function in Mathematics" we can say that, for this group and in these conditions, the introduction of a new tool like the computer ***did not*** help to accomplish the desired level of reification, although we can also say that most of them achieved a higher level of condensation than they had before. For example, when they recognize a larger scope of functions and do not say "parabola" anymore for any graph that is not a straight line; likewise, they recognize that $f(x)=\frac{2x+3}{3x-4}$ is not a first-degree polynomial function just because the denominator is a first-degree polynomial function; they also accept a function (as a mathematical model for something) with independent variable other than time *t*; they ask if it is possible to validate symmetry using tools from Cabri-géomètre II; and they say they will use trigonometric and rational functions with their students.

We can also bring up some questions about graphic registers of functions: "Why is it so difficult to break visual validation?"; "Why is it so difficult to understand and to assimilate global visualization of graphs?"; "Is it just a question of habit?".

We believe that more time and discussion about the subject function during undergraduate courses for future Mathematics teachers is necessary, because they must be in a certain level of reification in the subject if they are going to “teach” it, and we know from researchers in the area that reifying is difficult to achieve for many of us. In addition to this, the treatment of graphic registers must be stimulated, as well as conversions between graphic, algebraic and natural language registers, because they are not spontaneous and must be completely understood by future Mathematics teachers.

As regards our research question, *“Can a didactical sequence involving the use of graphic, algebraic and natural language registers help to achieve reification on the subject function?”*, we believe the answer is YES. In the case of the teachers in question, we encountered other difficulties that the three activities by themselves could not surpass, mainly because of the teachers’ lack of previous experience with graphs and natural language registers, as well as conversion of registers in the case of function.

It is not enough for a teacher to work with different registers in his or her classroom. It is necessary to promote the understanding of properties each register highlights, the relation between these properties and how to get from one register to another, in order to try to develop a more elaborate level of understanding functions, and then, reach a higher level of condensation, or even, reification, depending on the school level.

REFERENCES

Artigue, M. (1995). Ingeniéria didáctica. In: M. Artigue, R. Douady;, L. Moreno, & P. Gómez (Eds.), *Ingeniéria didáctica en educación matemática* (pp. 33–59). México, México: Grupo Editorial Iberoamérica.

Da Silva, B. A., Manrique, A. L., Bianchini, B. L., Dubus, M. T., & de Souza, V. H. (2002). Analysing functions behaviour in a computational environment. In *Proceedings of the International Conference on the Teaching of Mathematics.* 2nd Hersonissos: University of Crete. Retrieved on December 28, 2007, from http://www.math.uoc.gr/~ictm2/Proceedings/ICTM2_Proceedings_ Table_of_Contents.html

Duval, R. (2000). Basic issues for research in mathematics education. In M. J. Høines & A. B. Fuglestad (Eds.), *Proceedings of the 24th Conference of the International Group for the Psychology of Mathematics Education* (Vol. 1, pp. 55–69). Hiroshima, Japan: PME.

Duval, R. (1995). *Sémiosis et pensée humaine. Registres sémiotiques et apprentissages intellectuels.* Neuchâtel, Suisse: Peter Lang.

Duval, R. (1993). Graphiques et équations. L’articulation de deux registres. In Centre d’Etudes et de Recherches en Sciences de l’Education (CERSE) (Ed.), *Les Sciences de l’éducation - Pour l’ère nouvelle*, (Vol. 1--3), pp. 57--72).

Kieran, C. (1993). Functions, graphing, and technology: Integrating research on learning and instruction. In T. A. Romberg, E. Fennema, & T. P. Carpenter (Eds.), *Integrating research on the graphical representation of functions* (pp. 189--237). Hillsdale, NJew Jersey: Lawrence Erlbaum Associates, Inc.

Linchevski, L., & Sfard, A. (1991). Rules without reasons as processes without objects – The case of equations and inequalities. In F. Furinghetti (Ed.), *Proceedings of the 15th Conference of the International Group for the Psychology of Mathematics Education* (Vol. 2, pp. 317–324). Assisi, Italy: PME.

Sfard, A., & Linchevski, L. (1994). The gains and pitfalls of reifying – the case of Algebra. *Educational Studies in Mathematics, 26*, 191–228.

Sierpinska, A.nna. (1992). On understanding the notion of function. *Mathematical Association of America Notes, 25*, 25–58.

Rosana Nogueira de Lima
Programa de Pós-Graduação em Educação Matemática
Universidade Bandeirante de São Paulo

Tania M. M. Campos
Programa de Pós-Graduação em Educação Matemática
Universidade Bandeirante de São Paulo

Vera Helena Giusti de Souza
Programa de Pós-Graduação em Educação Matemática
Universidade Bandeirante de São Paulo

MIRCEA RADU

SYMBOLIC LANGUAGE VERSUS UNDERSTANDING IN MATHEMATICS EDUCATION

A brief archaeological investigation of mathematics education discourse

Over the past decades, a considerable amount of literature has emerged in which the distinction between understanding and abstract symbolic expression in respect to mathematics education is addressed.[1] The various studies often rely on different conceptual frameworks and it has become increasingly difficult to find a common denominator that would be helpful in fostering the communication between the promoters of the various theoretical views, and between these promoters and other important groups also dealing with mathematics education such as teachers, teacher trainers, mathematicians, policy makers etc. The objective of the present paper is to propose a framework that in my view is fruitful for comparing a wide range of conceptions. This framework is not intended primarily as a means for criticizing the various positions but rather as a means for identifying their central concerns and/or certain tensions between these concerns. I begin with a concise presentation of the framework. I then use it to explore the tension between formalism and understanding in Bertrand Russell's philosophy of mathematics. Next I explore the way in which this tension manifests itself in Paul Cobb's account of mathematics teaching. I close with a brief exposition of some of the dangers associated with a philosophical and/or didactic discourse which relies on a two strict distinction between understanding and symbol manipulation.

FORMALISM VERSUS MEANING IN MATHEMATICS

This paper is built around two statements expressing antithetical concerns about the use of signs in mathematics.

Statement (I)
If I am given a *formula*, and I am ignorant of the *idea* it expresses, then the *formula* cannot teach me anything, but if I already know the *idea,* what does the *formula* teach me?

Statement (II)
If I am given an *idea*, and I am ignorant of the *formula* expressing it, then the *idea* cannot teach me anything, but if I already know the *formula*, what does the *idea* teach me?

L. Radford, G. Schubring, and F. Seeger (eds.), Semiotics in Mathematics Education: Epistemology, History, Classroom, and Culture, 249–261.

Both statements are constructed so as to express as boldly as possible a distinction between two opposite positions concerning the role played by formal, symbolic language in mathematical work. The first statement reflects a Platonic view, giving *ideas* absolute priority over mathematical inscriptions. It comes close to a general statement about the use of signs that can be found, for instance, in Augustine's *De Magistro*.[2] It ammounts to saying that ideas are irreducible to symbolic mathematical language.

The second statement reverses the planes. Symbolic language is assigned priority over whatever kind of language-independent entity we may be looking at. Such a view was advocated, for example, by Robert Graßmann,[3] and it found its way into Alfred North Whitehead's *Universal Algebra.* Whitehead explains that mathematics relies on "substitute signs" and then draws a distinction between words and substitute signs in the following terms:

> A word is an instrument of thinking about the meaning which it expresses. A substitute sign is a means of not thinking about the meaning it symbolizes.[4]

The two statements address the relationship between the symbolic language of mathematics, on the one hand, and a type of extra linguistic entity that might be called "idea", "meaning", "taken-as-shared-meaning", "situation", "cognitive schemata", "implicit knowledge" etc. on the other. Both statements can be easily rephrased, replacing "idea" by "extra linguistic entity". I find these statements useful in classifying philosophical views concerning mathematical knowledge in general, and theories concerning the teaching/learning of mathematics alike.

The objective of this paper is to show how these two statements can be used to characterize the fundamental concerns about the relationship between symbolic language and meaning making in mathematics and mathematics education theory. I argue that they make it possible to trace down and emphasize certain fundamental tensions characteristic of texts that try to capture the nature of mathematical thinking.

Stated briefly, the thesis that I am proposing in this paper is that today we are (willingly or not) contributing to the emergence of a *discourse* which is unilaterally dominated by concerns about understanding, a discourse which emphasizes (I) to the detriment of (II). More specifically, I argue that many theories en vogue today in mathematics education, despite the wealth of fruitful ideas about the teaching/learning of mathematics they provide, contribute to the emergence of this type of discourse.

This discursive prevalence of (I) over (II) has, in my opinion, detrimental consequences for the development and implementation of school mathematics curricula. Among other things, it may lead to the emergence of symbolic illiteracy and fear of symbolism among students, and a whidespread skepticism on the part of teachers and political agents with respect to the importance of formal sign manipulation in mathematics. In my view, an approach to mathematics teaching/learning that deemphasizes (II) does not support the emergence of relational thinking, and risks annihilating whatever benefits the methods developed in order to cope with statement (I) might bring.

The path that I am taking in order to explore these issues goes along the lines of Foucault's idea of *Archaeology*. I take two different areas (philosophy of mathematics and mathematics education theories) and try to uncover points of convergence.[5]

UNDERSTANDING VERSUS FORMALISM: BERTRAND RUSSELL'S CASE

Bertrand Russell's "supreme axiom in scientific philosophizing" goes like this: "Whenever possible, substitute constructions out of known entities for inferences to unknown entities". Speak in terms of "inferences to entities", rather than in terms of "constructions out of entities". This sounds just like Ernst Cassirer's broader claim according to which a shift has taken place from designing theories in terms of substances (Substanzbegriffe) to designing theories in terms of relations (Funktionsbegriffe). The object (entity) is no longer the given (das Gegebene), but rather the goal (das Aufgegebene). Relations should (whenever possible!) define the relata and not conversely.

The relational approach to science dominates Russell's philosophical work. It finds a good expression in his "supreme axiom" and it owes a lot to ideas accumulated in the course of mathematical work. Russell's supreme axiom sounds very much like David Hilbert's "At the beginning is the sign", and to Paul Bernays' more explicit version of the same: "Whenever a concept is missing a sign jumps in to fill the gap." It is also reminiscent of Samuel Johnson's "I have found you an argument but I am not obliged to find you an understanding". Russell himself seems to have thought that the meaninglessness of mathematical signs is the foundation of their unique meaningfulness. This point of view seems to capture the core of the position shared by many philosophers and scientists alike in respect to mathematics and natural science: semantic bonds have betrayed us far too often, hampering the evolution of our theories, be it in science or in philosophy. The harsh words towards semantics, towards "construction out of entities" used by Russell, testify to our feeling that our love for semantic roots has been betrayed.[6] Formalism was—and to many still is—the new love. As Bernays said, signs bypass missing meanings, and in that way, they shape meaning in a particular fashion.[7] And Russell adds:

> For a non-mathematical mind, the abstract character of our physical knowledge may seem unsatisfactory. From an artistic or imaginative point of view, it is perhaps regrettable, but from a practical point of view it is of no consequence. Abstraction, difficult as it is, is the source of practical power. A financier, whose dealings with the world are more abstract than those of any other 'practical' man, is also more powerful than any other practical man. He can deal in wheat or cotton without needing ever to have seen either: all he needs to know is whether the prices will go up or down. This is abstract mathematical knowledge at least compared to that of the agriculturist. (Russell, 1969, p. 138)

These are clear words. In a sense, Russell's "financier" knows more about beans than the peasant cultivating them. Presumably, the pure mathematician, dealing with his meaningless signs has an even greater power over nature than the natural scientist or the financier. Thus, the formal seems to be assigned the greatest potential, and the expert, in dealing with the formal, the greatest power. Ancient cultures thought that being able to name something gives us power over that very thing. The previous formalist attitude seems to go in the same direction.[8] So far, therefore, we may expect Russell to adhere to (II) rather than to (I). But is this really so? Does Russell think that formalism is our only safe haven in mathematics? Russell's answer is, strange as this may perhaps seem, a categorical "No". This complementary attitude is perfectly illustrated by the ideas advanced by Russell in his famous *Introduction to Mathematical Philosophy* of 1919.

The first chapter of this book, "The series of Natural Numbers", deals with the foundations of the arithmetic of natural numbers. After having become acquainted with Russell's *supreme axiom*, the shift towards statement (I) is indeed bewildering, for most of the 19 pages of the first chapter of Russell's book represent an attack aimed at axiomatics. After a few remarks about the history of the subject, Russell reproduces Giuseppe Peano's axioms and then writes:

> Peano's three primitive ideas - namely, '0,' 'number,' and 'successor' – are capable of an infinite number of different interpretations, all of which will satisfy the five primitive notions (Russell, 1970, p. 7).

Fine, we might think. The power of formalism resides, after all, precisely in this. Does not Russell's own *supreme axiom* state that in order to increase generality one needs to exorcise content? Here, however, Russell writes these words not in order to state the mathematical counterpart of his *supreme axiom*, but in order to … criticize Peano! "Peano's treatment is less final than it appears to be", he writes (Russell, 1970, p. 7). After a wealth of examples aimed at revealing why Peano's approach is not good enough, Russell concludes:

> This point, that '0', 'number', and 'successor' cannot be defined by means of Peano's five axioms, but must be independently understood, is important. We want our numbers not merely to verify mathematical formulae, but to apply in the right way to common objects. (...) A system in which '1' meant 100, and '2' meant 101, and so on, might be all right for pure mathematicians, but would not suit daily life. (...) We have already some knowledge (though not sufficiently articulate or analytic) of what we mean by '1' and '2' and so on, and our use of numbers in arithmetic must conform to this knowledge. We cannot secure that this shall be the case by Peano's method; all that we can do, if we adopt his method, is to say 'we know what we mean by '0' and 'number' and 'successor', though we cannot explain what we mean in terms of other simpler concepts'. (Rusell, 1970, p. 9)[9]

It is indeed extraordinary, given what was said previously, to read lines such as these. The entire point of axiomatics, and both Peano and Russell were fully aware of it, was to get rid of the idea that our mathematical constructions must necessarily be linked to some particular set of entities, to be understood "independently" of axiomatics. According to a modern conception of axiomatic theories, theories must be developed so as to exclude any significant debt to "known entities". There is nothing that comes prior to whatever the axioms state. It is rather the axiomatic texture that constitutes the *a priori* of whatever can be derived from it. This, however, does not mean that formal axiomatics is completely cut off from all semantics, but only that it gives us a structural description—not of the entities in case as such—but rather of the net of relations linking whatever entities can be made to inhabit our symbolic nets. Russell does admit that this latter approach is useful to the pure mathematician. He argues, however, that formal-axiomatics leads to an infinite regress of reference, and this seems to be Russell's main concern in his attack on Peano.

Russell is looking for a foundation of the natural numbers themselves, whatever this may mean, and not just for a foundation able to describe our handling of arbitrary "progressions" such as, for instance, 100, 101, 102 etc. "Natural Number" is thus assigned a privileged ontological statute. The axiomatician disregards the distinction *arbitrary progression/natural number*. He approaches the matter by simply saying: even though I cannot show you any extralinguistic objects called numbers, I am able to show how I put certain signs in relation to each other; if you will watch me closely, you will soon find entities of your own with which you can operate in the same way. Russell agrees but he is not satisfied.

Adopting a term en vogue in current debates on mathematics education, Russell can be said to be arguing against a fully decontextualized mathematics. Formal games of signs are fine—no, they are absolutely necessary—but only as long as we have a firm ground on which to put them. Moreover, this cannot be some *particular* semantic universe that may be different for you and me, although perhaps isomorphic at the formal level; it must be universal.

In a certain sense, Russell's objections to axiomatics are objections against the point of view he himself advocates in respect to natural science. His work vindicates both those who emphasize that in mathematics education we need to focus on procedural skills, for there is no mathematics outside procedural skills, as well as their opponents who insist that there cannot be any sense in dealing with technique and formalism unless semantic contexts have already been taken into account.

We thus realize that despite his emphasis on statement (II) linked to his supreme axiom, at the end of the day, Russell returns to (I) and in this way he expresses a type of concern about the relationship between sign and meaning-making that comes close to the fundamental belief in the priority of understanding over language often invoked in texts dealing with mathematics education. Of course, Russell is no constructivist in the sense in which this term is used today in theories of mathematics education, and it is certain that his views do not endorse most of the fundamental views put forward in research on "situated cognition", "ethnomathematics" and the like. For Russell, the true foundation is logic, a logic which, however, is assumed to be irreducible to Peano's formalism. This important difference must however not make us blind to the fact that Russell, just like many mathematics education theorists, moves inside a discourse that ultimately embraces statement (I) as the most important.

What does the previously revealed tension stand for? Russell wants to build up a logical approach to arithmetic that does not require formal axiomatics and that captures the logical essence of number, as it were, independently of any formal mathematical expression. On this issue, Russell tries to rescue Gottlob Frege's dream. Freges's project was brought down by Russell, and Russell's own attempt to rescue it on a higher level failed, as Russell himself acknowledges by the end of his book.[10]

Is there a lesson to be learned from this? Peano's symbolic language in arithmetic cannot be reduced to anything more elementary or more transparent. There does not seem to be any absolute semantics that can be used to define our mathematical models. If we wish to understand what "0", "number", and "successor" mean outside the framework of Peano's axiomatics, logic and set-theory are not enough. Russell's thinking is haunted by the problem stated in (I), bridging the gap between formula and idea through a universal language uniting both, and at the same time, by the awareness of the reality and power of (II). In Russells work both (I) and (II) are alive and well, even though Russell is, in respect to arithmetic, looking for a solution to (I) that would bypass (II). Thus, Russell's attitude towards the formal language of mathematics is, in a sense, split. The fact that Russell did not find the true definition of the primitive ideas of arithmetic in the terms desired by him, the fact that he did not manage to resolve the contradiction involved in statement (I), does not exclude the possibility of an alternative path for explaining the primitive ideas in terms other than logic. We can move away from purely logical investigations and shift towards anthropological studies, examining language games played by people using the formal models of arithmetic. This was the path pursued by Ludwig Wittgenstein, a path that Russell never accepted.[11]

This sort of dilemma has its counterpart in the debates over the relationship between understanding and mathematical technique in mathematics education. The question is, how do these theories deal with statements (I) and (II)? Do they, like Russell, at least accept (II) as a necessary level, and then look at the way in which mathematical models can be used to support rich semantic language games that may be productive for the learning of mathematics? Or do they rather tend to reject (II) and look for a solution for the problem of learning mathematics, which in a sense marginalizes procedural skills?

LOOKING FOR THE BEST DIDACTIC SETTING

One particular approach to teaching/learning mathematics, constantly criticized by mathematics education theorists, is sometimes referred to as "traditional instruction". One of its most outspoken critics, Paul Cobb, calls it the "internalization account of learning" (Cobb, 1991, p. 11).[12] The backbone of the approach criticized by Cobb is roughly the assumption that it is possible to hand-down a bundle of abstract symbolic mathematical techniques (simple algorithms, for instance) to the students and expect them to be able to operate with these techniques, to make sense of them, to be able to use these techniques in a flexible way while solving a variety of problems, including meaningful problems related to daily life situations:

> as a consequence of traditional instruction in the early grades, children learn to rely on instrumental procedures at the expense of sense making. (Cobb, 1990, p. 129).

If "traditional teaching" or "teaching to the formal model", as I would rather call it, is so bad, then why is it still around? And what is the alternative? The crux of the matter is that there seems to be a gap separating abstract symbolic mathematical technique on the one hand, and the meaning attached to it by the learner, on the other. In a sense, the two appear to behave like oil and water. How then should we teach abstract symbolic mathematical technique in school settings in a way that will eventually lead to a significant enrichment of the semantic universe of the learner? This is indeed an important question.

In order to solve this problem, we need an account of what sort of a "solution" it is, that we have in mind. Such "solutions" depend on certain epistemological choices that need to be clarified beforehand.

If we feel attracted to a strict behaviorism, then the best we can do to solve our problem is to make a detailed catalogue of all the elementary skills that our students are expected to exhibit (including prescriptions for the way to apply each tool to each type of problem that the learner is supposed to be able to solve) and then design drill programs aimed at making sure that these skills are implemented.[13]

The other option is to emphasize that it is not enough to hand-down standard abstract mathematical tools and explain how such tools apply to standard situations. What students need to learn depends not only on the explicit rules of the symbolic language-game involving this or that individual mathematical technique, or this or that closed range of applications. There seems to be more. The trouble is that it is not easy to characterize this additional element in a way that can be stated in operational terms as easily as an algorithm can be.

Mathematical knowledge, if it is to be meaningful to students, cannot be reduced to the narrow context of procedural/axiomatic/logical/formal justification used by the practicing mathematician. The teaching of mathematics, so it would seem, cannot be reduced to procedural knowledge. It must take into account a variety of other aspects such as "implicit knowledge", "higher thinking skills", "mathematically rich environments" etc. Solving the above problem then means engineering learning situations in which procedural skills and higher thinking skills can grow.

Paul Cobb explains that mathematics should be regarded as a "normative conceptual activity" (Cobb, p. 1991, p. 12), meaning, among other things, that learning mathematics should be regarded as a process of "acculturation" into a given normative conceptual practice. What happens when we learn mathematics is not a piori determined by some universal process of cognitive development passing through an invariable hierarchy of alleged cognitive stages. Neither is it determined by any set of abstract symbolic mathematical tools as such. On the contrary, as we move on from one community of mathematical practice to another or from one situation to another, we are somehow able to grasp, not just technical skills, but also meaning. Learning mathematics means constructing "the conceptual actions that constitute the taken-to-be-shared mathematical practices of our culture" (Cobb,

1991, p. 12). What the learner grasps/constructs are both the observable actions as well as the implicit frames connected with these actions inside a given microculture.

This may sound like a plausible approach to mathematics teaching/learning, but putting it into practice is not so simple. Cobb is certainly aware of this, and the work developed by him and his collaborators is an interesting attempt at coming to terms with the difficulties involved. It is, I think, possible to agree with many aspects which Cobb derives from his conception. He is clearly right when he points out that children do not discover mathematics on their own, that is, that the process of mathematical "acculturation" cannot merely be a passive or receptive one. Mathematics is not learned by "internalizing" it either from objects, pictures or other didactic gimmicks, nor can it be learned by parroting this or that piece of mathematical symbolism. Neither standard mathematical technique nor didactic aid materials are enough.

Cobb believes that the foundation of learning is providing students opportunities to participate in situations "that will be personally meaningful to students in terms of their lived experiences" (Cobb, 1991, p.11). Thus, the entire burden of the "acculturation" process is attached to the "personally meaningful experiences". There can be no mathematical enculturation without personally meaningful situations. This last element, however, is not easily pinned down. Cobb's papers provide various examples which may be used as an illustration of the sort of thing that he has in mind. I will present this in a simplified form.

Assume that you confront a primary school student with the following two problems, (a) as a part of a worksheet containing a list of several routine computation exercises and (b) in the context of a discussion related to a situation which is meaningful to the student.

$$\begin{array}{r} 16+ \\ 09 \\ \hline \end{array}$$

(b) You have 16 apples, and your mother gives you 9 more. How many apples have you got?

Assume further, as Cobb does, that one of your students (Cobb calls her Auburn) proceeds like this. In solving (a), she adds 9+6, gets 15, writes down 5 and gets confused about the remaining "1". This can perhaps be described by saying that the student does not know "how to carry" (Cobb, 1991, p. 5). The student disregards the "1" in 15 but takes the "1" in 16 into account, getting 15 as a final result. In solving (b) the student counts on and reaches the correct answer 25.

Assume now that you want the child to realize that her solutions contradict each other. How do you go about that? In Cobb's example, the teacher engages in a dialogue with the student trying to clarify the issue:

> I: So when we count, we get 25 and when we do it this way [as in (a) – M. R.] we get 15. Is that okay to get two answers or do you think there should be one?
> A: (Shrugs her shoulders).
> I: Which do you think is the best answer?

A: 25.
I: Why?
A: I don't know.
I: If we had 16 cookies and 9 more, would we have 15 altogether?
A: No.
I: Why not?
A: Because if you counted them up together, you would get 25.
I: But is this (points to the answer 15 on the worksheet) right sometimes or is it always wrong?
A: It's always right (Cobb 1991, 5)

Cobb comments:

> Think about Auburn's last comment, 'It's always right.' What she meant was that 15 is always right in the school worksheet setting and that 25 is always right when she thinks about real world situations such as those involving collections of cookies. In other words, doing school-like worksheets had nothing to do with common sense or solving arithmetic problems that arise in the course of daily life. (Cobb, 1991, p. 5).

Cobb interprets this as a symptom of a radical rupture between "worksheet mathematics" and "daily life meaningful mathematics". The child lives in two worlds and does not feel compelled to coordinate the two. This situation in return, or so we are told, points to a fundamental deficit of the school mathematics to which the student had been exposed. What may this deficit consist in? A great deal depends on our answer to this question. If I am not mistaken, Cobb can be read as claiming that the "worksheet mathematics" alledgedly characteristic for "traditional instruction" is the source of the evel.

The prevoius example is not as extreme as Morris Kline's famous "3 is the right half of 8", which was used by him to illustrate the absurdity of the unilateral emphasis on artificial symbolism assumed to be typical for "New-Math" mathematics teaching (Kline, 1973), it is, however even more disturbing because it indicates how difficult it is to integrate everyday experience and standard mathematics. How do we deal with this?

Cobb's answer is: avoid letting mathematics appear to the students as a "highly ritualized, symbolic activity divorced from the rest of their lives" (Cobb, 1991, p. 6).

This point is certainly correct. The answer then is to create didactic settings which are "personally meaningful" to the students by bringing them closer to "their lived experience". This may perhaps prove a practicable recommendation for primary school mathematics. But the issue is really not so simple. Do Cobb's case studies really manage to prove that meaningful daily-life contexts can provide a sufficient background for productive mathematical teaching? Do they really prove that abstract symbolic mathematics is by itself so artificial?

It seems to me that, even though it is possible to interpret episodes such as the previous one in terms of Cobb's acculturation model of thinking, at the same time, his treatment of the problem raises serious questions as to his other claim according to which meaningful daily-life contexts provide a sufficient semantic background for learning.[14]

CONCLUSION

On the surface, both Russell and Cobb care just as much for (I) and (II). A closer look at the details of their conceptions reveal the fact that, despite the fundamental differences between them, in the end, both tend to unilaterally emphasize the importance of (I). Meaning making appears then as a process that does not necessarily benefit from formal sign manipulation within the formal mathematical models themselves.

From my point of view, highly ritualized symbolic activity is just as much an irreducible part of genuine mathematical experience as meaningful daily-life situations are. In some cases, the student may benefit from daily-life experience in order to make sense of a piece of abstract mathematics, in others not. As far as I can see, Cobb takes a stronger stand on this. But does he manage to *prove* that this sort of experience is a sufficient (I am not convinced that it is even necessary) prerequisite for productive school mathematical work? Paraphrasing Russell's axiom, I am inclined to say that, even in mathematics teaching, if we in fact wish to help students practice self-reflection and experience mathematical generalization, we cannot avoid substituting constructions out of known entities for inferences to unknown entities. Indeed, in some cases, daily-life issues or more generally "applications" become meaningful only in connection with the ritualized symbolic activity of the mathematician. On the whole, it is important to keep in mind the strict distinction between statements (I) and (II). There is no straight-forward natural link between "applications" and "situations" on the one hand and "symbolic methods" on the other, between the logic of substances and the logic of relations. Everything needs to be constructed: the applications, the symbolic methods, and the relationship between the two. Statements (I) and (II) emphasize the fact that there cannot be any "understanding" in respect to any field whatsoever that is either sufficient or necessary for learning and transfer, as long as the symbolic culture of the field, the culture of the formal models of thinking in the field is not a part of the work. Acculturation is required all the way. I suggest that understanding is something that cannot be provided to the student while disregarding the structure of the abstract symbolic mathematical technique. The problem is that this view has almost become taboo, with the consequence that many students never get the chance to experience the close link between generalization, symbolism and every-day situations, thus being hampered from experiencing relational thinking. We are writing and speaking a bit too much about meaningful applications and too little about generalization and its debt to symbolic language. We are too much involved in advocating the Egyptian mathematical style in schools. I am afraid that we are currently embarking on a trip to a kind of Disneyland of meaningful applications to the detriment of the Cartesian mathematical style. Are we trying to help students become autonomous self-reflecting subjects or are we mainly concerned about helping them become jolly consumers of allegedly meaningful applications, diagrams, applets, experiments etc.? We should of course try to prevent creating addiction to blind formalism, but should we not, as well, carefully consider the limitations involved in the view that meaningful every-day situations are the only path to a mathematical experience that is significant for the students?

Let me close by saying that in my view, even though (I) expresses an important and indeed legitimate concern, this concern cannot be dealt with without seriously taking (II) into account as well. This may seem obvious, but it seems to me that in many cases research aimed at exploring this problem ends up by adopting a rhetoric overemphasizing (I) to the detriment of (II). Russell and Cobb and their preference for statement (I) are everywhere. This may be fine for the purposes of certain narrow investigations, but on the whole, it is problematic because it leads to the emergence and reinforcement of a way of speaking about mathematics that lives independently of the results of these investigations themselves, shaping the way mathematics education is viewed within society at large. Insufficient attention to the distinction between the positions expressed in (I) and (II) have serious consequences not just for academic debates, but for the more down-to-earth everyday decisions involving practitioners.

NOTES

1 Compare for instance Ernest, 1999; Wu Hung-Hsi, 1999; D'Ambrosio, 2001; Begg, 2001; Radu, 2002.

2 Aurelius Augustinus' *De Magistro* contains one passage which comes close to statement (I). Compare Radu, 2003a, p. 172.

3 Compare Radu, 2000 and Radu, 2003c p. 345.

4 Compare Whitehead, 1898, p.3-4.

5 Cf. Lemke, 1997, p. 38.

6 One may perhaps be tempted to restrict this statement to the relationship between an abstract mathematical theory and its applications to practical matters. This, however, would miss the point, for the fact of the matter is that formalism appealed to pure mathematicians precisely because it provided a productive answer to a question that was the subject of heated debate: why is it possible to use the results of a given, purely mathematical, theory dealing with one distinct set of mathematical objects in another mathematical theory dealing with a completely different kind of entities? Many discussions were fuelled by the statute of combinatorics and they gradually paved the way for the view that mathematical theories do not have objects of their own, but rather deal with various ways in which we operate with signs. Russell himself, even though he is fully aware of this power of mathematical formalism, is still on the search for an absolute "logical" foundation of mathematics that will go beyond formal changes.

7 Compare Rotman, 1987.

8 For a more detailed historical account combined with a general semiotic analysis, consider Rotman 1987.

9 It is, of course, possible to argue that if Russell's critique of Axiomatics is correct, then Peano's Axioms mark the borderline separating the things we can speak of (i.e., the relationships between abstract but undefined entities) and a kind of experience that we cannot speak of, our experience with the entities themselves, prior to their axiomatic description. We would then not be able to address Rusell's concerns by trying to speak further, but rather by immersion in the practice of the various arithmetic language games played. Russell does not pursue this path. He is looking for a truly universal language.

10 Compare Otte 2002.

[11] Michael Otte explains Russell's attitude by emphasizing Russell's lack of appreciation for the fact that, in modern science and mathematics, the objects of a theory cannot be dealt with independently of some given theoretical framework. The objects of the theory, the theoretical symbolic generalizations used to describe these objects, and a certain social practice in which the two are embedded are linked together in a chain of reference, each pointing to the other two (Otte 2002, LIV ff.). On this account, one may perhaps be led to regard the tension between statements (I) and (II) as caused by precisely this type of lack of appreciation for the fact that neither the objects of a theory nor the symbolic generalizations of the theory as such can be understood independnetly of a wider unifying social practice (compare Otte 1994, p. 407 ff.). I will not pursue this issue any further in this paper. My objective here is not so much to resolve the tension (indeed, I believe that it cannot be resolved), but rather to show various contexts in which it comes to the fore, and indicate some of the diffculties and dangers linked to that.

[12] Similar ideas can be found in Cobb & Bauersfeld (Eds.), 1995.

[13] A good illustration of this kind of solution can be found in Thorndike's work (cf. Radu, 2003a). Another illustration of this is the frequently encountered expectation of a considerable number of students, parents, teachers and teacher trainers concerning homework. Homework should consist in drill exercises designed to apply skills presented in class to problems that do not significantly differ from those treated in class.

[14] In a recent paper, Radford offers a critical appraisal of constructivism in education and stresses some of the difficulties that result from constructivism's overemphasis on the individual and its weak concept of culture and the social realm (see Radford, 2008).

REFERENCES

Begg, A. (2001). Ethnomathematics: Why and what else? *ZDM, 33*(3), 71–74.

Cobb, P. (1991). Reconstructing elementary school mathematics. *Focus on Learning Problems in Mathematics, 13*(2), 3–32.

Cobb, P., & Bauersfeld, H. (Eds.). (1995). *The emergence of mathematical meaning: Interaction in classroom cultures*. Hillsdale, NJ: Lawrence Erlbaum.

D'Ambrosio, U. (2001). *General remarks on ethnomathematics. ZDM, 33*(3), 67–69.

Ernest, P. (1999). Social constructivism as a philosophy of mathematics. *ZDM, 31*(2), 71–73.

Kline, M. (1973). *Why Johnny can't add*? New York: St. Martin's Press.

Lemke, T. (1997). *Eine Kritik der Politischen Vernunft*. Berlin, Germany: Argument Verlag.

Otte, M. (1994). Das Formale, das Soziale und das Subjektive. Frankfurt a.M., Germany: Suhrkamp.

Otte, M. (2002). Introductory essay to Betrand Russell's *Einführung in die mathematische Philosophie*. Hamburg: Meiner Verlag, p. I–LXII.

Radford, L. (2008). Di sé e degli altri: riflessioni su un problema fondamentale dell'educazione [The self and the other: Reflections on a fundamental problem in education]. *La Matematica e la sua didattica, 22* (2) 185-205.

Radu, M. (2000). Justus Graßmann's contributions to the foundations of mathematics: Mathematical and philosophical aspects. *Historia, 27*, 4–35.

Radu, M. (2002). Basic skills versus conceptual understanding in mathematics education: The case of fraction division. A reply to Hung-Hsi Wu. *ZDM, 34*(3), 93–95.

Radu, M. (2003a). Als das Zeichen nicht am Anfang stand: Eine Untersuchung der amerikanischen Grundschularithmetik zwischen Colburn und Thorndike. In M. H. G. Hoffmann (Ed.), *Mathematik verstehen* (pp. 114–159). Hildesheim: Franzbecker Verlag.

Radu, M. (2003b). Peirces Didaktik der Arithmetik: Möglichkeiten ihrer semiotischen Grundlegung. In: M. H. G. Hoffmann (Ed.), *Mathematik verstehen* (pp. 160–194). Hildesheim: Franzbecker Verlag.

Radu, M. (2003c). A debate about the axiomatization of arithmetic: Otto Hölder against Robert Graßmann. *Historia Mathematica, 30*, 341–377.

Rotman, B. (1987). *Signifying nothing – The semiotics of zero*. New York: St. Martin's Press.

Russell, B. (1970). *Introduction to mathematical philosophy*. London, England: George Allen and Unwin Ltd.
Whitehead, A. N. (1898). A treatise on universal algebra. Retrieved February 20, 2008, from http://cdl.library.cornell.edu/cgi-bin/cul.math/docviewer?did=01950001&seq=32&frames=0&view=50
Wu, H.-H. (1999, fall). Basic skills versus conceptual understanding – A bogus dichotomy in mathematics education. In *American Educator*, Fall.

Dr. Mircea Radu
Oberstufenkolleg des Landes Nordrhein-Westfalen
Germany

INDEX

D

N

O

P

R

T

U

V

W

Z

ABOUT THE CONTRIBUTORS

Adalira Sáenz-Ludlow teaches mathematics and mathematics education courses at the University of North Carolina at Charlotte. She has conducted longitudinal teaching experiments with elementary school students and teachers and with pre-service high school teachers. Her research interests include the evolving and co-constructed processes of semiosis and inter-semiosis in the teaching and learning of mathematics.

Angel Contreras is Senior Lecturer at the University of Jaén, Spain. His research is focused on advanced mathematical thinking, mathematics teacher training and the application and development of the onto-semiotic approach to mathematics cognition and instruction.

Anna Athanasopoulou is a part-time instructor of mathematics at the University of North Carolina at Charlotte. She also develops the mathematics curriculum in Greek at the Socrates Academy, an elementary charter school at Charlotte North Carolina. She got her PhD in Curriculum and Instruction in June 2008. Her research interests focus on the investigation of the teaching and learning mathematics from socio-cultural and semiotic perspectives.

Cristina Sabena holds a post-doctoral position at the University of Torino, Italy. Her research focuses on the role of different kinds of signs (gestures, in particular) in the teaching-learning processes in mathematics, as well as on the role of new technologies.

Falk Seeger is retired from a position of Senior Lecturer at the *Institut für Didaktik der Mathematik* at Bielefeld University, Germany. His research interests include the analysis of mathematical learning from a cultural-historical perspective. In his current work, he puts a focus on the semiotic roots of intentionality in human development, drawing heavily on Vygotsky's work and on Peirce's ideas.

Gert Schubring is department head at the *Institut für Didaktik der Mathematik* at Bielefeld University, Germany. His research interests focus on the history of mathematics and the sciences in the eighteenth and nineteenth centuries and on their systemic interrelation with socio-cultural systems.

Juan D. Godino is Full Professor at the University of Granada, Spain. His research is focused on the theoretical foundation of mathematics education. In this field, he is developing an onto-semiotic approach to mathematics cognition and instruction that is being applied to different mathematics themes.

Luis Moreno-Armella has been a Full Professor at Cinvestav and at the University of Massachusetts Dartmouth, and a Senior Executive Research Associate of the Kaput Center for Research and Innovation in Mathematics

Education. His research interests are focused on the impact of digital technologies on cognition and epistemology, and on the evolution of representation infrastructures.

Luis Radford is full professor at Laurentian University in Canada. His research interests include the investigation of mathematics thinking and knowing from a cultural-semiotic perspective. For many years he has been conducting classroom research with primary and high-school teachers about the teaching and learning of mathematics as well as the historical and cultural roots of cognition.

Michael N. Fried is a senior lecturer in the *Program for Science and Technology Education* at Ben Gurion University of the Negev. His research interests are eclectic and include mathematics pedagogy, mathematics teacher education, sociocultural issues, semiotics, history of mathematics, and history and philosophy of education.

Michael Otte is Emeritus Professor at the University of Bielefeld, Germany. He has published articles and books in the areas of pure mathematics (complex analysis), mathematical education, epistemology and the history of mathematics. A recent book published in his honor (*Sign and Activity*, 2005, published by Springer) recognizes Otte's important contribution to the theoretical foundations of mathematics education.

Mircea Radu teaches Mathematics at the Oberstufenkolleg des Landes Nordrhein-Westfalen in Germany. His research interests include the history of the foundations of mathematics, the history of abstract algebra, and the relationship between mathematics, semiotics, and pragmatism in Charles S. Peirce's writings. He is also interested in the implications of certain theories about human cognition (constructivism, activity-theory, ethnomathematics, and semiotics) for classroom work, textbook design, curriculum development and educational policy.

Norma Presmeg is full professor in the Mathematics Department at Illinois State University, and editor-in-chief elect of *Educational Studies in Mathematics*. Her research interests include visualization in mathematics education, ethnomathematics, metaphor, metonymy and imaginative rationality, and semiotic theoretical frameworks for research on the teaching and learning of geometry and trigonometry.

Raymond Duval is Emeritus Professor at the Université du Littoral Côte d'Opale, France. After studying philosophy and psychology, he worked at the IREM of Strasbourg from 1970 to 1995. His research focuses on issues relating to the understanding of proof, the various kinds of reasoning, the cognitive functioning of visualization in geometry, with graphics, etc., and the use of auxiliary representations in problem solving. He has published *Semiosis and human thought* (1995), now translated into Spanish.

Rosana Nogueira de Lima is a Lecturer at the Post-Graduate Programme in Mathematics Education at the Bandeirante University of São Paulo in Brazil. Her research focuses on understanding relationships between the embodied, symbolic and formal aspects of algebraic thinking.

Stephen Hegedus is a Full Professor at the University of Massachusetts Dartmouth and the Director of the Kaput Center for Research and Innovation in Mathematics Education. His research focuses on the large scale impact of digital technologies on learning and motivation and the co-evolution of representation and communication infrastructures.

Tania M.M. Campos is the coordinator of the Post-Graduate Programme on Mathematics Education at the Bandeirante University of São Paulo in Brazil. Her research interests centre on investigations of all aspects of teaching and learning algebra in school mathematics and in teacher education.

Vera Helena Giusti de Souza is a Lecturer at the Post-Graduate Programme in Mathematics Education, Bandeirante University of São Paulo in Brazil. Her research interests are mainly directed towards exploring the possibilities associated with connecting graphical and algebraic approaches to function in school and university mathematics.

Vicenç Font is Senior Lecturer at the University of Barcelona, Spain. His research interests include the epistemology of mathematics education, advanced mathematical thinking and the development and application of the onto-semiotic approach to mathematics cognition and instruction. He has a great deal of experience as a secondary school teacher and has written textbooks for secondary education and teachers' training.

Wolff-Michael Roth is Lansdowne Professor of Applied Cognitive Science at the University of Victoria, Canada. He researches knowing and learning in a number of fields across nearly the entire life span, from kindergarten schools to professional practice and retirement. He integrates ethnographic, linguistic, and sociological studies of communication with cultural-historical activity theory in numerous fields of inquiry including education, pragmatics/semiotics, social studies of science, and sociology.

If you have any questions regarding this title, please contact:

Koninklijke Brill BV
Plantijnstraat 2
2321 JC Leiden
Email: info@brill.com

Batch number: 10723294